T0304511

Enterprise Integration and Information Architecture

A Systems Perspective on Industrial
Information Integration

Advances in Systems Science and Engineering

Series Editor: Li Da Xu

PUBLISHED

Enterprise Integration and Information Architecture:
A Systems Perspective on Industrial Information Integration
by Li Da Xu
ISBN: 978-1-4398-5024-4

Systems Science: Methodological Approaches
by Yi Lin, Xiaojun Duan, Chengli Zhao, and Li Da Xu
ISBN: 978-1-4398-9551-1

Enterprise Integration and Information Architecture

A Systems Perspective on Industrial Information Integration

LI DA XU

CRC Press
Taylor & Francis Group
Boca Raton   London   New York

CRC Press is an imprint of the
Taylor & Francis Group, an **Informa** business

AN AUERBACH BOOK

CRC Press
Taylor & Francis Group
6000 Broken Sound Parkway NW, Suite 300
Boca Raton, FL 33487-2742

Printed on acid-free paper
Version Date: 20140402

International Standard Book Number-13: 978-1-4398-5024-4 (Hardback)

Library of Congress Cataloging-in-Publication Data

Xu, Li Da.
    Enterprise integration and information architecture : a systems perspective on industrial information integration / Li Da Xu.
        pages cm. -- (Advances in systems science and engineering (ASSE) ; 2)
        Summary: "This book provides a detailed description of enterprise information integration, from the development of enterprise systems for individual enterprises, to the extended enterprise information integration in supply chain environment, through discussing enterprise architecture, information architecture for enterprises, business process/work flow modeling, and enterprise information integration. The author clearly explains how industrial information integration can be more successful through the integration of systems approach as well as how systems science will profoundly impact and change the research in industrial information integration. "-- Provided by publisher.
    Includes bibliographical references and index.
    ISBN 978-1-4398-5024-4 (hardback)
    1. Management information systems. 2. Business enterprises--Computer networks. 3. Software architecture. I. Title.

HD30.213.X8 2015
658.4'038011--dc23   2014009434

Visit the Taylor & Francis Web site at
http://www.taylorandfrancis.com

and the CRC Press Web site at
http://www.crcpress.com

Contents

Preface

Enterprise systems (ES), also known as enterprise information systems (EIS), have transformed our economy in recent years and presents itself as an engine of change for the future. In the past decade, ES has emerged as a promising tool for integrating and extending business processes across the boundaries of business functions at both intra- and interorganizational levels. This emergence of ES has fueled an information technology revolution in the global economy. The development of information technology and the technological advances in ES have provided a viable solution to the growing needs of information integration in both the manufacturing and service industries. This is evidenced by the fact that a growing number of enterprises worldwide have adopted ES, such as enterprise resource planning (ERP), to run their businesses instead of using functional information systems, which were previously used for partial functional integration within many industrial organizations.

We have witnessed that, in the global economy and in global business operations, there has been a need for ES, such as ERP, to integrate extended enterprises in a supply chain environment. The core objective of this initiative is to achieve efficiency, competency, and competitiveness. For example, global operations have forced enterprises such as Dell and Microsoft to adopt ERP in order to take advantage of a global supply network. Today, companies both large and small are quickly learning that a highly integrated ES is a requirement to operate in a rapidly changing business environment. For instance, business-to-business (B2B) integration generally comprises connections to ES. ES has become a basic information processing requirement for many industries. Thus, the ES market is one of the fastest growing and most profitable areas in the software industry.

It is well recognized that ES has an important long-term strategic impact on global industrial development. Due to the importance of this subject, there has been a growing demand for ES research to provide insights into the issues, challenges, and solutions related to the design, implementation, and management of ES. At a June 2005 meeting of the International Federation for Information Processing (IFIP) Technical Committee for Information Systems (TC8) held in Guimarães, Portugal, the committee members intensively discussed the important role played by ES in the global economy and the innovative and unique characteristics of industrial information integration engineering (IIIE) as a scientific subdiscipline. Broadly speaking, IIIE is a set of foundational concepts and techniques that facilitate the industrial information integration process. Specifically speaking, IIIE comprises methods for solving complex problems when developing IT infrastructure for industrial sectors, especially with respect to information integration. It was decided at this meeting that the IFIP First International Conference on Research and Practical Issues of Enterprise Information Systems (CONFENIS 2006) would be held in Vienna, Austria, during 2006. In August 2006, at the IFIP 2006 World Computer Congress that was held in Santiago, Chile, the IFIP TC8 WG8.9 Enterprise Information Systems was established, based on the endorsement of the IFIP. In 2007, the Enterprise Information Systems Technical Committee was established within the IEEE SMC Society. In the meantime, it became apparent that there was a greater need to respond to the needs of both academicians and practitioners for communicating and publishing their research outcomes on ES. Therefore, the science and engineering journal *Enterprise Information Systems*, which devoted itself exclusively to the topics of ES and IIIE, was launched in 2007 in order to link the world's greatest minds on the topic of ES.

Concurrent with the ES and IIIE progress made in the IFIP and the IEEE SMC Society, new opportunities in the field of industrial informatics were discovered by Wilamowski, Kaynak, and various other researchers. They found that, driven by information technology, there has been a paradigm shift from industrial electronics to industrial informatics. In the current business environment, the industrial application of IT has been emphasized more than ever before; furthermore, one of the new trends in critical infrastructures demonstrates the crucial interplay of information and communications technology. Experts in the field realized how important industrial informatics could be for various industry sectors. The logical response to this realization was to launch the journal *IEEE Transactions on Industrial Informatics* in order to directly respond to the ever-growing impact of ES and industrial information integration.

This book intends to provide readers with guidance to the ever-expanding field of ES in order to gain a new perspective. This information technology for industrial organizations requires greater collaboration among various fields in order to maximize its potential. The topics covered in the book include, but are not limited, to the cross-disciplinary study of enterprise integration, enterprise information architectures, and enterprise information integration techniques. The objective of this book is to introduce the current development and future opportunities that exist in the exciting field of ES. While it is a roadmap to the subject of ES, it is by no means meant to be an exhaustive narrative on the topic.

This book provides a detailed description of enterprise information integration ranging from the development of enterprise systems for individual enterprises to extended enterprise information integration in supply chain environments. In order to achieve the aforementioned task, it provides a discussion of enterprise architecture, information architecture for enterprises, business process/work flow modeling, and enterprise information integration. It attempts to explain how industrial information integration can achieve greater success through the integration of a systems approach while addressing how systems science will profoundly impact and change the way research is approached in industrial information integration. It chronicles the intellectual history of the emerging field of industrial information integration and deals with many critical issues of implementing industrial information integration in a professional setting.

This book offers comprehensive coverage of important topics that have recently received a great deal of press. Furthermore, there is in-depth discussion regarding new trends in enterprise integration and its information architecture, with an emphasis on technologies and methodologies. However, the book also provides interested *outsiders* with a clear and comprehensive introduction to the field while remaining accessible to novices.

Chapter 1 is designed to provide an introduction that discusses various views on modern enterprise solution and traces the emergence and growth of enterprise systems. Chapter 2 describes enterprise integration relative to manufacturing integration, engineering integration, and customer integration. The critical concepts of ERP, industry-oriented enterprise resource planning (IERP), as well as entire resource planning are introduced. Chapter 3 details real-world examples of extended enterprise integration in a supply chain environment. Chapter 4 provides a transition from extended enterprise integration in a supply chain environment to a methodologically based enterprise architecture, enterprise modeling, and enterprise modeling in a supply chain environment. In Chapter 5, a new information architecture for enterprise and supply chain is proposed. This information architecture is proposed as a new discipline in the context of industrial information integration. Chapter 6 elaborates the enterprise process modeling into detailed work flow modeling and management. Chapter 7 presents modeling and integrating information flows for enterprise information integration with the Internet of Things (IoT), the emerging revolutionary technology for global Internet-based information architecture. Chapter 8 probes the theory and methods of industrial information integration, including integration approaches and enterprise application integration (EAI). As the book attempts to elucidate how industrial information integration can be more successful through the integration of systems approach, Chapter 9 integrates the contents discussed in previous chapters from a systems perspective and illustrates how a systems approach will impact existing and future research in industrial information integration. Finally, Chapter 10 overviews current research and discusses the future evolution of ES and IIIE. The time has come for ES research to expand in new directions while building on the current base founded by the pioneers in the field.

During the preparation of the manuscript, I benefited considerably from discussions and correspondence with colleagues in the United States, the

United Kingdom, Sweden, China, and Thailand. They included Dr. Zhuming Bi, Dr. Wattana Viriyasitavat, Dr. Wu He, Dr. Shancang Li, Dr. Zhibo Pang, Dr. Wenan Tan, Dr. Nan Niu, Dr. Li Wang, and Dr. Huimin Liu. I would like to especially acknowledge Liuliu Fu, who prepared most of the graphics, and Dr. Alison Schoew, who proofread the entire manuscript.

Author

Li Da Xu serves as the founding chair of IFIP TC8 WG8.9 and the IEEE SMC Society Technical Committee on Enterprise Information Systems and as the founding editor in chief of the engineering journal *Enterprise Information Systems*. He is an endowed Changjiang Chair Professor by the Ministry of Education of China. He has been affiliated with the Institute of Computing Technology of the Chinese Academy of Sciences, the University of Science and Technology of China, Shanghai Jiao Tong University, the China State Council Development Research Center, and Old Dominion University, United States. Professor Xu participated in early research and educational academic activities in the field of systems science and engineering. He collaborated and worked extensively with pioneering scholars such as West Churchman, John Warfield, and Qian Xuesen. Furthermore, he has spearheaded early research and educational academic activities in the field of information systems and enterprise systems, which started in the early 1980s. He is a coauthor of the recent book *Systems Science: Methodological Approaches* published by the Taylor & Francis Group. His work has been cited by Qian Xuesen and other well-known scholars.

Chapter 1

Introduction

1.1 Modern Enterprise Solution

During the last two decades, the complexity of enterprises has grown dramatically. In particular, increasing product varieties, market globalization, and the expected quick responsiveness to market changes all require enterprises to be more agile and responsive. Today's enterprises are struggling to meet those challenges. One approach to meeting such challenges is the leveraging of advanced information technologies as global economic integration requires more rapid information sharing.

Information technology is revolutionizing information systems within enterprises, particularly the networking, communication, and control systems. It provides the means for enterprises to better integrate their intra- and interorganizational business processes, activities, and resources. Intraorganizational integration requires the enterprise to build a set of consistent and mutually supportive practices that can support the enterprise's goals and objectives. Interorganizational integration matches those goals and objectives with the needs of the market and with the enterprise's competitive needs (within the existing environmental constraints). Intra- and interorganizational integration support and complement each other. In many enterprises, both intra- and interorganizational integration are needed to realize the enterprise's objectives.

To achieve integration, efficiency, competency, and competitiveness, many enterprises have a strong need to build enterprise systems (ESs) that can manage their existing business processes and operations. By definition, ESs are software packages that are developed to support many aspects of an enterprise's information needs (Davenport, 2000). "Enterprise systems" refer to large-scale software systems that are developed for the seamless integration of material and information flows within an organization. Qi et al. (2006) define an enterprise information system (EIS) as an integrated information system that seeks to integrate every single business process and function in the enterprise to present a holistic view of the business within a single information technology architecture.

In general, an ES is composed of a suite of different modules. Typical modules include executive direction and support, customer integration, engineering integration, manufacturing integration, and support service integration (Langenwalter, 2000). An enterprise can make its ESs available by integrating a number of modules.

It is well recognized that ESs have an important long-term strategic impact on the organizations that adopt them. Many large companies have adopted ESs to support their global business operations and supply chain management (SCM). For example, global operations have forced enterprises such as Dell and Microsoft to adopt ESs in order to take advantage of the global supply network. Meanwhile, many small enterprises have also adopted ESs to facilitate their business operations. So far, ESs have become a basic information processing requirement for many industries. As a result, the ESs market has become one of the fastest growing and most profitable areas in the software industry.

1.2 Emergence of ESs

Today's business environment is a collaborative one. Due to the intense competition in the business environment, enterprises need to establish relationships with various stakeholders and collaborate with their business partners and customers in order to survive and stay competitive in the marketplace. Collaboration can take place in a variety of forms. In general, collaboration includes intra- and interenterprise business collaboration. However, supporting intra- and interenterprise business collaboration is difficult, since enterprises and their partners may be located in different places. To effectively support collaboration, many enterprises have recognized the need to build an integrated ES by integrating their processes within the enterprise and also with external enterprises in order to conduct business (Browne et al., 1995). Thus, integrated ESs such as enterprise resource planning (ERP) have been used by many enterprises to support intra- and interorganizational business collaboration. By definition, integrated ESs are designed to integrate different enterprise functions and are used by different user groups and departments in and across organizations. They have the capability to significantly improve the operational performance of enterprises and enterprise collaboration. By linking databases, applications, and systems across enterprise networks, integrated ESs can provide real-time and accurate enterprise-wide or supply chain–wide information to enterprises.

Due to the benefits brought by integrated ESs, many enterprises have been using various information technologies to implement integrated ESs and to link disparate data sources, applications, and systems within their enterprise and beyond their enterprises. Many successful ESs have been implemented and used in manufacturing, healthcare, transportation, telecommunication, logistics, and SCM.

However, realizing these benefits requires a lot of effort. Some enterprises have had very successful implementation, while others have not achieved the results that they anticipated to gain from ESs (Li et al., 2008a). It is well recognized that implementing an integrated ES is a complex process. It often takes years and a large team of people to implement such systems. There can be many challenges during the implementation process. For example, many enterprises use a large number of disparate heterogeneous applications to run their businesses. To share information between these heterogeneous systems, technical work is required to create an integrated ES so that different applications can interact with each other. As different applications often use different programming languages, protocols, and operating systems, integrating these applications can be time-consuming and technically challenging. Furthermore, enterprise collaboration involves more than technical integration and often requires process integration between enterprises in order to realize effective communications and collaborations. During the process of implementing an ES, enterprises often need to change their business processes and tasks, change hierarchical levels of their organizations, and build further relationships with clients and suppliers. For example, enterprises in supply chains often need to collaborate with customers, suppliers, distributors, financial institutions, and government agencies and seek their support to implement a successfully integrated ES.

Research indicates that many strategic, financial, managerial, and technical benefits can be leveraged if stand-alone and disconnected information systems are replaced by integrated information systems (Xu, 2011). ESs can be used to support different types of collaboration in a typical business life cycle. The following are some examples (Ho and Lin, 2004):

- Product life cycle collaboration
- Engineering project collaboration
- Customer order and inventory collaboration
- Distributor–reseller collaboration
- Supplier and procurement collaboration
- Demand planning collaboration
- Warehouse management and freight collaboration

As noted earlier, different enterprises often have different information systems and applications built upon different technologies (Liu et al., 2008). Due to the constant evolution of information technologies, many enterprises have installed hardware and software from third parties (such as software vendors) and have used them to build their systems. Many applications are also distributed across a number of business units and are deployed on heterogeneous platforms. These hardware parts, software parts, and applications need to be updated regularly to ensure that they are secure and efficient. The vast varieties of hardware and software and their upgrades can often complicate the integration problem and can make integration challenging because there is always a complexity about whether an integrated system will be able to communicate with the existing systems of the enterprise or with the systems of its partners. ESs also have to be implemented in environments with certain existing legacy applications. To bridge the so-called information islands seamlessly, ESs need to be integrated with other application systems or commercial software packages. However, many commercial software packages hide their data formats and processing transactions completely behind the interfaces. To integrate these applications, ES developers need to have sophisticated knowledge and a solid understanding of the information infrastructures. This can make integration a challenging task (Wang et al., 2005).

Despite the difficulty in developing integrative ESs, research and development on integrated ESs has received a lot of attention and has been growing, as a field, in recent years. Developing integrated information systems has attracted a lot of attention in industries, as well. Especially since 2000, there have been increasing efforts and resources devoted to developing integrated systems in industries, notably in the manufacturing and healthcare sectors. An increasing amount of literature has been written on the integrated systems. In fact, an engineering journal specifically focusing on ESs was launched in 2007 by the Taylor & Francis Group.

1.3 Growth of ESs

An ES is an integrated information system for streamlining business processes. It is able to facilitate the flow of data and information among all of the partners in the supply chain. ESs are also called enterprise information systems. In the past decade, ESs have emerged as a promising tool used to integrate and extend business processes and functions across the boundaries of corporate walls at both the intra- and interorganizational levels. The emergence of ESs has been fueled by the global economy, by intense competition, and by the rapid development of information technology. In particular, the advances in information technology have provided a valuable tool to help information integration in industries, including manufacturing and service industries. A growing number of enterprises worldwide have adopted ESs such as ERP to run their businesses, instead of continuing to use functional information systems for only partial functional integration. As a type of ES, ERP facilitates transaction processing in a distributive environment and has been used to manage all of the resources that an enterprise has (Zhang and Liang, 2006). Theoretically speaking, ERP can optimize all of the resources in enterprises, supply chains, and value chains. Through implementing a CAD/CAE/CAM/PIMS (C3P) ES, the Ford Motor Company has realized that its C3P system is a strategic initiative for achieving its business targets of maintaining high product quality, shortening cycle times, and reducing costs. Boeing, GM, and Chrysler have made similar efforts. In the late 1980s, GM started its "C4" ES. Through implementing ERP in the coal mining industry, once separately managed resources can be integrated so that enterprises can make management more efficient and business operations more profitable by reducing costs, increasing sales, and monitoring the entire operation process in real time (Zhang and Zhao, 1999).

There has been a growing need for in-depth research about ESs. The development of networking technologies and distributed computing, especially Industrial Information Integration Engineering (IIIE), has had a significant impact on ESs. According to Camarinha-Matos et al. (2014), "Developments in Enterprise Systems, namely from the Industrial Information Integration Engineering and Information Systems for Collaborative Networks, can inspire the development of more advanced platforms in this domain, similar to what is happening with health information systems." Networking technology, Internet of Things (IoT), and other new technologies not only help to make information and knowledge more widely available, sharable, and retrievable, but also provide the infrastructure for distributed and collaborative knowledge acquisition and sharing. As the amount of processed and stored information in enterprise computing is growing quickly, requiring advanced technologies to deal with the large amount of data, using new technologies in ESs has become an important research topic. In particular, more research on the enabling technologies related to ESs' design, development, and applications is needed.

1.3.1 Brief History of ESs

In the 1960s, centralized computing systems were designed and developed for automating inventory control systems. These legacy systems were written in COBOL and other programming languages. In the 1970s, material requirements planning (MRP) appeared. It was developed to manage inventory control, bill of materials (BOM), and elementary scheduling. MRP is still used to plan manufacturing, purchasing, and delivering activities and to help organizations maintain low inventory levels. MRP is material-centric. Specifically, MRP is mainly used to manage (1) manufacturing-related activities such as fabrication, assembly, inventory, and purchasing; and (2) delivery activities related to suppliers and customers. In the 1980s, MRP was further extended to manufacturing resource planning (MRP II), which is a method for the planning of all of the resources of a manufacturing enterprise. MRP II can apply to many business processes including manufacturing, planning, finance, order management, inventory, and distribution. MRP II can apply to business processes at the intraorganizational units within an organization, including plants and distribution centers. MRP II was a step further moved from its predecessor MRP. MRP II is manufacturing-centric. Specifically, MRP II applies to (1) demand management activities such as sales forecasting and sales order processing; (2) resource planning activities such as master production scheduling, capacity requirements planning, and MRP; (3) production control activities such as purchase order processing, production order processing, shop floor control, inventory control, and distribution management; and (4) cost management activities (Shtub and Karni, 2010). Both MRP and MRP II were mainly developed for the manufacturing sector. Starting in the 1980s, the software industry started to develop integrated information systems in which multiple functional applications shared a common database. Through sharing a common database, a single transaction (such as a sales order) could flow through the entire system, automatically updating related financial and inventory records without additional actions. Such systems are designed to integrate the information used by all of the functional areas of an enterprise into a single system, in order to streamline business processes for an enterprise. Due to this apparent benefit, many enterprises adopt integrated systems and use them to manage the physical assets and information flow within the enterprise. As the software industry continues to enhance the functionality of its integrated systems, these systems evolve and become known as ERP systems, which typically include administrative (financial and human resources) functions and nonadministrative functions, such as materials resource planning in manufacturing. A main characteristic of the early ESs is to ensure the integration of the intraorganizational business processes for an enterprise's coherent internal operation. The tools for bridging the various isolated systems include common databases and intraorganizational coordination. These ERP systems were beneficial to the enterprises that adopted them, since they efficiently managed the enterprises' information processing needs. Some enterprises such as the Ford Motor Company and Wal-Mart achieved business success by adopting ERP systems. Learning of this, more and more enterprises have adopted ERP systems. As ESs provide a platform that enables industrial organizations to integrate and coordinate their business processes, ESs are considered a revolutionary advance in the continuous evolution of computer applications for business and industry. Nowadays, the ESs' market is one of the fastest growing and most profitable areas in the software industry.

ESs have become increasingly popular since 2000. However, for industries within specific sectors, those with specific needs and tasks, customized ESs are generally preferred. For example, in the textile and apparel industry, the BOM for textile production has a "one-to-many" characteristic. Although the production of yarns with different counts, colors, and warp types can be made from the same raw material, the difference in the BOM creates a long list of requirements that can span almost all aspects of development including the planning, production, dyeing and finishing, quality control, and sales and distribution specific to the textile industry. Under such circumstances, extraordinary customization efforts are required in order for general ESs to meet an industry's specific requirements and needs. Since many industries have their own processing requirements, general-purpose ESs may not exactly fit a specific industry's requirements for a variety of possible reasons: limited flexibility for customization or the inability to address real-world problems in specific industry sectors; in these circumstances, the scale of business process reengineering (BPR) and the customization tasks involved in the software implementation process are considered to be the major insufficiencies of general-purpose ERP (Wu et al., 2009). On the contrary, an industry-oriented ERP (IERP) system designed for enterprises belonging to a specific industry sector can support specific business needs that are not covered by existing general-purpose systems. Implementing an IERP can tackle problems such as low adaptation to specific business processes, the complexity of configuring thousands of parameters, and especially low industry relevance. IERP is mainly characterized by more adaptability for a specific industry and by a software componentization approach. Due to the fact that the processing requirements in a specific industry can be closely examined and continuously refined, it is relatively easy to make continuous improvements to IERP software. As a result, the software modules of IERP systems are able to incorporate more industry-specific requirements, to eliminate redundant modules and functions, and to maintain a moderate scale and size; meanwhile, they can deliver with both a much shorter implementation cycle and a higher success rate. Besides, explicit industry orientation is also helpful for ES vendors to benchmark the best practices within a specific industry sector. Thus, the concept of ERPII was proposed in 2000. It emphasizes two specific aspects: business coordination and industry orientation, along with the roles and functions of a system that is expected to expand to fit the needs required by an industry sector or a particular industry (Xu, 2011).

Since the 1990s, ESs have been mainly developed and implemented for managing the physical assets of an enterprise. To survive in today's highly competitive and ever-expanding global economy, there is a need for ESs to support the management of corporate knowledge. It is well recognized that knowledge is a compilation of an enterprise's invisible assets. Increasing requirements for extended enterprises have also stimulated the integration of the knowledge management function into ESs for use in knowledge asset management. Lorincz (2007) indicated that the real assets of any organizations are actually its knowledge resources. It was also predicted that the next stage of the coevolution of business and ESs will be the development of ESs with knowledge management capability (Lorincz, 2007). Due to the fact that both types of assets need to be properly managed, the integration of knowledge management with ESs has become a strategic initiative for providing competitive advantages to enterprises. In 2006, the concept of deploying knowledge management and ESs concurrently within the framework of ESs was proposed, and thus, the concept of ERP III was developed (Xu et al., 2006). ERP III enables ES applications to transform an enterprise into a knowledge-based learning organization and to capture know-how in order to develop business solutions and to create real competitive advantages.

In 2008, a new theory about material flows, called the Comprehensive Material Flow Theory (CMFT), revealed the essence of material flow objects and phenomena (Xu, 2008). In this theory, material flows are specified in terms of the material flow in the economic dimension, the material flow in the social dimension, and the material flow in the natural dimension, as well as their interrelationships. It was pointed

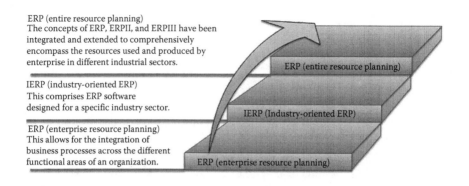

ERP (entire resource planning)
The concepts of ERP, ERPII, and ERPIII have been integrated and extended to comprehensively encompass the resources used and produced by enterprise in different industrial sectors.

IERP (industry-oriented ERP)
This comprises ERP software designed for a specific industry sector.

ERP (enterprise resource planning)
This allows for the integration of business processes across the different functional areas of an organization.

ERP (entire resource planning)

IERP (Industry-oriented ERP)

ERP (enterprise resource planning)

Figure 1.1   The evolution of enterprise systems.

out that material flow is not only an economic phenomenon but also a social and a natural one. In other words, there exists not only economic material flow but also social and natural flows. The economic material flow is the core of the material flow, whereas the social and natural material flows are the basis of the material flow. This theory provides extremely important insights into future enterprise architecture and integration, in considering both sustainable economic growth and societal development. Based upon this, the concept of next-generation ESs has been proposed. The next generation of ESs is called Entire Resource Planning (ERP) or Complete Resource Planning (CRP), as Figure 1.1 shows. In Entire Resource Planning or CRP, not only have the concepts of ERP, ERPII, and ERPIII been integrated, but also the design has been extended to comprehensively encompass the resources used and produced by enterprises in different industrial sectors, all within the context of economic and societal development. In Entire Resource Planning, not only is economic material flow included, but social and natural material flows are included as well. Entire Resource Planning is considered to be a significant step forward in the evolution of ESs. The major factors that have contributed to the birth of Entire Resource Planning are obvious. The past decade has brought fundamental changes to the global economy and to global business operations. The main challenges facing the enterprises are (1) both the globalization of operations and implementation of SCM are moving toward a deeper level and (2) the expectations for sustainable economic growth and global environmental protection are rising. Figure 1.1 shows the evolution of ESs in terms of the three stages of development efforts.

Due to the importance of this subject, there has been a growing demand for research about ESs, in order to provide insights into the issues, challenges, and solutions related to the design, implementation, and management of ESs. In June 2005, at a meeting of the International Federation for Information Processing (IFIP) Technical Committee for Information Systems (TC8) held at Guimarães, Portugal, the international committee members intensively discussed the important role played by ESs in the global economy and the innovative and unique characteristics

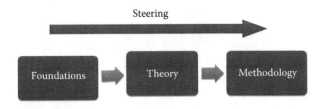

Figure 1.2 Hierarchical structure of science.

of IIIE as a scientific subdiscipline. IIIE is a set of foundational concepts, theories, and applications that facilitate the industrial information integration process. IIIE comprises methods for solving complex information integration problems. At the historical Guimarães meeting, it was decided that the IFIP First International Conference on Research and Practical Issues of Enterprise Information Systems (CONFENIS) 2006 would be held in 2006 in Vienna, Austria. In August 2006, at a continent other than Europe (in fact, at the IFIP 2006 World Computer Congress held in Santiago, Chile), the IFIP TC8 WG8.9 Enterprise Information Systems was established. In 2007, the Enterprise Information Systems Technical Committee was established within the IEEE SMC Society. To further respond to the needs of both academicians and practitioners for communicating and publishing their research outcomes on ESs, a new science and engineering journal entitled *Enterprise Information Systems*, exclusively devoted to the topics of ESs and IIIE, was launched in 2007.

According to Warfield (1990), a science can be structured in the form shown in Figure 1.2. In this hierarchical structure, foundations steer theory, and theory steers methodology. Figure 1.3 further illustrates the relationships between foundations, theory, methodology, and applications in industrial information integration. When IIIE was proposed in 2005 (Xu, 2011) as an interdisciplinary scientific subject, industrial information integration became described as having three integrated components: foundations, theory, and methodology. The products of IIIE are applications such as ESs' applications in business and industry, which provide corrective or supportive feedback to IIIE. Chapter 5 introduces the foundations, theory, and methodology of IIIE. Since IIIE was formally proposed, research on ESs has entered a new era—an era in which more research efforts have been devoted to the foundations, theory, and methodology of industrial information integration, in addition to the applications of ESs in business and industry.

1.3.2 Characteristics of ESs

Some general characteristics of ESs are as follows:

Modular design: ESs are composed of numerous modules such as engineering, manufacturing, financial, and distribution modules, which use centralized databases.

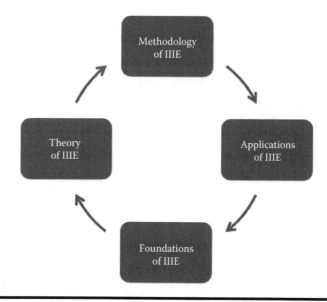

Figure 1.3 The relationships between foundations, theory, methodology, and applications in industrial information integration.

Integration: ESs are intended to seamlessly integrate all of the information flows within an enterprise, including customer information, financial information, human resource information, manufacturing/service information, and supply chain information, in order to increase operational transparency through standard interfaces. There have been numerous vendors who specialize in developing ESs. As software packages, ESs enable the integration of transactions-oriented data and business processes throughout an organization. By providing a single system that is central to the organization, ESs ensure that information can be shared across all of the functional areas within the organizations. According to Langenwalter (2000), ESs such as ERP include related modules as executive direction and support, customer integration, engineering integration, manufacturing integration, and support services integration. Since ESs include related modules, organizations can have choices in adjusting which modules will be included for the integration objective. In 1996, Ford started two vehicle development programs using the new C3PES. In 1999, there were 40 programs; in 2000, there were 60 programs. Organizations can implement ESs via a wide range of options including doing it themselves, doing it through selective external assistance, and outsourcing.

Adaptation and modification for application purposes: ESs can be acquired rather than developed in-house from scratch. The legacy ESs that are developed to integrate the intraorganizational functions within the walls of an enterprise have been expanded to extended enterprise or supply chain enabling ESs. ESs can be classified into a number of categories according to their application, for example, general-purpose systems and industry-oriented systems (Wu et al., 2009). A general-purpose ES such as SAP R/3 is a software package designed for a wide range of enterprises. The benefits of a general ES include integrated data and applications, implementation of a generic business model (BM) based on the best practices, standardized solutions of business problems, and opportunities for customizations. Such systems have been categorized into systems specifically designed for traditional or selected nontraditional industries.

The adopters of an ES, either a general-purpose or an industry-oriented system, can either adjust the organization's ways of working to fit the package or modify the package to fit the organization's way of working. During the implementation process, configuration is often needed to adapt the generic functionality of a package to meet the needs of a particular organization.

Organizational efforts: Enterprises often face challenges in integrating their ES software with existing operating systems, database systems, and networking systems to suit their particular organizational structures and requirements. In many cases, organizations need to do some technical development work in order to effectively interface the system with the enterprise's own proprietary legacy systems, with which the ESs are required to work.

Rapid evolution: ESs are rapidly evolving from MRP to Entire Resource Planning. Many ESs have evolved from original ESs to modern ESs. The functionality of ESs is also evolving over time. Many enterprise software vendors have been releasing new products designed for specific purposes, including SCM or specialized vertical industry solutions. New technologies are also constantly incorporated in order to make ESs more efficient and effective.

Complex systems: ESs are complex systems involved with high complexity and multidisciplinary design characteristics.

The characteristics related to the application and implementation of ESs are as follows:

Business planning: Business management must deal with various kinds of planning, such as production planning, inventory planning, and workforce planning. To develop a quality plan, a tremendous amount of information on demand, production, capacity, lead time, and inventory is needed. An ES has the capability to provide quality information to support business planning. For example, demand patterns can vary dramatically and can be difficult to capture. At the same time, it is very important that demand forecasts reflect customers' true needs. The customer relationship management (CRM) module of an ES can be used to record information about the types of products that customers purchase, to analyze demand patterns, and to project targeted forecasts. ESs add enormous value to customer relations management and, as such, are an essential input to a quality plan as described earlier.

E-Procurement: ESs have become a prevalent e-procurement method. For example, GM, a giant in the auto industry, spends more than $80 billion on purchasing. It uses its channel power to coerce its suppliers to adopt ESs and to move their trading to the Internet (Lucas, 2005). In e-business environments, customers and suppliers are demanding access to some information that are related to e-business. Research suggests that, assisted with advanced ESs technology, successful collaboration among supply chain echelons can improve company's market performance. After conducting a Delphi study with 23 Dutch executives on ESs, Akkermans et al. (2003) point out that ESs could potentially enhance transparency across the supply chain by eliminating information distortions and information delays. Furthermore, ESs can help integrate supply and demand data and can provide real-time information, thus effectively controlling routine production operations (Mandal and Gunasekaran, 2003).

Business process management and business process redesign: One of the primary objectives of an ES is to help integrate an organization's business operations and processes effectively and efficiently. To ensure a successful implementation of ESs, enterprises often need to redesign their business processes to some degree. In particular, the introduction of an ES requires a number of changes and adaptations. An enterprise may need to adapt to the new requirements. It is recognized that computerized data transactions differ considerably from paperwork-based data processing methods. Shifting from manual operations to computer-aided operations is—by no means—merely making electronic duplication of paperwork and the processes handling them. It is likely to be necessary for enterprises to discard some of their traditional business processes that have become obsolete. Enterprises that decide to implement ESs must create and follow new business processes that accommodate general ES requirements. To achieve the desired ES performance, the enterprise may need to incrementally or radically redesign its business processes. ES implementation promotes enterprises to reengineer their business processes (Wang et al., 2005). In a certain sense, ESs implementation indeed provides an impetus to enterprises to update their business processes.

BPR generally starts with identifying and explicitly documenting the existing business processes. Then, the existing redundant and inconsistent activities are analyzed against performance goals and specific ES requirements. Based upon such analysis, new business processes that appropriately align with the ES operations are developed. The business processes need to be hierarchically decomposed into differing levels of detail. It is commonplace that enterprise organizational structures need to be changed to comply with the reengineered business processes (Wang et al., 2005). BPR often fails because of the difficulty and the huge expense of reprogramming enterprises' core transaction processing systems to support the new processes. To support BPR, workflow products can enable enterprises to reengineer and streamline their business processes using workflow recommendation systems. In business and industry, both frequent changes of custom demands and the specialization of the business process require the capacity of modeling business processes for enterprises effectively and efficiently. Traditional methods for improving business process modeling, such as workflow mining and process retrieval, still require much manual work. To address this, based on the structure of a business process, a method called the "workflow recommendation technique" is proposed by Li et al. (2014). It provides process designers with support for automatically constructing the new business process that is under consideration. Furthermore, it helps to develop a recommendation system for improving the modeling efficiency and accuracy. The system is designed to select most probable activities or tasks for recommendation. Their workflow recommendation system can (1) speed up business process design work by reducing the time that is needed when domain knowledge is inadequate or missing; and (2) provide guidance for choosing the most likely tasks by minimizing the errors that could possibly be made in business process design.

Implementation preparations: ES implementation requires a well-designed team structure that organizes and coordinates team members' work and synchronizes various interactive activities (Wang et al., 2005). The steering group, led by top management, consists of key staff members and is responsible for making corporate decisions on ESs. The steering group develops the strategic goals of ES implementation in alignment with the enterprise's business goals. It evaluates and ratifies the ES requirements, application scopes, solutions, and estimated budgets. It makes trade-off decisions on operation stability versus the improvement of business processes. Commitments from the steering group are crucial for motivating the project team and discouraging latent resistance. The steering group periodically holds sessions to review project progress, to discuss emerging problems, and to decide further steps.

The project manager is delegated by the steering group to lead, organize, and coordinate the ESs' implementation process effectively. In addition to a thorough knowledge of the nature of the enterprise's business, the project manager needs to have solid knowledge of ES technologies. To ensure smooth ES implementation, enterprises normally seek advice from external experts. The experts may be invited from academia or from enterprises that have previously implemented ESs. The experts make presentations on the state of the art, best practices, and lessons learned from previous experiences.

They are often asked to help with the evaluation and appraisal of phased and final project outcomes. Subsequently, some task forces are established with members from both the enterprise and the ES vendors, who jointly carry out implementation. The requirement analysis task force identifies and specifies user requirements through interviews or related documents. They analyze and refine requirements in the context of ES functionalities and business scope, which will be the basis of system planning and implementation. The system planning task force is responsible for building the "as-is" and "to-be" models of the business processes using different approaches. The to-be models help determine how business will be conducted and how information will be processed within the ES. System planning activities further specify what organization units and business processes will be affected by ES implementation. The system implementation task force actually installs the software package and customizes interfaces and some modules. The training and support task force is responsible for training users at different levels and for providing support services during system operation.

1.4 ES Examples

An ES offers a set of integrated solutions that cover many of the units and tasks of an enterprise, such as manufacturing, engineering, human resource management, customer service, and SCM. Examples of ESs are ERP systems, product data management and product life cycle management (PDM/PLM) systems, CRM systems, SCM systems, and various advanced collaboration tools that support collaborative work. In this section, some successful examples of ESs and applications in various industries are introduced.

1.4.1 ES Applications in the Manufacturing Industry

ESs have become an increasingly critical part of any organization seeking to play a part in today's business world. During the past 20 years, impressive computing technology developments have changed the way in which enterprises manufacture products and deliver services. Many manufacturing enterprises have begun to recognize ESs' potential to create a broader source of competitive advantage and have capitalized on the increasingly interorganizational span of ESs. This has motivated manufacturing enterprises to invest heavily in ESs.

1.4.1.1 C3P System

In the mid-1990s, the Ford Motor Company found itself struggling to increase its product development efficiency and found that, over many years, different units of the organization had installed design automation software that would increase local productivity, but in relative isolation. As the company looked to take engineering productivity to the next level, which involved the sharing of product information more seamlessly across disparate company functional units, it discovered that these multiple software systems were difficult to link together. Like most of the automotive industry, Ford was faced with the challenge of matching the complexity of transitioning from an analog world of drawings and clay models to a digital world in which product designs exist in the computer only—with no physical presence.

To meet the challenge, a large cross-functional team of Ford engineers and system developers created an enterprise-wide information system known as the C3P system. The acronym combines the three "Cs" of computer-aided design, computer-aided manufacturing, and computer-aided engineering (CAD/CAM/CAE) with yet another acronym: product information management (PIM).

Staley and Warfield (2007) state that the C3P system is a large system that has been carefully designed and highly integrated to serve a very large enterprise extending around the globe. It is intended to be a single integrated system for the design and manufacture of vehicles using the concepts of ESs and IIIE. From a technological perspective, the system is an integration of subsystems through which product information management system (PIMS) data are integrated and can be shared. A single CAD system was chosen, per the strategic design developed using interactive management, the system was linked with CAM and CAE systems through data standards and protocols managed by the PIMS. Considered to be one of the world's largest and most comprehensive PDM implementations, the C3P ES manages data, coordinates processes and workflows, and integrates the widely dispersed groups, divisions, and facilities of Ford and its suppliers in the supply chain, making collaboration in vehicle design and engineering possible. This system was applied to design, engineer, and manufacture automobiles and was further used to provide product information across and beyond the entire enterprise, extending into Ford's supplier and customer base. The architecture of the system can be seen as generic and applicable to a wide range of industries. The system implementation is considered to have been very successful. It is still in use today, largely unchanged in concept and application. The system is now used by the following groups: design, body engineering, power train (engine, transmission, and drive line), vehicle and package engineering, interior styling, interior layout and test development, manufacturing plant layout, work cell design, tool design, and assembly line development. In short, it is used over the entire life cycle of product design, development, and service. Close to 1000 suppliers have been using the system at various development stages. An ES such as C3P enables more efficient reuse of vehicle designs and manufacturing tooling and allows for better relying on an integrated system to integrate subassemblies, in order to improve collaboration throughout Ford and its huge network of suppliers in the supply chain. C3P is a typical example of a manufacturing ES that is being used to support and fully optimize the Ford product development process. It has provided an industrial example of an integrated ES for vehicle manufacturing and has provided global access to related engineering and manufacturing information. In addition, it provided an early example of the ability to facilitate multidisciplinary engineering design through ESs. Figure 1.4 shows the relationships between C3 (CAD, CAM, and CAE) and P (PIM). Figure 1.5 shows how C3P contributes to digital manufacturing.

1.4.1.2 Reconfigurable Manufacturing Systems

As a type of ES in the manufacturing industry, manufacturing systems must be cost-effective and responsive to market changes. Reconfigurable manufacturing systems (RMSs) provide a cost-effective solution to mass customization. As the core of RMS, reconfigurable machine tool (RMT) is constructed from a set of

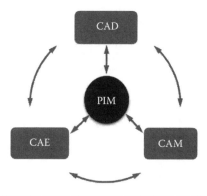

Figure 1.4   Relationships between C3 and P.

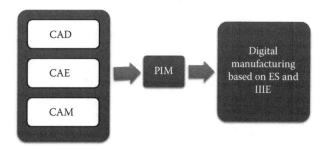

Figure 1.5   How C3P contributes to digital manufacturing.

standard modules and is designed to process a given family of machining features. As a result, RMSs have emerged as an advanced system for applying ESs in manufacturing enterprises to dynamically adapt to the changing market. Although it has been recognized that human-machine interaction plays a significant role in configuration design, especially for RMTs, at the core of RMS, there are no platforms available to enable harmonious interaction between human and computer, mainly because of the limitations of the traditional artificial intelligence frame. Yin et al. (2012) have developed a human–machine design methodology for the configuration of RMT. By using their design methodology, both human decision-making ability based on knowledge and experience and the computer's high-speed logical computation can be exploited to full extent. Furthermore, their methodology can be extended to other layers of the manufacturing execution system, the planning and scheduling layer, and the control layer. Moreover, information processing in other parts of ESs can also employ the methodology. This method narrows down the gap among different parts of an ES to ensure maximum resource utilization.

Yin and Xie (2011) have also developed a reconfigurable manufacturing execution system (RMES) for pipe cutting. The RMES for pipe cutting is mainly responsible for the implementation of pipe production and for the scheduling of cutting tasks, which fills the gap between ESs and the resource layer, provides a consistent information flow for enterprises, and improves the agility of enterprises. This system consists of ERP, a manufacturing execution system (RMES), and computer numerical control (CNC) pipe-cutting machine tools as a reconfigurable pipe-cutting system (RPCS). So far, more than 100 companies have effectively implemented RMES for pipe cutting, which proves that RMES for pipe cutting is a cost-effective solution to aid in mass customization and rapid adaptation to a dynamic market.

1.4.1.3 Aero-Engine Pipe Routing

Due to the complexity of aircraft manufacturing, the development of ESs has been growing in the field of aerospace manufacturing. One ES application is related to aero-engine pipe routing, which is one of the sophisticated design activities in aircraft design. To improve the safety of air transportation, it is very important to design reliable pipe systems for aero-engines. The problem in designing an aero-engine piping system is finding satisfactory connections between the sets of pipe outlets and inlets in a narrow space with the consideration of spatial obstacles. As a requirement, an ES for aero-engine pipe routing design should be able to process and share massive data efficiently and coordinate the tasks and resources at different levels, in order to achieve system-level optimization. However, existing systems have their limitations in collecting and integrating experts' knowledge in decision making. This is particularly true since human interventions are an essential part in the design evaluation and selection.

As a result, it is advantageous to include a human–machine interface, in order to integrate human qualitative analysis capability with the computer's quantitative analysis capability, when developing an ES. An ES should allow design experts to guide the design process at the critical decision-making points in order to accelerate the design process and improve the reliability of the design. Yin et al. (2013) have proposed a methodology for designing aero-engine pipe routing by integrating human visual perception and knowledge, and the computer's superior capabilities of graphic modeling and simulation. The proposed methodology integrates experts' knowledge effectively with both design systems and ESs.

Thus, ESs are being successfully used for aero-engine pipe routing. Such ESs integrate designers' knowledge and experience into decision making and provide an effective collaboration between designers and computing machines. The design methodology and algorithms have been applied and validated in the design of aero-engine pipe systems. This can be further extended to the designs of other complex systems in which human knowledge and reasoning are essential to the success of the design process.

1.4.1.4 Assembly Planning

Assembly planning is an important stage in the product life cycle and is crucial to the success of a product. As industrial products are becoming more sophisticated, more and more enterprises are adopting technology for product assembly. As a result, the importance of digital assembly as a key component of digital manufacturing for assembly purposes is increasing. The adoption of a digital manufacturing approach for the assembly planning of complex products makes it possible to generate, analyze, and evaluate feasible assembly designs in a short time. Digital assembly has the potential to greatly improve the success ratio for onetime assembly, reducing the cost of the product by making an optimal assembly plan. Although many CAD/CAM software tools are available to support design activities in the product life cycle, none of them specializes in assembly planning.

As a result, ESs have emerged as a promising tool for integrating and coordinating digital manufacturing. Xu et al. (2012) have recently developed an integrated assembly planning system named AutoAssem to automate assembly planning. AutoAssem is dedicated to assembly planning for complex products. The novelties of AutoAssem include implementations of some key technologies including the automatic generation and extraction of multiple relational matrices, assembly sequence planning with multiple algorithms, automatic generation of the exploded view and the assembly sequence, and the generation of interactive 3D assembly documentations. AutoAssem has been developed as an integrated CAD system for assembly modeling; assembly planning for sequence, paths, and processes; assembly evaluation; and simulation. In comparison with existing CAD/CAM software tools with their certain level of support for assembly planning, the advantages of the AutoAssem are as follows:

1. For assembly relation matrices, they can be generated automatically and can be updated dynamically to respond to changes in a product model. Values for matrices' elements correspond to the nodes in the tree structure of the product. Matrices' elements are editable.
2. Other software tools are not able to find a solution automatically for the sequence and paths of assembly. On the other hand, AutoAssem has implemented four heuristic algorithms that are able to find and compare the solutions with minimal manual input.
3. For the representation of an assembly sequence, AutoAssem has adopted multiple representations including an assembly sequence table, an assembly tree, and assembly fish bone graphs. Designers are able to use different representations based on their needs.
4. For the assembly technical graphs, exploded views generated by AutoAssem have a compact structure. They can be used to track the assembly processes. Annotations on the technical graphs are automatically aligned based on the given assembly sequence. Directions of annotations indicate the assembly directions of parts. The graphs are concise and easy to understand.

5. For simulation, based on the given configuration of the product, the system can automatically learn positions, orientations, and assembly relation data for the simulation. Additional data can be inserted into the system and can be cited in technical documentations, which, in turn, can be generated automatically from the system.

AutoAssem has provided a comprehensive solution to the assembly planning of complex products—in particular, complex aircraft products. Its uniqueness is its higher level of automation in assembly planning. It offers a theoretical and practical value to ESs and IIIE in the aspect of the assembly of complex products. As a completed ES, it has been successfully employed in the assembly planning of many complex products including aircraft engines, car molds, naval power valves, CNC machine tools, and the general assembly of automobiles.

1.4.2 ES Applications in Healthcare

ESs have been widely used in the area of healthcare. Many medical information systems have been adopted by healthcare organizations. By sharing information through an integrated system, a patient's medical information can "flow through" the entire system, with medical and financial records being automatically updated without additional data entry. The following are some examples.

IHHS: The IoT integrates all kinds of sensing, identification, communication, networking, and information management devices and systems and seamlessly links all the things based on needs, at any time and anywhere, through any device or medium, allowing access to any information or allowing any object to obtain any service more efficiently. The impact of the IoT is significant as use of the Internet has grown astronomically in recent decades. As such, the IoT is recognized as the future generation of the Internet. A part of the enabling technologies include the wireless sensor network (WSN), mobile Internet, cloud computing, radio frequency identification (RFID), machine-to-machine communication, human–machine interaction, middleware, web service, information systems, data mining, and distributed intelligences. Currently, the IoT revolution is sweeping across all sectors, in which the in-home healthcare service based on the IoT is one of the most attractive potential applications. In-home healthcare services based on the IoT are promising to resolve the challenges caused by the aging of the population. Pang et al. (2014) propose an ES solution for the cross-boundary integration of in-home healthcare devices and services. In the proposed ES, key elements of the solution, such as the healthcare process model, device, service integration architecture, and information system integration architecture (ISIA), are fully integrated. In particular, a cooperative Health-IoT ecosystem is formulated, and the information systems of all stakeholders are integrated in a cooperative health cloud, which is even extended to patients' homes through an in-home healthcare station (IHHS). The design principles of the IHHS include the reuse of the 3C platform, certification of the Health Extension, interoperability and extendibility, convenient and trusted software distribution, standardized and secured electrical healthcare record handling, effective service composition, and efficient data fusion. These principles are applied to the design of an IHHS solution called iMedBox. Detailed device and service integration architecture (DSIA) and hardware and software architecture are verified by an implemented prototype. A quantitative performance analysis and field trials have confirmed the feasibility of the proposed design methodology and solution. That means that the future of healthcare services can be changed from career-centric to patient-centric. Moreover, the evolution of the healthcare delivery model has been boosted from the present hospital-centric, through hospital-home-balanced in 2020, to home-centric in 2030, as technology will eventually transform the medical system by bringing technology directly into the home.

The ES based on the Health-IoT technologies and services is promising to address the challenges faced by the healthcare sector. First, the ever-larger aging population requires much larger spatial and temporal coverage in its healthcare service. In response to this requirement, the Health-IoT technologies can provide suitable solutions. Powered by the ubiquitous identification, sensing, and communication capacities, all of the objects in the healthcare service can be tracked and monitored on a 24/7 basis. Enabled by the global connectivity of the IoT, all of the healthcare information can be collected, managed, and utilized. Second, the IoT technologies can enable the transformation of healthcare service from career-centric to patient-centric. Healthcare information and service can be accessed by personal computing devices through mobile Internet access such as Wi-Fi, 3G, LTE, and under the authentication of the individual. That is, the Health-IoT service will be mobile and personalized. Third, the Health-IoT services are feasible in technology and promising in business. In recent years, technical feasibilities have been proven by a number of technical solutions.

In the past decade, ESs have emerged as a promising tool for integrating and extending business processes across the boundaries of business functions at both the intra- and interorganizational levels. From an enterprise point of view, the IoT system will be the fundamental infrastructure of future ESs. According to Pang et al. (2014), recent progress in ESs and IIIE has offered a number of interdisciplinary and powerful approaches for this purpose, for example, business process management, workflow management, and service-oriented architecture. In the context of ESs, the integration of healthcare devices and services includes the integration at all the layers, that is, business logic integration, data integration, communication layer integration, and presentation layer integration. Meanwhile, noticeable progress has been made in ESs for healthcare. The system proposed by Pang et al. (2014) applies the latest theory and techniques of ESs to the architectural design of Health-IoT so that the proposed solution can be easily integrated to existing medical ESs.

In summary, an ES solution methodology is proposed and applied to in-home health solutions (IHHS). The core of the methodology is the alignment among the three key elements: the BM, the DSIA, and the ISIA. In particular, as the primary task of BM design, a cooperative Health-IoT ecosystem can be formulated by the deconstruction and reconstruction of traditional healthcare and mobile Internet value chains. To accomplish this business ecosystem, the information systems of all stakeholders are integrated in a cooperative health cloud and are extended to patients' homes through the IHHS. In order to meet the requirements of BMs and ISIA, the design principles of the IHHS solution are derived including the reuse of the 3C platform, certification of the Health Extension, interoperability and extendibility, convenient and trusted software distribution, standardized and secured electrical healthcare record handling, effective service composition, and efficient data fusion.

Then, a concrete IHHS solution, the intelligent medicine box (iMedBox) system, can be designed based on the proposed design principles. It is an intelligent medicine box, based on a high-performance and open platform–based tablet PC. Figure 1.6 shows the in-home extension and integration of hospital information systems (Pang et al., 2014). Figure 1.7 shows the hardware and software architecture of the proposed IHHS solution (Pang et al., 2014).

Environmental public health system: In the United States, environmental health data have been collected at a national level by government agencies for

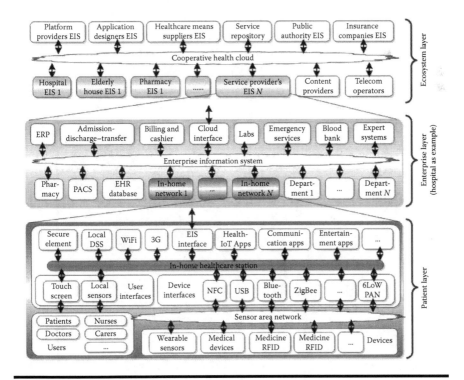

Figure 1.6 In-home extension and integration of hospital information systems.

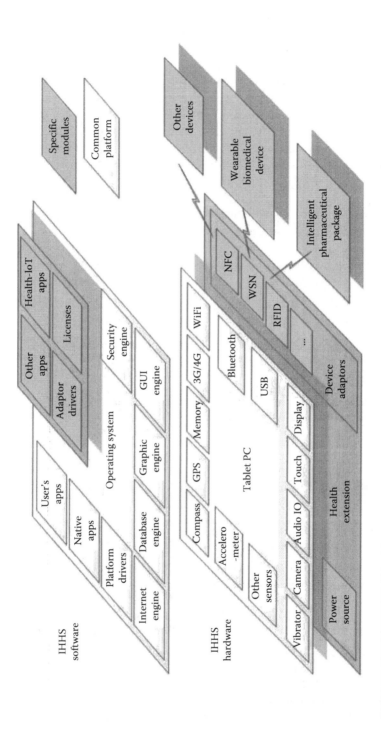

Figure 1.7 Hardware and software architecture of the proposed IHHS solution.

several decades. All state and city governments are required to report communicable diseases such as smallpox and tuberculosis to a federal agency. There is a need to establish a national environmental public health tracking network to link fragmented data or neglected information in order to respond adequately to environmental threats. An ES-based environmental health information infrastructure has been developed (Li et al., 2008b). The developed environmental health system is a web-based platform that integrates databases, geographic information systems, and other systems in order to support public health service and policy making.

The environmental health information system is a robust web-based portal/platform for environmental health tracking in Virginia. This web-based system provides a vast repository of information and can be used for a variety of purposes such as sharing environmentally related data, searching for information, providing healthcare service and support, and facilitating collaborative work among various healthcare providers and stakeholders. The system provides a variety of functions including web-based data entry, secure and automated exchange of data between agencies, data visualization, automated data analysis and decision support, environmental health information dissemination, and environmental health information infrastructure. Data can be sent electronically to a system where it can be analyzed, explored, and mined for environmental healthcare tasks. In addition, the system is able to provide information useful to emerging infectious disease surveillance in order to help prevent bioterrorism.

Integrated medical supply system: Xu et al. (2011) developed an integrated medical supply information system for healthcare clinic medical supply management. The system is a robust web-based portal/platform for managing medical supply. The system provides a variety of functions including web-based data entry, secure and automated exchange of data, data visualization, automated data analysis, automated decision support, and healthcare information infrastructure development. Data can be sent electronically to the system, where it can be analyzed, explored, and mined using decision support techniques for medical inventory management tasks. Thus, the proposed integrated system can serve as the basis for building a more robust and comprehensive information infrastructure for efficient healthcare delivery. The system consists of eight major components: a service demand module, an appointment/scheduling module, a patient record module, a service supply module, a doctor record module, an inventory record module, an external medical supplier module, and a decision support module.

The web-based system provides a vast repository of information and can be used for a variety of purposes such as demand projection, patient appointment scheduling, inventory replenishment schedule, patient service analysis, and collaborative work among various healthcare providers and medical supply vendors. The integrated medical supply inventory control system is a hybrid system that is shaped by the nature of medical supply, usage, and storage capacity limitations of healthcare facilities. The system links demand, service provided at the clinic, healthcare service provider's information, inventory storage data, and decision support tools into an integrated system.

The ABC analysis method, the economic order quantity model, the two-bin method, and the safety stock concept are applied as decision support models to tackle inventory management issues at healthcare facilities. In the decision support module, each medical item and storage location has been scrutinized to determine the best-fit inventory control policy. The pilot case study demonstrates that the integrated medical supply system holds several advantages for inventory managers, since it offers the benefits of deploying ESs to manage medical supply and to better patient services. In summary, the integrated system helps facilitate material management at healthcare clinics and generates values that support public health decision making. Better tracking and monitoring of healthcare medical supplies can have wider benefits beyond the immediate goals of the application involved.

1.4.3 ES Applications in Managing Dams

ESs have been used for managing dams such as China's Three Gorge Dam, which is the largest in the world. To manage the dam effectively, numerous advanced systems have been successfully developed including the process control system, analytics system, project cost management and quality control system, and the comprehensive supervision system for concrete production, transportation, and molding.

As a part of the ES, a particular system is a knowledge based subsystem developed for Three Gorge Dam (Wang et al., 2009). This system adopts a multi-layer architecture and includes a knowledge portal based on web services and the Java 2 Platform, Enterprise Edition (J2EE) while offering a method for integrating existing resources. The portal system is logically divided into four layers: a resource layer, a function layer, a support layer, and a portal layer. The portal site provides a unified entry to knowledge resource platforms for users by utilizing the service from the service layer. The service layer provides a unified presentation and makes access to resources available by extracting the service from the support layer. The support layer provides a connection to the resource of the related layer. The resource layer includes all of the resource information from all of the subsystems.

The knowledge portal system, which is an important component of the decision support system, is the platform for knowledge creation, sharing, and reuse. Through the interoperability in a distributed environment, software integration technology, and heterogeneous system integration, the interoperability of heterogeneous systems is realized by the enterprise application integration (EAI). In this system, an open and unified integration environment is provided by the analytic system, making all kinds of software resources readily accessible. A platform-free EAI environment improves the interoperability of heterogeneous systems and provides an open and unified environment; it is realized by message processing mechanisms, interoperability in distributed environments, software combination technologies, heterogeneous system integrations, and application middleware specifications. In the EAI environment, data, information, and knowledge from all of the branch divisions are integrated effectively.

1.4.4 ES Applications in the Telecommunication Industry

ESs have been used in the telecommunications industry. The telecommunications industry is a very specific high-tech service industry. The main feature of the telecommunications industry is its tight integration of business process and information technology applications. An example ES is the operations support system (OSS), which is becoming increasingly popular in the telecommunications industry. OSS is a mainstream technology that supports large-scale network operation, maintenance, and management. OSS is a common term for the collection of all of the support systems required to run a telecom operator's business. OSS consists of four subsystems: an OSS, a business support system (BSS), a resource support system (RSS), and a system support system (SSS). The functions of OSS consist of activation, inventory management, fault management, and workforce management, among other projects. BSS includes customer care, multiservice provisioning, service assurance, and billing. RSS handles network resource management, operation information management, customer basic information management, and customer service information, among other things. SSS deals with log file, system parameters, etc.

The main functions of OSS include the following:

- Customer care: It provides an interface to customers for all issues related to customer order, sales, billing, and problem handling.
- Multiservice provision: It activates instances of service for particular customers.
- Service assurance: It monitors and upholds the quality of the delivered services.
- Billing: It charges for the service.
- Planning and administration: It plans, designs, and administers the services and infrastructures.
- EAI: It automates the exchange of data between internal applications.
- Activation: It executes a service in an optimal and well-defined order.
- Inventory management: It keeps track of the equipment such as where it is, how it is configured, and its status.
- Fault management: It can handle alarms.
- Workforce management: It manages and schedules teams of technicians, installers, and engineers.

OSS has been increasingly adopted by the telecom industry with New Generation Operations and Software Systems as its next-generation product. As ESs help the manufacturing industry to achieve a competitive edge in the global market, OSS plays a similar role in the telecommunications industry.

OSS has been acquired by many telecommunications enterprises, since OSS is considered to be a basic ES to support value chain management for the telecommunications industry. OSS also provides the enabling tools required to fulfill effective knowledge management. Qi et al. (2006) proposed a knowledge management system for the OSS in the framework of ESs. They proposed that each subsystem of the OSS should be equipped with knowledge management capacity, and the knowledge management of the OSS should be realized through its subsystems. An integrated OSS is a combination of applications that interact with each other to enable support and management of services for the telecom industry. It includes systems that manage the networking infrastructure, planning tools, billing systems, customer care, trouble management tools, and the like. Integrated OSS has become the fundamental integrated platform for the telecom industry.

1.4.5 ESs in Transportation

There are also many ES applications in the transportation industry. For example, Chen et al. (2011) developed a computer-based interlocking system (CIS), which is a safety-critical ES. This CIS has been widely applied in the transportation industry for signaling. In a CIS, the relay logic is always used to describe the interlocking logic in CIS. Normally, all of the interlocking logic has been designed by experienced signaling engineers manually, which can lead to low efficiency and high cost. To address this issue, a novel computer-based approach has been developed, which uses a component-based model to represent the topology of the station layout and uses state charts to describe the interlocking logic. Then, the state charts' description is transformed to the relay logic. The entire procedure of interlocking logic development can be finished automatically, and a software toolkit can be implemented, according to this approach. The introduction of state charts also makes the formal verification of interlocking logic possible; this can guarantee the generated logic correction.

This proposed approach has been used for a railway interlocking system, which is a type of railway control system used in a railway ES to maintain traffic safety—for example, to prevent high-speed trains from colliding. The railway interlocking system is a critical safety system. It must guarantee the safe operations of those signaling apparatus, which means their actions must strictly follow the interlocking logic that is the core of the interlocking software. The interlocking logic is dependent on the station layout and the interlocking rules. The interlocking rules define the sequences of basic operations and general interlocking relationships between different kinds of signaling devices. These rules can be applied to a specific station layout. In summary, ESs are playing an important role in the transportation industry such as high-speed train operations.

1.4.6 ES Applications in Other Areas

There also many ESs applications in other areas. The following are some examples.

Managing reverse logistics: Shi et al. (2012) pointed out that with the coming of a low-carbon society, the reverse logistics of used batteries for lowering the carbon emission is becoming an important research topic. By definition, reverse logistics is the process of planning, implementing, and controlling the efficient and cost-effective flow of raw materials, in-process inventory, finished goods, and related information from the point of consumption to the point of origin to the recapture value or to proper disposal. Information integration of reverse logistics is the key for implementing reverse logistics systems. Currently, only a few enterprises are capable of using ESs to manage reverse logistics. Reverse logistics of used batteries not only protects the natural environment, but also promotes the important economic value of the used batteries. Reverse logistics activity involves wider regions, more partners, and more complicated operations than forward logistics. Therefore, in the entire management process, information sharing, transaction coordination, decision support, and the allocation of resources are inseparable and require support from the ESs. It is obvious that the integration of the information flow of reverse logistics is important, in order to realize the environmental and economic value of the used batteries. By implementing an ES for managing the reverse logistics of used batteries, processing enterprises can share information with other enterprises through an integrated platform. Processing enterprises are the main beneficiaries of information integration because they can adjust the various plans according to the information provided by ESs (Shi et al., 2012).

Plateau ecosystem management: Sustainability is a measure of the capacity of enduringness. Sustainability of ecology describes how biological systems remain diverse and productive over time. Due to various causes such as increasing population, pollution, and the inappropriate use of natural resources, the global ecosystem has deteriorated rapidly. Take as an example the native antelopes on the Qinghai-Tibet Plateau; the population was more than one million at the beginning of the century, but the number had been reduced to less than 75,000 in the mid-1990s due to poaching for their hair to be woven into shahtoosh, or shawls. Now, it is estimated that there are merely 70,000–100,000 antelopes inhabiting the Qinghai-Tibet Plateau, which is less than 1/10th of the population of 100 years ago. Great efforts are in demand to restore and preserve the ecosystem. An ecosystem is extremely complex, since a large number of system components are involved, and the relationships among these components are dynamic. Decision making for a sustainable ecosystem needs the scientific understanding of the system based on reliable real-time data collected from the environment. Therefore, it is essential to develop an ES for data management.

An integrated systems approach has been applied (Liu et al., 2014). A plateau ecosystem management system has been developed, and its system architecture is an extension of the integrated system approach developed in an Agriculture Ecosystem

Enterprise Information System (AEEIS) (Xu et al., 2008). The AEEIS provides a variety of the functions, including secure and automated exchanges of data, data mining, online analytical processing, knowledge management, and information dissemination. It allows the integration of disparate data sources into a single coherent framework. The concept of integrated databases is becoming commonplace as agricultural enterprise evolves to engage in digital agriculture that requires the integration of data of diverse sources. The system can be used to analyze, explore, and mine for scientific study and consultation. Its architecture has been developed based on the concept of IIIE. The integrated system for the plateau ecosystem consists of data acquisition systems, a database, and data mining, analysis, and optimization tools. The data acquisition system is responsible for collecting data from the ecosystem. Instruments for various purposes are required to collect different types of data. For example, remote sensors have been widely used to monitor and locate the objects in an ecosystem. The database is the core of the developed integrated system. All of the acquired data are preserved in the database; a massive amount of data has to be classified accordingly. Data can be classified and maintained with a hierarchical structure. The data are first classified into water/land systems, animal systems, and plant systems at the highest level; it can then be further classified into lower levels. For example, data in the animal system can be classified into the data for insects, retiles, mammals, birds, amphibian, and so on; the data for mammals can be further classified into yaks, sheep, antelopes, cows, pika, and so on. Note that the database for an actual ecosystem can be extremely comprehensive. Take the example of the Qinghai-Tibet Plateau; it has 9600 species of plants, 210 species of mammals in 29 families including 40 native animals, 537 bird species in 57 families, and 115 species of fish.

Analysis and optimization tools are an assembly of all of the required hardware and software tools and the implemented methodologies for data acquisition, filtering, manipulation and mining, and planning activities for ecosystem management. Let us consider, for example, the required tools for a scientific experiment; the instruments and tools are required to clone genes and quantify the level of STAT3 and identify the primary sequences of genes. The integrated platform for this application provides users with access to available data and tools in the system. Since an ecosystem is usually very complex, the integrated system must be capable of providing an appropriate service for various applications. Graphic user interfaces are designed to be tailored applications, so that users can retrieve and manipulate data and can apply the available tools effectively.

Product quality management: The popularity of ES applications in product quality management has attracted considerable attention. Beheshti and Beheshti (2010) have highlighted the way in which ESs help organizations achieve better performance and higher productivity (including in product quality management). Popular quality management methods such as total quality management provide enterprises with vital tools to ensure the expected product quality in production. One of the most important features of product quality is safety. With economic globalization, concern for the safety of consumer products is growing. However, the safety of a consumer product also relies on a number of factors involved in its life cycle, some of which are difficult to control. It is challenging to acquire and integrate the information from all of the sources involved in the supply chain to assess the risk level of a product. Several information systems for product risk management have been reported; however, those systems are not universally applicable since the safety standards as well as other important factors involved in the deployment of products vary from country to country. Wang et al. (2012) propose to use ESs as a solution to address this problem. They have developed an ES component for the quality and safety evaluation of consumer products. In the system, knowledge is acquired from groups of experts to assess product safety. The level of safety is then measured. The implementation of the system allows assessment and comparison of a broad range of risks and makes the process of risk assessment highly efficient.

1.5 Conclusion

Today, keen competition in the marketplace has challenged the survival of enterprises throughout the world. Many enterprises currently face the challenge of unprecedented and abrupt changes, including economic globalization, a saturated market, and ongoing revisions to the information technology infrastructure. To maintain a competitive advantage, many enterprises have recognized the importance of introducing advanced information technologies into their business processes and operations and adopted such advanced information technology to maintain their competitiveness in the marketplace. To facilitate cooperation and the coordination of work within the organization and across the organizations, ESs have been developed to integrate and support all phases of an organization's operations and to help align business strategy, processes, and information technology infrastructures. ESs are highly integrated systems that manage all of the aspects of the business operations of an enterprise, including executive direction and support, customer integration, engineering integration, manufacturing integration, and support integration (Langenwalter, 2000). In the last two decades, many enterprises have adopted ESs to integrate their business processes, in order to achieve both effectiveness and efficiency in their operations. Many ESs have been successfully introduced into industries. In particular, the adoption of ERP systems is one of the major endeavors. The implementation of ESs enables the firm to reduce the transaction costs of business and to improve its productivity, customer satisfaction, and profitability. For example, the CRM subsystem in the ESs is increasingly more popular today, since it allows enterprises to manage customer contacts, customer data, orders, and follow-ups. The system also allows for data analytics and tracking market activity, specific to a product or industry. In recent years, more enterprises have invested in ESs in order to integrate all of their business activities into a uniform system, and many enterprises have used such ESs to support and manage their mission-critical applications. ESs have opened up many opportunities for economic and social development.

The implementation of an ES is a major investment and commitment for any organization; involvement goes far beyond mere information technology applications in the enterprise. An ES involves whole picture-strategic goals, business processes, management, hardware, software, people, etc. When implemented correctly, the organization will have a powerhouse system to keep pace with the economic environment, the fast-changing market, and ongoing technological developments.

Even though ESs were traditionally focused on the manufacturing sector and intraorganizational functions, their expanding functionality continues to take an important place in today's business and industry. The evolution of ESs has spread outward from intra- to interorganizational functions. ESs have found use not only in manufacturing industries but also in service industries. Enterprises that implement ESs now span a wide range of industrial sectors, such as aerospace, shipbuilding, refineries, machinery, electronics, and healthcare. Some examples introduced in Section 1.4 are for individual industrial sectors. With advances in information and communication technologies, ESs offer a new level of service.

ESs have evolved from primarily manufacturing materials planning systems to all-around ESs such as Entire Resource Planning that allows for the planning of all resources. Since the limitations of existing systems may restrain their flexibility to cope with the challenges, and the simple modifications of and/or extensions to existing systems may not meet all of the requirements and/or address all of the challenges, new types of ESs are always needed to help enterprises cope with new challenges. The key characteristic of the new system such as Entire Resource Planning is the provision of comprehensive coverage of all relevant types of resource planning. The new terms "ERP" and "IIIE" have been coined to encompass the expanding definition of what ESs are today and will be tomorrow. Figure 1.1 not only describes the historical development of ESs in terms of three representative development stages but also implies that there are opportunities available for ES manufacturers, since many new technologies are ready to contribute to the new-generation ESs. The new technological innovations will ultimately take the lead in the ES market. The idea behind ESs is that, as an information technology infrastructure, ESs will eventually represent the whole of the supply chain.

References

Akkermans, H. A., Bogerd, P., Yücesan, E., and van Wassenhove, L. N. 2003. The impact of ERP on supply chain management: Exploratory findings from a European Delphi study. *European Journal of Operational Research*, 146(2), 284–301.

Beheshti, H. and Beheshti, C. 2010. Improving productivity and firm performance with enterprise resource planning. *Enterprise Information Systems*, 4(4), 445–472.

Browne, J., Sackett, P. J., and Wortmann, J. C. 1995. Future manufacturing systems-towards the extended enterprise. *Computers in Industry*, 25(3), 235–254.

Camarinha-Matos, L., Rosas, J., Oliveira, A., and Ferrada, F. 2014. Care services ecosystem for ambient assisted living. *Enterprise Information Systems*, DOI: 10.1080/17517575.2013.852693.

Chen, X., He, Y., and Huang, H. 2011. An approach to automatic development of interlocking logic based on Statechart. *Enterprise Information Systems*, 5(3), 273–286.

Davenport, T. 2000. *Mission Critical: Realizing the Promise of Enterprise Systems*. Boston, MA: Harvard Business School Press.

Ho, L. T. and Lin, G. 2004. Critical success factor framework for the implementation of integrated-enterprise systems in the manufacturing environment. *International Journal of Production Research*, 42(17), 3731–3742.

Langenwalter, G. 2000. *Enterprise Resource Planning and Beyond Integrating Your Entire Organization*. Boca Raton, FL: CRC Press.

Li, L., Markowski, E. P., Markowski, C., and Xu, L. 2008a. Assessing the effects of manufacturing infrastructure preparation prior to enterprise information-systems implementation. *International Journal of Production Research*, 46(6), 1645–1665.

Li, L., Xu, L., Jeng, H. A., Naik, D., Allen, T., and Frontini, M. 2008b. Creation of environmental health information system for public health service: A pilot study. *Information Systems Frontiers*, 10(5), 531–542.

Li, Y., Cao, B., Xu, L. et al. 2014. An efficient recommendation method for improving business process modeling. *IEEE Transactions on Industrial Informatics*, 10(1), 502–513.

Liu, F., Bi, Z., Xu, E., Ga, Q., Yang, Q. et al. 2014. An integrated system approach to plateau ecosystem management—A scientific application in Qinghai and Tibet plateau. *Information Systems Frontiers*, DOI 10.1007/s10796-012-9406-5.

Liu, X., Zhang, W. J., Radhakrishnan, R., and Tu, Y. L. 2008. Manufacturing perspective of enterprise application integration: The state of the art review. *International Journal of Production Research*, 46(16), 4567–4596.

Lorincz, P. 2007. Evolution of enterprise systems. *Proceedings of the International Symposium on Logistics and Industrial Informatics*, Wildau, Germany, pp. 75–80.

Lucas, H. 2005. *Information Technology: Strategic Decision Making for Managers*. New York: Wiley.

Mandal, P. and Gunasekaran, A. 2003. Issues in implementing ERP: A case study. *European Journal of Operational Research*, 146, 274–283.

Pang, Z., Zheng, L., Tian, J., Walter, S., Dubrova, E., and Chen, Q. 2014. Design of a terminal solution for integration of in-home health care devices and services towards the Internet-of-Things. *Enterprise Information Systems*, DOI: 10.1080/17517575.2013.776118.

Qi, J., Xu, L., Shu, H., and Li, H. 2006. Knowledge management in OSS-an enterprise information system for the telecommunications industry. *Systems Research and Behavioral Science*, 23(2), 177–190.

Shi, X., Li, L. X., Yang, L., Li, Z., and Choi, J. Y. 2012. Information flow in reverse logistics: An industrial information integration study. *Information Technology and Management*, 13(4), 217–232.

Shtub, A. and Karni, R. 2010. *ERP The Dynamics of Supply Chain and Process Management*. New York: Springer.

Staley, S. M. and Warfield, J. N. 2007. Enterprise integration of product development data: Systems science in action. *Enterprise Information Systems*, 1(3), 269–285.

Wang, C., Xu, L., Liu, X., and Qin, X. 2005. ERP research, development and implementation in China: An overview. *International Journal of Production Research*, 43(18), 3915–3932.

Wang, L., Shi, H. B., Yu, S., Li, H., Liu, L., Bi, Z., and Fu, L. 2012. An application of enterprise systems in quality management of products. *Information Technology and Management*, 13(4), 389–402.

Wang, L., Xu, L., Wang, X., You, W. J., and Tan, W. 2009. Knowledge portal construction and resources integration for a large scale hydropower dam. *Systems Research and Behavioral Science*, 26(3), 357–366.

Warfield, J. N. 1990. *A Science of Generic Design Managing Complexity Through Systems Design*. Salinas, CA: Intersystems Publications.

Wu, S., Xu, L., and He, W. 2009. Industry-oriented enterprise resource planning. *Enterprise Information Systems*, 3(4), 409–424.

Xu, E., Wermus, M., and Bauman, D. 2011. Development of an integrated medical supply information system. *Enterprise Information Systems*, 5(3), 385–399.

Xu, L. 2011. Enterprise systems: State-of-the-art and future trends. *IEEE Transactions on Industrial Informatics*, 7(4), 630–640.

Xu, L., Liang, N., and Gao, Q. 2008. An integrated approach for agricultural ecosystem management. *IEEE Transactions on SMC Part C*, 38(4), 590–599.

Xu, L., Wang, C., Bi, Z., and Yu, J. 2012. AutoAssem: An automated assembly planning system for complex products. *IEEE Transactions on Industrial Informatics*, 8(3), 669–678.

Xu, L., Wang, C., Bi, Z., and Yu, J. 2013. Object-oriented templates for automated assembly planning of complex products. *IEEE Transactions on Automation Science and Engineering*, DOI: 10.1109/TASE.2012.2232652.

Xu, L., Wang, C., Luo, X., and Shi, Z. 2006. Integrating knowledge management and ERP in enterprise information systems. *Systems Research and Behavioral Science*, 23(2), 147–156.

Xu, S. 2008. The concept and theory of material flow. *Information Systems Frontiers*, 10(5), 601–609.

Yin, Y. and Xie, J. 2011. Reconfigurable manufacturing execution system for pipe cutting. *Enterprise Information Systems*, 5(3), 287–299.

Yin, Y., Xie, J. Y., Xu, L., and Chen, H. 2012. Imaginal thinking-based human-machine design methodology for the configuration of reconfigurable machine tools. *IEEE Transactions on Industrial Informatics*, 8(3), 659–668.

Yin, Y., Xu, L., Bi, Z., Chen, H., and Zhou, C. 2013. A novel human-machine collaborative interface for aero-engine pipe routing. *IEEE Transactions on Industrial Informatics*, 9(4), 2187–2199.

Zhang, H. and Liang, Y. 2006. A knowledge warehouse system for enterprise resource planning systems. *Systems Research and Behavioral Science*, 23(2), 169–176.

Zhang, H. and Zhao, G. 1999. CMEOC-an expert system in the coal mining industry. *Expert Systems with Applications*, 16(1), 73–77.

Chapter 2

Enterprise Integration

2.1 Enterprise Integration

An enterprise can be a single enterprise, an extended enterprise, or a part of a larger enterprise, as it has been shown that the scope of enterprises has grown dramatically. An enterprise has the mission, goals, and objectives that allow it to provide an output such as a product or a service. This broad definition applies also to virtual enterprises. Enterprise architecture (EA) is an important tool for understanding and designing an enterprise, since an EA defines the scope of the enterprise, the structure of the enterprise, and the relationship between the enterprise and its environment. Thus, the concept of EA can help enterprises complete missions and achieve their goals and objectives to meet strategic and organizational challenges.

The scope of the structure of an enterprise (which EA can describe) includes those main enterprise components such as enterprise goals and objectives, organizational structures, business processes, and information infrastructures. As the information infrastructure is within the scope of EA, the term "enterprise" as used in EA generally relates to the information architecture in an industrial organization. EA is highly relevant to industrial information integration, as industrial information integration concerns itself with the information flow within the entire industrial organization.

Enterprise architects use a variety of conceptual models and analytical tools to describe the structure and dynamics of an enterprise. "Enterprise artifacts" are used to describe the logical organization of business processes and business functions, as well as information flow and architecture. The collection of these artifacts is considered to be an "EA." The landscape of EA can be divided into various domains that allow enterprise architects to describe an enterprise from a number of important perspectives. One of the main domains in EA is the information domain. The important components in this domain include information architecture. Database architecture, software architecture, and network architecture are components of information architecture. The other two domains with components that are also highly relevant to industrial information integration are the application domain and its component "interfaces between applications" and the technology domain with its components such as middleware, networking, and operating systems.

Architecture can be a description of the basic connectivity of parts of a system either at a conceptual or at a physical level. The word "architecture" conveys various meanings depending on its contextual usage: (1) a formal description of a system at the system and subsystem levels and (2) the structure of the system and subsystems, their interrelationships, the principles of evolution over time, and the guidelines for their design. During the 1980s, active research was conducted to develop EA frameworks. Among them, the most well known are Computer Integrated Manufacturing Open System Architecture (CIMOSA), which coined the term "enterprise architecture," Purdue Enterprise Reference Architecture (PERA), GIM architecture, and ARIS. The Zachman Framework is an example of such an initiative. It structures various enterprise modeling and engineering concepts according to the perspectives of various stakeholders involved in the enterprise engineering. This is because different stakeholders use different levels of abstraction to view an enterprise, and consequently, they expect different deliverables. Comparing these architectures, some researchers considered that CIMOSA and ARIS present strong similarity and are both process-oriented approaches that aim to integrate functions through modeling and monitoring. GIM is based on the GRAI model in which integration is seen as the coherence between global and local objectives, and the PERA and Zachman architectures mainly contribute to complex architecture frameworks. Although these architectures are heterogeneous, in general, they are complementary to each other. As such, complementarities can be identified and analyzed for formal modeling and for purposes of integration. Related work was done by the IFAC-IFIP Task Force and other researchers during the mid-1990s. In addition to the frameworks, during the last decade, active research has also been conducted to develop EA methods. EA methods are based on assumptions, the underlying principles within formal frameworks, and models. EA is considered by some researchers in both academia and industry as an established discipline.

Representing the architecture of an enterprise correctly and logically improves the performance of an industrial organization. This includes innovations in restructuring an organization, reengineering business processes, and ensuring quality and the timeliness of information flows that represent material flows (MFs). It also ensures that efforts generated toward industrial information systems are effective.

Based on the EA with the formal frameworks and methods, the concept of enterprise integration (EI) has emerged. EI is a research subject emerged in 1990s as an extension of computer integrated manufacturing (CIM). EI has now become a key issue for many enterprises and is considered to be a major challenge of these times. The idea of EI is to improve business processes through integrating and streamlining the processes both intra- and interorganizationally. EI is closely related to EA. It is the process of ensuring that the required interactions between enterprise entities achieve the appropriate enterprise goals and objectives. It consists of methods and techniques that aim to consolidate, coordinate, and integrate applications. EI can be approached in various ways and at various levels, for example, (1) physical integration (interconnection of devices), (2) application integration (integration of database systems and software applications in heterogeneous computing environments), and (3) business integration (coordination that control and manage business processes). Methods are available for integration for achieving objectives both intra- and interorganizationally. Methods are also available for integration through enterprise modeling. A suggestion for this is to use a consistent modeling framework.

In the earlier paragraph, we mention application integration under EI. Application integration that can help an enterprise achieve integration is referred to as enterprise application integration (EAI). In a broad sense, application integration mainly refers to the interoperability of applications on heterogeneous platforms. Application integration activities initially emerged in the mid-1980s. Originally, EAI was focused only on integrating ES with intraorganizational applications, but now it has been expanded to interorganizational integration. EAI facilitates the integration of both intra- and interorganizational systems. Ranging from electronic data interchange (EDI) to web services and XML-based process integration, EAI-enabling technologies provide flexible, adaptable, and scalable EAI frameworks. Solutions comprise the conversion of various data representations among the systems involved, the integration of diverse data and business processes across the enterprises, and the integration and interoperation of intra- and interorganizational enterprise applications, including the connection of proprietary/legacy data sources, processes, workflows, applications, and ESs interorganizationally.

The objectives of EAI are to facilitate information exchange among business enterprises in a timely, accurate, and consistent manner and to support business operations with the objective of seamless integration. EAI entails integrating enterprise data sources and applications so that business data and processes can be easily shared. The integration of enterprise applications includes the integration of data, business process, applications, and platforms, as well as the setting and maintaining standards for integration. In general, those enterprise applications that were not designed as interoperable can be integrated on an intraorganizational and/or interorganizational basis. Typically, an enterprise has existing legacy systems that are expected to continue in service while a new set of applications is migrated to or added. Thus, legacy and newer systems can be integrated to provide greater competitive advantages, meeting constantly changing business requirements and the need for adaptation to the rapid changes in the supply chain. This kind of adaptation requires that EAs provide newer functionality all the time. The integration of data and applications is expected to be accomplished without significant changes to existing data and applications. As such, EAI must be able to integrate heterogeneous applications that were created with different methods and on different platforms. By creating an integrative structure, EAI connects the heterogeneous data sources, applications, and systems intra- or interorganizationally. EAI provides a flexible and convenient integration mechanism. A complete EAI offers functions including better business process integration and information integration. Through coordinating the business processes of multiple enterprise applications and by combining software, hardware, and standards together, EAI provides the integration of intraorganizational systems and is moving toward integrating ESs both intra- and interorganizationally. As such, with EAI, intra- or interenterprise application systems can be expected to be integrated seamlessly, since EAI ensures that different divisions or even different enterprises can cooperate with each other, even with different systems.

2.2 Manufacturing Integration

2.2.1 Brief Description of Manufacturing Enterprises

A simplified model of a manufacturing enterprise is shown in Figure 2.1. A manufacturing enterprise is generally driven by customer requirements. Manufacturing covers basic activities such as designing, manufacturing, and assembling, including determining product components and assembling products. Manufacturing activities make parts from raw materials, and the assembly activities assemble parts to the subcomponents, to the components, and finally to the complete product. Hardware and control resources are required in manufacturing processes. A "manufacturing execution system" (MES) consists of required hardware and control resources related to the design, manufacture, and assembling activities. "Hardware resources" refer to the resources involved with the MF, and "control resources" refer to the resources involved with the information flow.

2.2.2 Challenges Facing Manufacturing Enterprises

The associations between a manufacturing enterprise and its business environment are shown in Figure 2.2. As mentioned earlier, manufacturing enterprises are driven by customer needs, which, in turn, are determined either by the market in the push mode or by customer orders in the pull mode. Within the life cycle of a product, a manufacturing system experiences a series of manufacturing activities including product design, manufacturing, and assembly. Interacting with the business environment, logistical activities are required to acquire raw materials and to deliver the final product to customers. A manufacturing activity gains its value

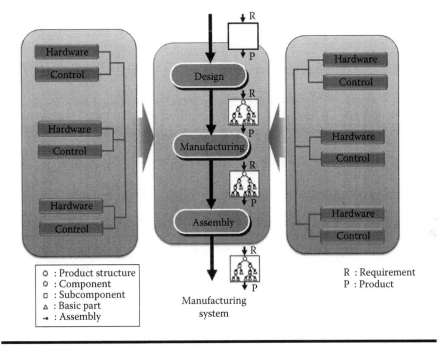

Figure 2.1 A simplified model of a manufacturing system.

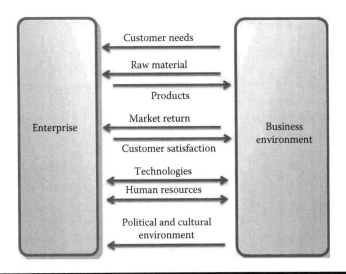

Figure 2.2 Association of a manufacturing enterprise with its business environment.

as the customers acquire the products. Meanwhile, manufacturing activities are constrained by manufacturing technologies, by rules and regulations, and by other resources (see Figure 2.2). In general, the business environment determines the operation domain of a manufacturing system; therefore, the changes in a business environment play a key role in the changes of activities of manufacturing systems. The challenges of the business environment have always been significant, and they are summarized as follows (Bi, 2002).

Short lead time: Due to the rapid development of technology and economic globalization, the market for manufactured goods has become more and more saturated with respect to the demand of the global market. Timing has become one of the most important factors in determining the performance of a manufacturing system. A short lead time is critical to enterprises at least in three aspects: (1) If a product can be introduced earlier, it is likely to be more competitive, as the time lag in matching or surpassing it is likely to be longer; (2) earlier product introduction can increase the duration of peak sales, since the earlier that a product is introduced, the brighter are its prospects for obtaining and retaining a larger share of the market (in particular for software and certain types of industrial machinery); and (3) profit margins, since an enterprise that is introducing a new product will have more pricing space to make a higher profit margin possible.

More variants: Products are becoming more and more versatile and customized. As a result, manufacturing enterprises are forced to produce more variants to meet customized and fragmented, but increasingly sophisticated requirements. Manufacturing enterprises are aware that customers are willing to accept products that cater to specific needs. A versatile product has many features, and those features have multiple characteristics, and different characteristics correspond to different materials or parts. Therefore, the number of product variants increases along with the increase in the number of product features. Customized products must be manufactured to satisfy individual customers; the more specifically individual customers are targeted, the more product variants must be provided. In general, an increase in the customer population has an impact on the increase in requested product variants, due to demands for customization.

Versatility also increases the complexity of the product: (1) if a product is versatile, each functional feature needs a component to perform the function; therefore, the increasing features of a product imply that more components must be considered for the product; (2) a product may embrace the components of mechanism, electronics, and control software, and the more components are involved, the more intrigue the interfaces of the product is.

Low and fluctuating volume: The volume of specific products can be low for several reasons: (1) Limited market niches are shared by numerous manufacturing firms; (2) the life cycles of new products may tend to be shorter while the durability of these products tend to be longer; therefore, products of different generations overlap with each other on the market; and (3) product customization has divided customers' demand into smaller partitions because of the customized features. Meanwhile, the uncertainty and turbulence of the global market, as well as the extension of the product service (such as the maintenance and warranty) can make the product volumes more fluctuated and unpredictable.

Pricing competition: Product pricing is the primary concern to most customers. On one hand, fierce competition among manufacturing firms provides more room for customers looking for a lower price for a certain product but demanding the same quality and delivery time. On the other hand, price is heavily time dependent, price can change abruptly along the product life cycle, and the price margin can reach its limit very quickly after a product is introduced to the market.

These four aspects (there can be many other aspects involved) have a significant impact on the shift of manufacturing paradigms and enabling technologies. As a result, today's manufacturing enterprises are characterized by economic globalization, advanced automation, virtual engineering, knowledge management (KM), and distributed computing. In such a competitive and highly dynamic environment, manufacturing enterprises need information integration to succeed.

2.2.3 Modularity and Integration to Meet Challenges

Quick changes in customers' requirements are the theme of today's manufacturing environment. Many companies face the challenges of providing as much variety in products as possible for the market, while keeping as little variety as possible among the products in order to maintain economies of scale. The topics of time-to-market and product variety have been discussed both in industry and in academia. The partial causes of the problem are the explosion of product variety and the globalization of the marketplace. The key challenge for most manufacturing enterprises is to improve flexibility, adaptability, and responsiveness in order to meet the quick changes in the dynamic environment.

The performance of manufacturing enterprises can be achieved in two general ways: (1) by preparing organizations and/or their systems to be capable of managing dynamic parameters. In manufacturing industries, this implies that they adopt advanced manufacturing technologies such as flexible manufacturing systems. The fundamental idea behind these technologies is that flexibility can be made available through parameter changes in planning, scheduling, and controlling aspects of these systems; (2) by modularizing system components and by providing the varieties needed by employing different modules and/or different connections among them. Implementing external dynamic parameters implies a need for modular architecture in a system at the hardware level. It also implies that the system topology should be able to be changed by adding or removing modules. As such, the flexibility, adaptability, and responsiveness of manufacturing enterprises can be enhanced significantly in order to deal with the complexity and dynamics that manufacturing

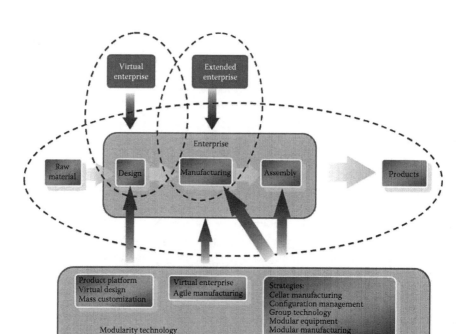

Figure 2.3 Applying modularity technologies to achieve flexibility, adaptability, and responsiveness.

enterprises are facing. As shown in Figure 2.3, quite a few techniques in manufacturing (including CIM, agile manufacturing, and virtual manufacturing) have adopted modularity technology and/or object-oriented methods, that is, an enterprise consisting of encapsulated objects with the defined functions.

From a systems point of view, in any manufacturing enterprise, entities can interact with each other. They can be coordinated and integrated as well, and then they can collaborate "as a whole" to achieve the system-level objective and performance. Based on this systems perspective, it has been suggested that modular components be integrated in manufacturing operations. As shown in Figure 2.4, the integration within manufacturing enterprise occurs at different levels based upon the scope of its business activities. The integration of manufacturing activities is called "manufacturing integration." It is related to the activities in the engineering design. In engineering design, the main integration activities are called "engineering integration." As customers play a critical role in the operation of an enterprise, "customer integration" takes care of the activities involved in customer relations. Finally, as an integration over the scope of an enterprise's or an extended enterprise's entire business process, the concept of "EI" is applied. Due to its high-level complexity, dynamics, constantly emerging new technologies, and global supply

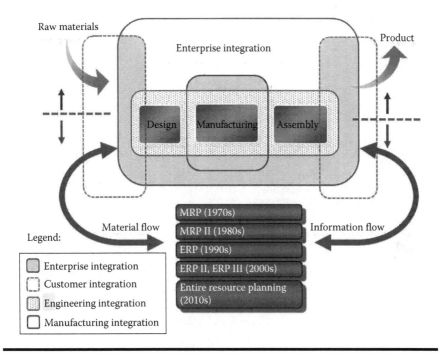

Figure 2.4   An example of EI—the relationship between manufacturing integration, engineering integration, customer integration, and enterprise integration.

chain, the boundaries of a manufacturing enterprise have become more and more extended (particularly during the most recent decade). As such, the level of complexity related to manufacturing integration (as shown in Figure 2.4) will dramatically increase. Section 2.1 introduced the concepts of EA, EI, and EAI, and Figure 2.4 provides a typical example of EI that includes complex interactions involved in manufacturing integration, engineering integration, and customer integration within an enterprise.

As mentioned earlier, as manufacturing enterprises face great challenges from constant market competition and changes, the ability to respond in a timely manner to rapid changes becomes one of the most significant competitive advantages. As such, the role played by information integration in manufacturing (IiM) systems has been highly emphasized.

2.2.4 Manufacturing Integration

In today's manufacturing atmosphere, ESs related to the manufacturing need to be constantly and smoothly reengineered in order to allow them to respond to the fluctuating market and to technological evolution.

Several computer-aided tools are available for designing and implementing manufacturing systems. These tools operate at different levels of abstractions for different stages of the life cycle (such as capacity planning, job shop scheduling, production design, and facility layout) and provide with different knowledge-level support from a variety of disciplines. Manufacturing integration has been the strength of traditional ESs such as ERP systems; however, complete integration across the entire manufacturing process is relatively new, as it was virtually nonexistent in early 2000. Manufacturing integration may include (1) integration covering materials and capacity planning, master production scheduling, rough cut planning, material requirement planning, capacity requirements planning, and inventory systems; (2) integration covering MESs, further covering planning systems interacting with MRP, CRP, cost accounting, inventory control, product data management, work order management, process control, workstation management, inventory tracking and management, material movement management, data collection, and exception management; (3) integration covering JIT; (4) integration covering advanced planning and scheduling (APS); (5) integration covering supplier integration; and (6) integration covering quality management systems and maintenances (Langerwalter, 2000). The main objective of manufacturing integration is to offer flexibility and adaptability and to meet productivity and quality requirements.

Information and communication technologies (ICT) play a key role in manufacturing integration. The following technological areas related to ICT have been identified as important in manufacturing integration (Panetto and Molina, 2008):

- Integrated equipment, processes, and systems that are adaptable and can be readily reconfigured.
- System-level functioning enabling through methods including systems synthesis based upon the modeling and simulation of manufacturing operations.
- Advanced human–machine interfaces.
- Intelligent systems and collaboration.
- Product- and process-design methods that address a broad range of product requirements.
- Innovative processes for designing and manufacturing.
- Manufacturing processes that can reduce energy consumption and are moving toward green manufacturing.
- KM. Recent advances in ICT allow manufacturing enterprises to move from high data-driven environments to a more cooperative information/knowledge-driven environment.

In the manufacturing integration process, significant considerations include supporting EI and interoperability, supporting distributed computing, supporting heterogeneous environments, supporting cooperation, accommodating dynamic structures, and promoting agility, scalability, and fault tolerance. A number of important enablers are needed to support the creation of successful collaborative networks, for example, common reference models, effective interoperability approaches and mechanisms, supporting infrastructures based on open architectures, design and engineering methodologies to instantiate/duplicate already successful cases, and standardized technologies and tools (Panetto and Molina, 2008). Active research has been conducted to develop new manufacturing integration methods, and many examples are available that extend the capacity of existing ESs.

As an integrated component in manufacturing integration, MES can provide real-time information integration support. For management, MES provides tactical level support; for operations, MES provides operations support by building a bridge between planning systems in ESs and manufacturing floor control supervisory control and data acquisition. MES also links the manufacturing information systems' strategic planning and direct execution layers by relying on online information. MES has been valued in ESs. In some CIM frameworks, the MES application development/integration cost and the cycle time can be reduced through the adoption of an integrated system architecture along with faster equipment and factory integration. Efforts have also been made based on systems perspectives. IiM is a systems effort to integrate humans and machines into a "whole" system, not only at the field level, but also at the management and corporate levels, in order to create an integrated and interoperable ES in which MES plays a significant role (Panetto and Molina, 2008). In MES, a reconfigurable MES for pipe cutting has been proposed (Yin and Xie, 2011). In the offshore oil industry, pipes are used to intersect with a variety of components such as other pipes, rectangular hollow section tubes, and frustums. Currently, pipe-cutting machine tools are used with open architecture computer numerical control (CNC). Although CNC pipe-cutting machine tools provide efficiency in pipe cutting, they are insufficient in their response and adaptability to the dynamic market and the current customer-driven manufacturing pattern. A partial reason for this is the lack of reconfigurable manufacturing execution system (RMES) to fill the gap between the existing ESs and pipe-cutting machine tools. In the RMES for pipe cutting, the concept of an integrated system called reconfigurable pipe-cutting system (RPCS) was defined as a system consisting of ESs, RMES, and CNC pipe-cutting machine tools (see Figure 2.5). This RMES for pipe cutting has been successfully implemented in hundreds of enterprises.

In automobile manufacturing, autobody assembly consists of more than 30% of the total manufacturing cost, and the assembly information of autobody plays an important role in developing competitive advantages. Much research has been conducted in order to facilitate the integration of autobody assembly system with ESs. These include assembly sequence planning, determining the precedence constraints among the components of the final product, graphical methods for representing the assembly sequences of a product, functional precedence relation, locating graphs, virtual link mating graphs, the predicate calculus method, and

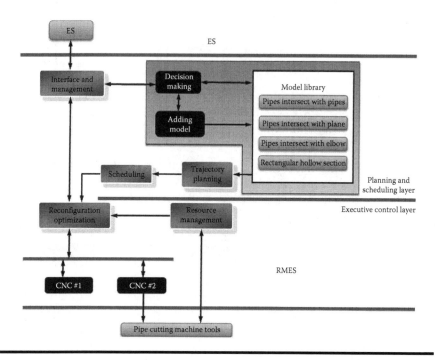

Figure 2.5 An example of integrated subsystem in ES involving ES, RMES, and CNC.

others. A polychromatic sets-based assembly information model for analyzing constraints between assembly relations has been proposed. In this model, an assembly algorithm is proposed, the polychromatic sets hierarchy structure graph can be built, and then, the feasible assembly sequences can be obtained.

The C3P system built by a large crossfunctional team of Ford engineers and system developers in the mid-1990s is a successful example of both manufacturing integration and the EI (Staley and Warfield, 2007). It covers the integrations included in Figure 2.4. The term "C3P system" refers to a CAD/CAE/CAM/PIMS system, in which PIMS stands for product information management system. This is an ES that can be applied to the design, the engineering, and the manufacturing aspects of automobiles. It is also a typical example of an ES for extended enterprises, as this system provides support across and beyond the entire enterprise, extending into SCM. For most industrial organizations, EI requires more integration than the integration of a single aspect. The key design idea of a C3P system is to make CAD, CAM, and CAE come together. Stemming from systems science principles, the implementation of the Ford C3P system began in 1996. Since then, there has been no change in the design strategy; only the technology has been changed (and it will continue to be upgraded along with the new innovations in ESs technology). Efforts to conceptualize the C3P system include three major issues and the focusing of six components (SC-1 to SC-6) toward the resolution of these three issues. The three issues are the product information issue, the technology issue, and the work practice issue. The six components are (1) SC-1 EI, (2) SC-2 process reengineering, (3) SC-3 methodology, (4) SC-4 product information management, (5) SC-5 software purchase, and (6) SC-6 technology awareness.

According to research, EI (including manufacturing integration) will likely be successful when systems approaches are applied, as systems approaches are good at managing multiple factors and complexities. According to Warfield, systems science broadly consists of two sequential components: discovery and resolution (Warfield, 2007). These two important components were reflected in the design of C3P system. The sequence carried out in the discovery portion of the system design of the C3P system include activities such as (1) generating a problem set; (2) filtering the problem set; (3) structuring the problematique; (4) computing the complexity metrics with data from the problematique; (5) placing problems in dimensions using Interpretative Structure Modeling (ISM); (6) delivering the results to the resolution group; (7) generation of options; (8) determining the interdependency of option categories; (9) forming options profiles independently; (10) reporting, comparing, and choosing an alternative; (11) arraying the options in dimensions; (12) reporting to the task sequence group; (13) generating task start sequences; (14) generating task end sequences; (15) constructing DELTA charts for tasks; and (16) preparing a work breakdown.

C3P is a successful example of the blending of manufacturing integration, engineering integration, and customer integration; overall, it is a successful example of EI. Its high-quality conceptual design based on systems thinking and methodology enables it to handle issues such as complexity in intra- and interorganizational EI.

2.3 Engineering Integration

In general, about 90% of a product's cost is determined during its design cycle; its quality characteristics are also determined during the product design stage. In a typical product development process (such as in plastic injection mold design), the design information flow may not be well supported by the existing systems. If associative relations among engineering features were not available through the system, data consistency and design changes would be difficult to manage. At different stages of a product's life cycle, from its requirement specifications to its conceptual design to its more detailed structure design and finally to its production, engineering knowledge must be integrated. A complete integration includes the design process, product data management, integration with customers, integration with suppliers, integration with the rest of the organization, and project management. The ways in which the engineering division integrates with the rest of divisions in an enterprise have been intensively researched.

According to Kulvatunyou and Wysk (2000), integration can be classified into three types:

1. Data-oriented integration, which integrates CAD, CAPP, CAM, and CIM
2. Structure-oriented integration, which is an implementation of team-oriented concepts, such as the use of a simultaneous engineering team, a concurrent engineering team, and an integrated product and process development team
3. Procedure-oriented integration, which refers to concurrent engineering-enabling technologies include QFD, the Taguchi method, axiomatic design

Traditionally, MRP II systems interface with engineering design systems to receive BOM and routing information. However, the interface is not always advanced, as it is unable to pass critical information back to the engineering design system. In concurrent engineering, all of the engineering processes should be supported by integrated computer-aided tools and should be based on a consistent set of data with different application views. Such applications include conceptual design, structural design, detailed design, design analysis for certain specific engineering aspects, computer-aided process planning (CAPP), and CAM tool path generation, etc. However, this desirable scenario has not been fully realized due to the interoperability limitations of different software packages.

A concurrent design process consists of many design activities that are interrelated with each other. Figure 2.6 introduces four types of concurrent design activity interactions (Gao et al., 2008). In sequential-type activities, a successive restriction exists in design activities, and the output of design activity is the input to the subsequent design activity (Figure 2.6a). In concurrent-type activities, two or more design activities can be executed synchronously. They have no information exchange and interaction on each other (Figure 2.6b). In the coupled-type and

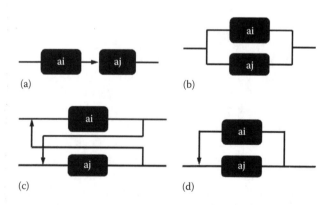

Figure 2.6 Types of concurrent design activities: (a) sequential type activities, (b) concurrent type activities, (c) coupled type activities, and (d) iterative type activities.

iterative-type activities, an information exchange occurs between two design activities (Figure 2.6c and d). Concurrent design has become increasingly important in designing complex products. When it is implemented in manufacturing ESs along with engineering integration, it is likely to generate better design. Numerous concurrent design techniques have been developed, such as Project Evaluation and Review Technique (PERT), ISM, Design Structure Matrix (DSM), Petri nets, and polychromatic sets. Each of these methods has some weaknesses. For example, PERT is useful for the design processes in which activities have a clear sequential relationship. It is inflexible and therefore unable to include feedback information and the iterative characteristics of the concurrent design. Using the adjacent matrix, ISM and DSM can apply partitioning algorithms and other algorithms in the concurrent design process. Although the Petri net is suitable for modeling concurrent processes, it does not have sufficient capacity to represent data flow or handle computational complexity. UML is a graphical and visual modeling language. Integrating UML with polychromatic sets provides a powerful tool for modeling and analyzing concurrent design processes. UML has been applied in concurrent design such that a UML model of concurrent design process has been developed and mapped into a polychromatic sets contour matrix model. Using this novel modeling and analysis method for a concurrent design process based on UML and polychromatic sets, the concurrent design process can be modeled formally and analyzed quantitatively, and the major factors that affect the concurrent design process can be considered.

In the CAD/CAM field, the comprehensive design of dimensional and geometric tolerances for mechanical products using computers is called Computer-Aided Tolerancing (CAT). This is a focal point of research in CAD/CAM. In the process of product design and manufacturing, the tolerance values of a mechanical part are closely related to its manufacturing process, which not only influences the quality of product but also affects the manufacturing cost. So far, considerable research has been conducted on CAT analysis and synthesis, tolerance information modeling and representation, concurrent tolerance design, dynamic tolerance control, and tolerance information verification. The research covers (1) the concept for determining the geometric shape and the dimensional and geometric tolerance of a part using a computer. Based on this, designed dimensions and tolerances of the part with a geometric shape can be described using mathematical formulae; (2) the method to control the tolerance of design and manufacturing using computerized dimension chain; (3) the theory of tolerance that defines the concept of tolerance according to the offset values of the real entity of a part and provides a theoretical basis for its CAT design; (4) the concept of virtual boundary requirements that describe tolerance and conditional tolerance; (5) topologically and technologically related surfaces (TTRS) theory, which establishes the important theoretical foundation for dimensional tolerance and geometric modeling in the CAD system; (6) the theory based on wavelet and fractal technology with application in designing the tolerance.

With the continuous development of CAT technology, a number of tolerance models have been proposed, such as attribute models, offset models, parametric models, kinematic models, and DOF models. In attribute models, a tolerance can be directly stored as an attribute of either geometric entities or metric relations. Offset models can obtain the maximal and minimal object volumes by offsetting the object by corresponding amounts on either side of the nominal boundary. However, they cannot distinguish the interactions of different tolerance types. Parametric models represent tolerances as ± variations of dimensional or shape parameters. In current CAD systems, the modeling method for parametric models has been widely applied. Kinematic models use vector additions to analyze tolerances. A kinematic link is used between a tolerance zone and its datum features. TTRS models have many similarities to DOF models.

With the development of three-dimensional (3D) CAD system, it has become urgent to construct a 3D dimension chain and to comprehensively design dimensional and geometric tolerances. Researchers have put forward the variational geometric constraint (VGC) theory, which can effectively handle the comprehensive design of dimensional and geometric tolerance, although with some difficulties in computation.

In CAD/CAM technology, the importance of CAT research has been emphasized by researchers. Consequently, CAT research is becoming more and more popular. On the basis of the studies mentioned earlier, Zhang et al. (2011) have introduced the polychromatic sets theory into CAT. Based on the VGC theory and TTRS theory, it develops a hierarchical reasoning model of tolerance information, applies the polychromatic sets theory to describe this model, optimizes computer-aided generation of tolerance types, and provides a basis for developing tolerance network and designing tolerance. Their research introduces the application of the hierarchical reasoning model as well as its reasoning method, based on assembly-oriented tolerance generation. Polychromatic sets theory is a mathematical theory tool that is regarded as a promising approach for many applications (Li and Xu, 2003). Due to the idea and the theory that were developed, namely, to use a standardized mathematical model to simulate different objects, the techniques of polychromatic sets have been widely applied to areas such as product lifecycle simulation, product conceptual design, concurrent engineering, and virtual manufacturing for product modeling, process modeling, and process optimization. Figure 2.7 shows a reasoning model used in VGC and TTRS research (Zhang et al., 2011).

In recent years, engineering design has required more and more multidisciplinary design activities. Engineering designers from a number of different disciplinary areas may interact and exchange views in the design process. Therefore, seamless integration and efficient processing of engineering data among numerous heterogeneous data sources play an important role in engineering design. Hence, engineering integration is assumed to support multidisciplinary engineering design activities throughout product development cycles. The ubiquitous characteristics of

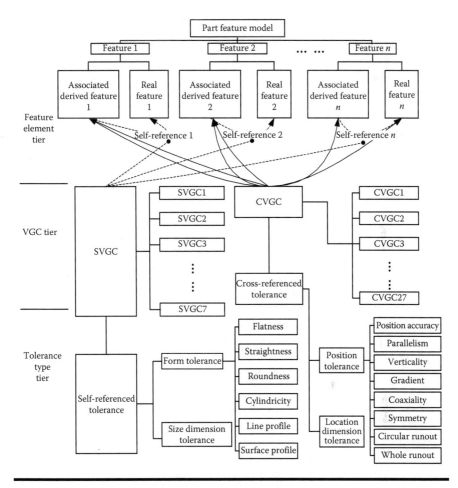

Figure 2.7 A reasoning model for hierarchical tolerance information.

data diversity, irregularity, and heterogeneity will distinctively differentiate engineering information integration from information integration in other domains. This poses a challenge to effective engineering integration. There has been much ongoing research in this area. The topics cover (1) the methodology for developing a virtualization-based simulation platform in support of multidisciplinary design of complex products; (2) approaches for engineering software integration and product data exchange to support interoperability among different engineering phases; (3) mathematical formulation and optimization method for engineering problems; (4) autogenetic design theory and distributed computing approaches and their applications to multidisciplinary design optimization; and (5) web services–based multidisciplinary design optimization frameworks that provide data exchange services and integration.

The research on engineering integration is becoming more prevalent now. Research has recently been conducted on the methods and models for establishing ESs for large-scale engineering projects. However, the attention paid so far to ways of establishing ESs for large-scale engineering projects is still not sufficient.

2.4 Customer Integration

As the competition among enterprises is becoming fiercer, satisfying customer needs becomes the key element to support an enterprise's effort to survive its competition. Customers are a potential goldmine of information for product and service development, not only in the innovation phase but throughout the development process as well. Edvardsson et al. (2012) reviewed and classified methods of customer integration and suggested four models of integration in which data are classified as *in situ* (data captured in a customer's use situation) or *ex situ* (data captured outside the use situation). The term "customer integration" describes a wide array of methods to import intelligence about customers' values and behavior. Integrating with its customers is absolutely essential for a business to thrive.

The most influential factors that affect corporate customers' satisfaction are the customer relationship and the service process. Both factors can serve as inputs to CRM, which can generate new ideas for improving customer services. Many organizations turn to a CRM solution when they are seeking better CRM. The concept of CRM assumes and requires the integration of all of the functions that affect the customer. CRM can be independent software or a module of ESs in which the ESs include CRM functionality that links to customers. Many enterprises seeking new approaches to maintaining or increasing market share have adopted the CRM system to improve their customer service. Many enterprises have constituted CRM implementation plans and now use CRM as an effective way to increase their competitive advantage. Through implementing CRM systems, enterprises are able to obtain their customers' information on a real-time basis to meet their customers' needs in a timely manner.

Over the last decade, the scope of ESs has enlarged to include CRM. As a module of ESs, areas of customer-focused information that are integrated with the rest of an ES include sales support, sales forecasting, order generation, order entry, quoting and promising deliveries, demand management, logistics and distribution, and field service (Langerwalter, 2000). The purpose of CRM is to capture, consolidate, and analyze customer data for two reasons: to retain existing customers and to attract potential customers. This process continues throughout the entire business process with the objective of better understanding customers and their interests regarding the offered products or services.

Customer integration mainly covers the exchanging and coordinating mechanisms of information flow. According to the concept of customer integration, realizing customization may require a variety of aspects to be considered. For example, it may include all of the internal factors of production (as long as they are related to customization). Examples of such internal factors may include tooling, raw and semifabricated material, personnel, technological know-how, and others. These factors of production need to be considered for customization purposes. Some internal factors may need to be carefully integrated to become part of a firm's internal resources for showing a competitive edge in customization. Customer integration can enhance an enterprise's innovation performance.

Customer-oriented enterprises engage in customer integration throughout the innovation process. A large body of literature indicates that innovation activities can benefit from customer integration, and thus, enterprises can seek opportunities through the accumulation of knowledge about current and future customer needs. Virtual customer integration (VCI) allows the exchange of information and the sharing of tasks between an enterprise and the potential customers of its products via the Internet within all stages of new product development (NPD)—from idea generation and conceptualization through to development and design and to testing and marketing (Bartl et al., 2012). Customers are willing to participate in and provide their know-how during innovation processes. But despite the high potential of VCI methods for NPD, practical use is still limited.

The integration of KM into CRM becomes a strategic initiative for providing competitive advantages to enterprises. KM has been integrated into CRM, as KM can optimize an enterprise's performance through creation, communication, and application of knowledge acquired when enterprises interact with their customers. KM is a process through which an enterprise can generate value from its knowledge assets such as customer management practices and experience. KM can also help enterprises in creating new business ideas that are generated from CRM, and in managing and improving the enterprise's relationship with its customers. Integrating KM into CRM becomes necessary as enterprises are interested in discovering the needs of customers and offering tailor-made products and services.

CRM integration is an important research topic in both intra- and interorganizational ESs. One of the major tasks is to facilitate communication between enterprises through underlying systems. One of the technical challenges is the lack of interoperation across the domains of the involved information systems. This is because interorganizational processes are often heterogeneous (especially multilingual business processes). Active research has been conducted in ensuring semantic consistency among multilingual business processes. It aims at automating a concept disambiguation process for the preprocessing of dictionary entries for collaboratively developing a multilingual vocabulary that is used for building semantically consistent interorganizational ESs, especially for multilingual CRM.

To enable information exchange across heterogeneous ESs without misinterpretation, research has been conducted in providing a collaborative conceptualization approach that resolves the semantic conflicts of information exchange based on a collaborative vocabulary editing mechanism. The main task of this approach is to create semantically consistent concepts across heterogeneous systems. Such concepts can be used to build consistent business communications for accurate information exchange since they are both syntactically and semantically consistent, for interorganizational ESs. Recent research has further improved a collaborative conceptualization approach by inserting a preprocessing mechanism that allows for the development of a collaborative vocabulary and maximally reduces the manual involvement. The main idea is this: all vocabulary entries that need to be collaboratively edited are preprocessed by a near-synonym finding process so that collaborative editors can resolve semantic conflicts between vocabulary entries by using sets of near-synonyms found in the preprocessing. To improve the existing solutions of finding and identifying near-synonym sets, particularly to avoid any translation ambiguity of a locally designed vocabulary, researchers have proposed a novel framework, called "near-synonym graph" (NSG). This framework assumes that a well-known multilingual dictionary is accurate in multilingual translations for its corresponding words and definitions. The NSG framework is illustrated in Figure 2.8 as a graph with examples of S_1 = {recognition}, S_2 = {actualization, realization}, and S_3 = {realization, realization, recognition}; W_1 = "recognition," W_2 = "actualization," W_3 = "realization," and W_4 = "realization"; $D_{1,1}$ = gloss 1 of W_1 in BD, $D_{1,2}$ = gloss 2 of W_1 in BD, $D_{2,1}$ = gloss of W_2 in BD, and etc. Technically, NSG framework is a seven-tuple of <S, W, D, R_1, R_2, R_3, P>.

In NSG framework, the solution is clearly divided into two stages: finding potential near-synonym sets and identifying final usable near-synonym sets.

In today's highly competitive global economy, enterprises are continuously enhancing their ESs, including their CRM subsystem. Much research has been conducted in this area.

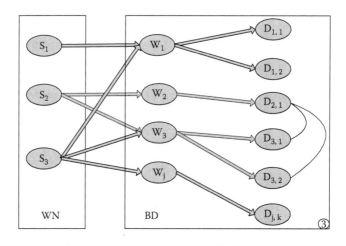

Figure 2.8   An overview of near-synonym graph framework.

2.5 ESs in Evolution

2.5.1 Design Considerations Changes

ES is a system that mainly concentrates on the realization of the concept of business integration. The key driving force appears more and more toward implementing the systems to meet the need of integrating processes and systems intra- and interorganizationally across the supply chain. The concept of the ERP was conceived in the early 1960s with the advent of MRP, which was then followed by MRP II. American Production and Inventory Control Society (APICS) defines MRP as a system that uses BOM, inventory, and open order data, and master production schedule information to calculate requirements for materials. As more functional modules have been added to MRP, the second-generation MRP system abbreviated as MRP II appeared. The original underlying mission of ERP is to achieve a capability of planning and integrating enterprise-wide resources. It is intended to support one-time entry of information at the point where it is created and to make it available to all the systems that need it. Early ERP systems mainly manage intraorganizational resources and operations. A successfully implemented ERP can link all areas of an enterprise including manufacturing functions, financial functions, CRM, and many other functions, forming a highly integrated system with shared data. Potential benefits include drastic declines in inventory, reduction in working capital, and abundant information about what customer wants and needs, along with the ability to manage the enterprise consisted of customers, suppliers, and alliances as an integrated whole. Today, ERP has become a basic business information processing requirement for many industries. ERP systems consist of numerous integrated applications including manufacturing, logistics, distribution, accounting, marketing, finance, human resources, and others. In an ERP system, software is available to support activities such as manufacturing, assembling, logistics, finance and accounting, sales and marketing, CRM, human resources, and SCM. ERP is intended to enable automation as it offers the availability of real-time data, improved process visibility, and an increased level of automation; as such, some benefits attributable to automation are credited to ERP.

There are numerous ways to describe ERP:

- The objective of an ERP system is to automate the business processes of the enterprise. The benefits resulting from this automation include supporting business operations leading to better enterprise performance.
- The core functions include financial planning, business process management, production management, quality management, and human resource management. The functions of an ERP system involve both primary business activities and supporting activities. Such integrated activities enable higher productivity and efficiency.

■ The architecture of an ERP system is of modular construction that makes it flexible for modification and expansion. Its components include APS, CRM, SCM, and others.

There are also numerous descriptions of ERP software modules; for example, for PDM and PLM:

PDM enables an enterprise to manage product-related information more effectively throughout the life cycle of a product. PLM software enables an enterprise to bring innovative products to market effectively. PLM incorporates Product Design Support, including cost estimation, product development, and prototyping.

Substantial research has been conducted on the most critical factors that affect the success of ERP. The identified factors include (1) inappropriate definition of requirements; (2) customizing ERP that is associated with an increase in cost, longer implementation time, and difficulties of software maintenance; (3) lack of commitment from senior management. Commitment from top management is critical to ensure the success of ERP project; it should be in line with the strategic direction of the enterprise. Senior leadership is critical to make an ERP successful; (4) lack of a clear strategy of improvement; (5) resistance to changes and a lack of participation; and (6) an inadequate level of qualified end-users. Theoretical models have been developed to explore the mediating effects of organizational culture and knowledge sharing on transformational leadership and ERP successes. Despite facing challenges, ERP has been successfully applied to many industrial sectors. So far, ERP systems have become the most significant information technology investment for many enterprises with positive outcomes. There are many applications provided by ERP. For example, with the economic globalization, the application of ESs has become more and more popular in the financial service sector. The price tag for ERP implementation ranges from millions to billions. According to research, in the past 10 years, the worldwide ERP market has been growing at an annual rate of more than 10%. Currently, the ERP market is one of the fastest growing and most profitable areas in the software industry.

In recent years, ERP has been increasingly applied to extended enterprises and global SCM. SCM is the integration of key business processes among a network of interdependent suppliers, manufacturers, distribution centers, and retailers, in order to improve the flow of goods, services, and information from original suppliers to final customers, with the objective of reducing system-wide costs while maintaining required service levels. SCM is a network of autonomous or semi-autonomous business entities collectively responsible for procurement, manufacturing, and distribution activities associated with one or more families of related products (Su and Yang, 2010).

A supply chain is a dynamic process that involves the constant flow of information, materials, and funds across multiple functional areas both within and between chain members. A supply chain is an instance of industrial networks. SCM can be digitally connected, and such digitization can provide real-time data to facilitate supply chain operations. Connection of SCM to the ESs enables an operation optimization level both intra- and interorganizationally. Research and practice in this area have covered numerous aspects such as planning/routing, real-time monitoring, and subject–infrastructure–enterprise (SIE) information integration. The SIE model supports industrial network flow and facilitates automatic MF systems. In summary, with the great challenges from both technology and the global economy, ESs are expected to play a more and more important role as an integral component of SCM; such systems must have the flexibility to adapt themselves to the rapidly changing supply chain environment.

Total EI (TEI) represents an early effort to integrate all of the subsystems within an enterprise into a larger scale. The idea of TEI was developed and built upon MRP, MRP II, and ERP (Langerwalter, 2000). It has mainly stayed at the conceptual level. In the year 2000, the concept of ERP II was proposed by the Gartner Group, and it emphasized the two specific aspects of business coordination and industry orientation, along with the role and function of the system that is expected to expand to those required by an industry sector or a particular industry. Since then, IERP has been given more attention from both ERP vendors and academic circles. Section 2.5.2 will introduce industry-oriented enterprise resource planning, that is, IERP. IERP represents another design idea change in the evolution of ESs.

In the 1990s, ESs were mainly developed and implemented for managing the physical assets of an enterprise. In the late 1990s, it became well recognized that knowledge is a compilation of an enterprise's invisible assets. Survival in today's highly competitive environment requires the efficient management of corporate knowledge. As KM has become increasingly critical for the success of enterprises, not only large-sized enterprises but also small- and medium-sized enterprises are positioned to tap into KM. Due to the fact that both types of assets need to be properly managed, the integration of KM and ERP has become a strategic initiative for providing competitive advantages to enterprises. Increasing requirements for extended enterprises have also stimulated the integration of the KM function into ERP systems for knowledge asset management. In 2006, Xu et al. proposed the concept of deploying KM and ERP concurrently in the framework of ESs (see Figure 2.9), and thus the concept of ERP III was developed. ERP III enables ES applications to transform an enterprise into a knowledge-based learning organization and to capture know-how for developing business solutions to create real competitive advantages. Since 2006, the importance of KM in ESs has been recognized, and KM in ESs has started receiving attention. In 2007, Lorincz also indicated that, in a broad sense, the real asset of any organization is its knowledge. Lorincz also predicted that the next stage of the coevolution of business and ESs will be developing ESs with KM capability.

The word "knowledge" has been defined differently by many authors. For example, knowledge is defined as a mix of experience, values, contextual information, and insights. Although knowledge is related to both data and information, it is neither data nor information. It is important to note that the terms "data," "information,"

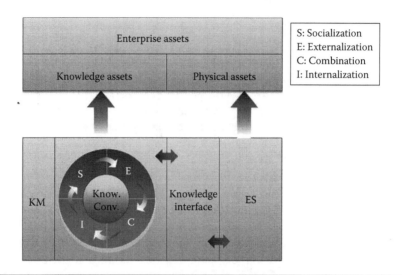

Figure 2.9   Integrating KM and ES for managing knowledge and physical assets.

and "knowledge" are not interchangeable. In decision-making processes, decision makers usually apply a combination of data, information, and knowledge. Knowledge is typically embedded in organizational processes. KM is a process used for acquiring, organizing, and communicating both explicit and tacit knowledge so that users of the knowledge can be more productive and effective. KM systems are a type of information systems that are specifically designed for collecting, coding, and disseminating knowledge. During the last decade, organizations have made increasing efforts in KM, since efficient KM can lead to superior business performance such as better operational effectiveness, higher-quality products and service, and greater organizational creativity. Many examples demonstrate that many of the world's most successful organizations are those that are best at managing their knowledge.

There has been an increasing research interest in applying KM to ESs as research has shown that KM is useful in ESs. There is a tremendous amount of knowledge that enterprises can tap into for their planning and management processes. Such knowledge is available at strategic, tactical, and operational levels and can be utilized to a competitive advantage if the knowledge can be tapped to develop a knowledge-based ES. Several different models have been developed for KM in ESs.

KM can support ESs across its entire life cycle. For manufacturing enterprises, the key facilitators for KM in an ES environment have been identified, and the relationship between KM and manufacturing enterprises' competitive advantages has been examined. Research indicates that organizational preparation for knowledge sharing, KM, and other factors are facilitators that can have a systematic and synergistic effect on a firm's competitive advantages. Currently, the research emphasis in the area has been on KM frameworks and approaches.

Xu et al. (2006) originally introduced ways to deploy KM and ESs concurrently within the framework of ES, with an emphasis on the interaction between KM and ESs in a systems perspective. Within this framework, a variety of research has been conducted. One example is the building of a KM system for managing knowledge in an ES implementation process. Its main tasks are to identify the types of knowledge needed in ES implementation, to summarize various KM activities, and to develop a KM system that consists of cooperative working platform, an organizational KM platform, and a knowledge transfer platform. This proposed system then targets the management of knowledge and provides support for the successful implementation of an ES. KM-specific data warehousing models for ESs have also been proposed. Traditional data warehouses have provided data to decision support systems (DSSs) or to online analytical processing (OLAP) systems, but not to knowledge-based systems. KM-specific data warehousing proposes that existing data warehouses can be extended to create knowledge warehouses for KM. A knowledge warehouse mainly manages the knowledge assets of an enterprise. Both tacit knowledge and explicit knowledge can be processed and analyzed. New knowledge can also be created through synergistic interactions within the knowledge warehouse. Substantial work, based on ESs, has been done on knowledge warehouse architecture. Applications of KM-oriented ESs have appeared. OSS, which are becoming increasingly popular in the telecommunications industry, are one kind of ESs. A KM system for OSS has been developed. Each subsystem of the OSS can be equipped with KM capacity, and the KM of the OSS can be realized through its subsystems.

Implementing KM and enterprise systems concurrently: Although KM and ESs emphasize different characteristics, the primary goal of each system is to improve the competitiveness of an enterprise. From a practical point of view, KM and ESs need to be implemented simultaneously within the framework of integrated ESs. As such, KM systems are best when they are not implemented in isolation but are concurrent with a variety of subsystems, among them PDM, CAD, and SCM subsystems. Simultaneous implementation of KM and ESs facilitates the incorporation of both KM and ESs into entire enterprise business process. Researchers have studied organizational mechanisms that are considered to be important for achieving simultaneous efficiency and flexibility. Researchers have also examined the combined effect of improving organizational efficiency and flexibility by concurrently implementing both ESs and KM in an enterprise. Research indicates that there is a large potential for applying KM to the ESs life cycle. KM can be employed through the entire ES life cycle, including feasibility analysis, user requirement analysis, system design, systems development, and systems maintenance, to support ERP implementation. A four-phase ES improvement model has been proposed. It incorporates KM into each major implementation phase. In this framework, ESs implementation allows knowledge to be captured, assimilated, refined, documented, and leveraged for the next project phase. This cumulative process builds a knowledge base that can be used to support the ESs through the entire life cycle.

Relationships between KM and enterprise system: KM and ES are the two philosophies for managing business enterprises. The knowledge-based view of enterprises argues that the KM system is the focus of enterprises and that an enterprise's competitive advantage depends upon the effective management of its knowledge assets. From another angle, the information processing view considers that ES is the focus of enterprise management, since ES enhances business performance through managing material and information flow. Although either view has its limitations, from a systems perspective, they are complementary to each other. As such, although KM and ESs are based on different management philosophies, KM and ESs complement each other. Also, despite the different focus of KM and ESs, the two systems have common goals. Both KM and ESs aim at improving business processes in order to achieve better business performance, with tasks based on data, information, and knowledge. KM systems are devoted to the knowledge processes of enterprises (knowledge creating, storing, and sharing). ESs emphasize the efficiency of business processes in enterprises. To achieve the operations goals of the organizations, ESs maintain enterprise data and information. In terms of corporate assets management, ESs and KM systems manage the physical and the knowledge assets, respectively. From the perspective of enterprises, the ultimate goals of the two systems are to help enterprises to improve performance. With a proper framework in which KM and ESs can cooperate with each other, an enterprise can benefit from the advantages of both KM and ESs and can be successful. Sometimes, an ES provides a platform for capturing knowledge, since both tacit and explicit knowledge may be acquired through ES. Such knowledge involves business transactions, individual experiences, and documents; this kind of knowledge will improve the performance of the entire enterprise, including the ES. In such a framework, ESs provide a platform for knowledge capturing, storing, and sharing; and KM integrated with ESs can help improve the business processes managed by ESs since it offers better management of knowledge assets. As such, the relationship between the KM and ESs is synergistic.

To continue the earlier-provided discussion, in a given enterprise in which KM and ESs are implemented concurrently, the interaction between KM and ESs can be studied from a variety of perspectives, including the effects of ESs on KM and vice versa. For example, ESs can provide a large volume of enterprise data and information for possible use as knowledge (after appropriate processing and integration). In this way, ESs can become a channel for capturing, exploring, and sharing knowledge. ESs can change the way of organizational learning by enabling and facilitating organizational innovation. Implementing ESs can also result in major changes as organizations learn more about their business and business processes, which may make related knowledge available for capturing, utilizing, and sharing. As such, knowledge about organizational business processes becomes more readily available through the use of ESs.

However, there are a number of challenges. For example, knowledge may be captured and processed by a transient process; as such, it will vanish quickly.

Traditional practices do not have an explicit process to ensure that the knowledge being captured will be stored for future use. The sheer volume of knowledge being captured may prevent it from being condensed into a single deliverable. As such, a KM system, which can be implemented with ESs, is able to support ESs efficiently only as numerous technical issues are resolved.

Integrating KM and enterprise systems: There are two ways to integrate KM and ESs. Either they can be integrated on the basis of existing KM and ESs, or a newly developed KM system can be integrated into an existing ES. In the first approach, the relationship between KM and ESs is cooperation. In the second approach, a KM system is integrated into an ES in terms of modules. The first approach is generally considered a more common approach.

Usually, a business's KM and ES are provided by different vendors and are stand-alone systems. Integrating these two separate systems can provide an enterprise with better business performance. With the integration, the system can manage physical as well as knowledge assets for achieving a more competitive advantage than ever before. From the enterprise's point of view, managing both types of enterprise assets is highly desirable. Although ES and KM emphasize different types of assets, the integration would satisfy the requirements of systematic management.

Knowledge interface between KM and ES: According to the functions performed, there exist two types of knowledge interfaces between KM and ESs. Knowledge interfaces are the software brokers that transfer knowledge from KM to ESs or from ESs to KM. In other words, knowledge interfaces are the channels though which knowledge flows between KM and ESs. Therefore, knowledge interfaces play important roles in incorporating KM and ESs. Also, there are two knowledge circles (the learning circle and the innovation circle). As ESs request knowledge for business processes, ESs employ the methods provided by the knowledge interface to obtain knowledge. As the KM requests operation knowledge, KM uses the knowledge interface to interact with ESs. In general, according to their function, knowledge interfaces can be classified into K-Discovery, K-Classifying, K-Storage, K- Identifying, and K-Indexing (see Figure 2.10):

K-Discovery: It is equipped with the methods to access the data from ESs. It will discover knowledge from ES process data with an information context.

K-Classifying: It is equipped with the methods to categorize knowledge according to the type of knowledge and the domain context to which the knowledge belongs.

K-Storage: It is equipped with the methods to save knowledge and knowledge context to the knowledge base. It also maintains the linkage between knowledge and its context.

K-Identifying: It provides users with the methods to send knowledge requests and transfers the requests with their contexts into queries on knowledge base.

K-Indexing: It consists of methods for interacting with the knowledge base engine based on the query commands given by K-Identifying.

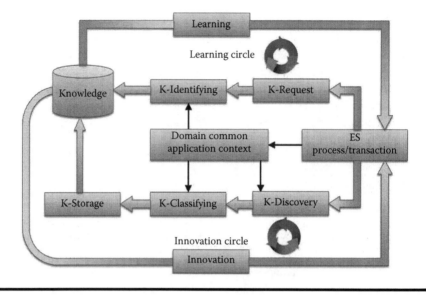

Figure 2.10 Knowledge transfer process in KM and enterprise systems.

In recent years, the concept of next-generation ES as well as their strategies and applications has been proposed by Xu et al. (2009). The system is called entire resource planning (ERP) or complete resource planning (CRP), in which the concepts of ERP, ERP II, and ERP III have been integrated and extended to comprehensively encompass the resources used and produced by enterprises in different industrial sectors in the context of societal and economic development. Figure 2.11 shows the relationship between MRP, MRP II, ERP (TEI), and ERP. Section 2.6 will introduce ERP. Figure 2.11 shows a brief history of ESs:

    1960s–1970s MRP
    1980s MRP II
    1990–2000 ERP
    2000 TEI
    2000–2009 IERP
    2009-ERP

2.5.2 Industry-Oriented Enterprise Resource Planning

General-purpose ERP systems usually have problems such as complex configuration processes, a low ability to adapt to specific industrial sectors, and an extensive implementation time period. In order to alleviate these issues, the concept of IERP and a component-based approach to IERP development were proposed in 2009 by Wu et al. (2009). This section introduces the framework of developing an IERP in which business process modeling and software reuse are employed as primary

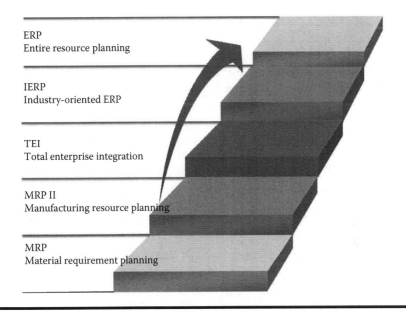

Figure 2.11   MRP, MRP II, ERP (TEI), and ERP (entire resource planning).

methods to improve the operability of component-based IERP. The framework consists of five layers including the server layer, the teamwork supporting layer, the IERP construction and customization layer, the reusable assets and toolset layer, and the IERP system instance layer. These layers, as well as their relationship with each other, are also discussed.

2.5.2.1 Introduction

From a technological perspective, ERP systems are viewed as configurable multi-module software packages. However, implementing ERP systems can pose problems (such as an extensive implementation time and a low ability to adapt to specific industrial sectors), which cannot be simply ascribed to the enterprise that implements the ERP system. The factors related to ERP software, such as the complexity of configuring thousands of parameters and low specific industry relevance, have been intensively researched. Although ERP systems have been categorized into systems specifically designed for traditional and nontraditional industries (Langerwalter, 2000), the issue of their low ability to adapt to specific industrial sectors has not been adequately analyzed until recently. According to Wu et al. (2009), ERP systems are classified into two categories according to their application, that is, general-purpose systems and industry-oriented systems. A general-purpose ERP system such as SAP R/3 is a software package, which is generally designed for a wide range of industrial sectors, while an IERP system is designed for enterprises belonging to a specific industrial sector and often supports specific business needs that are not covered by general ERP packages. As many industries have their own processing requirements, general-purpose ERP packages may not fit exactly to a specific industry; for example, the SAP R/3 solution comprises more than 5000 parameters. As a result, the complexity in implementing SAP R/3 solutions is obvious. For industries with specific needs and tasks, customized ERP packages are generally preferred, although the expenses incurred through development and maintenance can be relatively high. For example, in the textile and apparel industry, the BOM for textile production has a "one-to-many" characteristic. Although the production of yarns with different counts, colors, and warp types is made of the same raw material, the difference in the BOM creates a long list of requirements in almost all aspects including the planning, production, dyeing and finishing, quality control, and sales and distribution that are specific to the textile industry. Under such circumstances, extraordinary customization efforts are required to meet specific requirements and needs. As a result, adopting a textile and apparel IERP package, such as Datatex TIM, is obviously a better choice than using a general-purpose ERP system. The sheer number of customization tasks including the relevant business process reengineering involved in the software implementation process is considered the major challenge for general-purpose ERP. Research shows that many enterprises have modified ERP software in various ways to meet essential business needs. In order to better satisfy the requirements of specific industries and to reduce the configuration workload during the implementation of ERP software, more and more general-purpose ERP vendors have begun to take steps to redesign or reengineer existing software to provide industry-specific solutions. For example, SAP, a leading general-purpose ERP vendor, now offers business solutions to over 20 industries by tailoring its existing ERP package. Research also indicates that ERP systems' functionality and integration have been greatly improved over the last decade by incorporating specific industry solutions. However, in academic communities, there has been limited research on IERP to support ES industry.

2.5.2.2 IERP

An IERP system is designed for a specific industry sector and thus can satisfy most of the processing requirements of the target industry or industries. Since an IERP system has an explicit target industry, it is different from a general-purpose ES. Software modules of IERP systems usually incorporate more industry-specific requirements. Consequently, a shorter implementation cycle can be expected. The processing requirements in a specific industry usually can be relatively conveniently refined. Therefore, it is relatively easy to make continuous improvements to IERP. In addition, such explicit industry orientation is helpful for vendors to benchmark best practices within a specific industry sector. An IERP system is a version-based customizable system that can add or remove software components to support the continuous modification of business processes and is capable of integrating with other systems.

In general, an IERP system is a system that aims to satisfy the requirements of the target industry; meanwhile, it is quite different from traditional fully customized ERP. The term "version" is an important concept in software evolution and corresponds to those errors found in previous releases that will be corrected and those processing requirements not satisfied that will be considered in future versions. Generally, there is less room for version evolution in fully customized software, which means that it becomes obsolete more quickly, especially in a dynamic environment. It is commonly recognized that there are a significant number of similarities in business processes among different industries. Software reuse methodology and platform are included in IERP software. The earlier-mentioned similarities can be considered in IERP, and systematic software reuse (SSR) methodology provides guidelines and techniques for practice. Meanwhile, the capability of component-based variation design for individual processing characteristics is emphasized for efficiently constructing version-based IERP software.

2.5.2.3 IERP versus General-Purpose ERP

The position of IERP is that it stands between general-purpose ERP and fully customized ERP systems, as illustrated in Figure 2.12. On one hand, IERP software inherits all the merits of version-based evolutionary mechanisms from general-purpose ERP software. On the other hand, IERP software offers more adaptability for a specific industry.

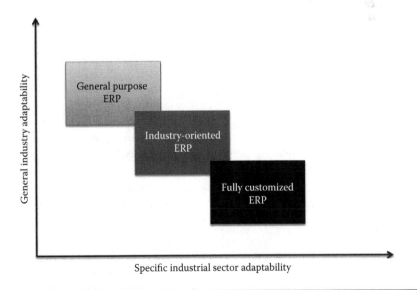

Figure 2.12   Industry-oriented enterprise resource planning (IERP).

In a traditional environment, a software component can be defined as a self-contained unit of software with well-defined interface or set of interfaces, which can be independently delivered, installed, and deployed. Such components can be easily combined or can collaborate with other components to provide useful functionality. Nowadays, component-oriented software technology has become a major approach in facilitating the development and evolution of software systems. The objective of this technology is to take elements from a collection of reusable software components and build applications. The component-based software development approach enables the practical reuse of valuable software assets and amortization of investments over multiple applications. In addition, the concept of software reuse has been extended to encompass all of the resources used and produced during the development of software. Accordingly, the concept of the software component has been extended to encompass all kinds of intellectual artifacts such as domain-specific software architecture (DSSA), application frameworks, design patterns, class libraries, and even test data, which can be produced and potentially reused at every stage of the software life cycle. At present, software reuse techniques can effectively support the expression, encapsulation, and reuse of enterprise knowledge, which offers the technology required for IERP componentization.

2.5.2.4 Connotation of IERP-Oriented Componentization

The connotation of IERP-oriented componentization mainly includes three aspects. First, IERP is involved with a component-based SSR process. Via SSR componentization in IERP software development process, organizations can identify common functionalities among applications within a domain and can build reusable components that can benefit future development efforts. In order to facilitate the assembling of components from different providers and the evolution of IERP systems, the quality of component documentation is usually emphasized. In most cases, documentation is used to access information regarding the applicability and quality of a component. In the documentation model defined by Forsell (2002), a software component is divided into two parts: the reusable part and the part that supports reuse. The reusable part defines the objectives and the design rationale for the component and the result of the production and tests used to certify that the component works correctly. The reuse support part includes maintenance information. The idea is to map the roles of the component provider, maintainer, developer, and reuser to all of the pieces of information produced during the development and use of a component.

Second, it is a domain-specific componentization process that involves acquisition and reuse of reusable ones such as DSSA. DSSA is the most important outcome of domain engineering (DE), which is a reuse-based approach to define the scope (domain analysis), specify the structure (domain design), and build the assets (domain implementation) for a class of systems, subsystems, or applications.

DE analyzes all of the systems in a specific domain, generalizes their functionality, and represents the requirements for the whole domain in a domain model. It involves a paradigm shift from system development, which emphasizes the development and maintenance of a family of software systems within the same domain, rather than the development of a specific system. As mentioned earlier, IERP strategy aims at developing version-based customizable ERP systems in a specific industry or industries that share a significant number of similarities; unquestionably, DE plays a vital role in acquiring domain models, DSSA, and other reusable assets.

Third, business components are the most important reusable artifacts. A business component is the software implementation of an autonomous business concept or business process. It consists of all the software artifacts necessary to represent, implement, and deploy a given business concept as an autonomous, reusable element of a larger distributed information system.

Unlike traditional software modules and technical components, business components should support some business functionalities. Examples include inventory transaction processing, accounts payable, and order processing components. The features that allow business components to help realize the encapsulation of business concepts and functionalities at a high level make them ideal places to deposit and reuse industry solution knowledge. An effective way to obtain business components of all kinds is to take a business-driven approach to an IERP system-oriented componentization.

2.5.2.5 Business-Driven Approach to IERP System-Oriented Componentization

Generally, manufacturing industries involve several business domains, each with a different function, such as purchasing, manufacturing, logistics, and distribution. Each business domain is composed of business functions, which are performed by executing particular business processes of the domain. For example, some typical business functions found in a purchasing domain are supplier management, purchasing requirement management, and purchasing order/contract management. There can be significant similarities in business domains, business functions, and business processes among industrial sectors. On the other hand, business processing in the same business function may vary depending upon their industry, and the requirements of specific business processes in enterprises within the same industry may even be somewhat different. In fact, it is possible to adopt a systems approach to analyze, obtain, encapsulate, and reuse business components of all kinds by systematically analyzing the business processes of enterprises. It is called a business-driven approach to IERP-oriented componentization.

In this approach, common business processes are a starting point. Each business process is described as it is abstracted from enterprises belonging to the same industry; then the business process is decomposed, hierarchically, into more detailed subprocesses. In each hierarchy, relevant business objects participating in certain task fulfillments will be identified, which will then be further elaborated at subsequent decompositions. The decomposition activities will continue until the point is reached at which the business object class can be realized directly by one of existing component models. The distinct feature of this approach focuses on the business processes, which is further driven by the activities from requirement collection, component acquisition, and the IERP system construction. Generally, this approach includes three aspects: (1) business-driven modeling, (2) acquisition of business components in the form of one or several available component models, and (3) IERP system construction with reusable business components. The first two aspects are related to the acquisition of reusable business components, and the last is about IERP software construction with existing reusable components.

2.5.2.6 Levels of Business-Process-Driven Modeling

There are three modeling levels in this approach. The first is the business function level, which describes what to do for an enterprise from the perspective of a corporate strategy. This level most deals with the corporate goals, external environments, competitive products, and business management characteristics. At this level, the business domains, functions, and relevant vocabularies are identified and documented. It is common that there will be a need for further decomposition into subindustries. The functional requirements collected and refined at this level can be guided by DE methodology. Based on the analysis of the existing business systems in a target industry, features such as commonality and variability can be obtained and then be generalized to create the artifacts of domain analysis models. If necessary, these models can then be further developed into submodels of a subindustry (or industries). At this time, the common components are abstracted and shared among submodels. This is crucial for building and maintaining several IERP versions of subindustries. In general, the typical outcomes include industry definition, an industry lexicon, and concept models. The "industry definition" defines the scope of an industry and characterizes its contents via a typical business processing domain. An "industry lexicon" defines a specific vocabulary of a target industry. "Concept models" describe the main concepts in business domains expressed in some appropriate modeling formalism.

The second is the business process level, which describes what should be done from a business process perspective to enable an enterprise to support the fulfillment of business functions. What this level is concerned about is how specific business functions can be accomplished by a sequence of logically related business tasks and events. The modeling process at this level provides the specification of tasks and task-related business objects, events, and resources. Business processes can be expressed using certain techniques such as UML or Petri nets. The initial business process model can be optimized through the elimination of non-value-added or invalid activities and by rearranging the possible sequence of the business tasks.

Considering reusability, more possible branches or subbranches in every business process should be carefully considered and documented in a formal way.

The third is the business task level. At this level, business tasks are further analyzed and documented. For every business task, every related business object and agent can be identified, and methods of business objects can be further outlined.

2.5.2.7 Metamodel for Business-Process-Driven IERP Componentization

There are many methods and tools available for modeling business processes, such as flowcharts, event-driven process chains (EPCs), and UML. An ideal modeling approach should provide an effective mechanism to specify the business processes, business objects, and resource requirements related to the tasks. It is also helpful to encapsulate related business object classes into business components. As a visual and extensible modeling language, UML provides several mechanisms for extension and customization purposes. In addition, the metamodels of UML can also be extended and modified. Therefore, UML is an ideal candidate tool for modeling business processes in IERP componentization.

The metamodel can be described as follows: a business function is achieved by executing a specific business process. One business function may associate with many other business processes, and one business process may participate in the realization of many business functions. Business objects provide methods to support the realization of a specific business function. A business process, which can be further divided into several business tasks, is generally triggered by external events. A business task may either be an atomic task or a compound task. An "atomic task" is an atomic unit; it is not necessary to divide it any further and it can be executed directly. A "compound task" can be further divided into smaller subtasks. The introduction of a compound task enhances task granularity and thus simplifies the management of the tasks. During the execution of business tasks, one or several business events can be generated, which, in turn, can further trigger the execution of new tasks. Trigger rules determine the conditions and mode (e.g., executing sequentially or simultaneously) of executing business tasks and serve as the agents responsible for the execution of business tasks. Business tasks can be completed by executing the relevant functions of business objects; and, in turn, the business objects are responsible for dealing with business events. The states of business objects can be affected by business events and business tasks.

When this metamodel is adopted in the componentization of IERP systems, the standard UML can be extended because key business concepts are not directly supported. In general, several types of diagrams can be employed together in order to model the multiple aspects of business components. Accordingly, related diagrams may be required to be extended at the same time. As an example, in order to model business processes in a more natural way, the activity diagram of a UML can be extended at the metamodel level by adding metaclasses.

2.5.2.8 Category of Business Components

For business processes, tasks and objects are identified, analyzed, and documented with the employment of the above-mentioned metamodel. By doing so, business components at each modeling level can be obtained. The candidate objects, which can be encapsulated into business components, include business tasks, business object classes, and even business processes. There are several ways to categorize business components. Generally, business components can be classified into two categories: atomic components and compound components. An "atomic component" is the unit of component reuse. A "compound component" is made up of smaller member components, which can be either atomic or compound components. Compared with an atomic component, a compound component, in general, has higher granularity. According to the level of generality, business components are categorized into industry-neutral components and industry-specific components. According to component granularity, business components can be categorized into business process components, business task components, and business object components. According to the level of generality, business object components can be further divided into basic business object components and supporting business object components. "Basic object components" are used to encapsulate basic business entity sets, which generally provide direct support for business tasks. "Supporting business object components" are generally a family of tool components. In general, business process components are compound components, that is, reusable assemblies made up of logically collaborated business task classes. Business process components are building blocks with low granularity used for constructing business process frameworks. Logically related business task classes can be encapsulated into business task components. Similarly, logically related business object classes can be packaged into one or several business object components. In order to satisfy the needs of different industries, the interfaces of business components may have different implementations. For example, the BOM component may have a different implementation to support business functions for discrete and process manufacturing.

Business software platform framework for constructing IERP systems: Developing business software from scratch is a hard task. Typically, a business software platform is used both to facilitate the construction and maintenance of business software and to separate business knowledge from specific technical details. An effective way to develop IERP software is to construct a component library-based business software platform by using SSR and software product line techniques. In an attempt to improve the processes of development and operability in IERP systems, a hierarchical business software platform framework has been proposed that takes advantage of software reuse ideas.

Research has shown that the use of a framework for software development has many advantages, such as modularity, reusability, and extensibility. This framework consists of five layers: the server layer, the teamwork supporting layer, the IERP system construction and customization layer, the reusable component library and toolset layer, and the IERP system instance layer. The reusable component library plays a major role in this framework. The IERP system construction and customization layer is the key to effectively constructing an IERP system, and the teamwork supporting layer provides a toolset for harmonizing the development efforts of multiple teams. The IERP system instances layer is made up of industry-specific ES applications that can be directly deployed in the enterprises of target industries. The other four layers are described in the following text.

Server layer: The server layer is composed of two sublayers: the application server and operating system and the database management system. The application server provides fundamental supporting services, such as load balancing, database connections, session management, HTTP request, and response management, which can be directly used in constructing IERP systems.

Teamwork supporting layer: This is a software abstraction layer that sits immediately above the server layer. The typical functions provided by this layer are task decomposition, resource allocation, access control, and communication support. It also provides software tools that are used to interface with other layers. The primary objective of this layer is to harmonize the activities of multiple teams in the entire IERP software life cycle.

IERP system construction and customization layer: This layer is used to effectively support construction and customization of IERP software for a target industry or industries; a group of correlated productive software utilities are integrated at this layer. Among these utility components, the enterprise workflow modeling and executing component is at core status, which is used to model, customize, simulate, and optimize the business processes of target industries. The model mapping and transforming component makes an enterprise model–driven IERP system development pattern possible, through which the framework of target IERP software can be generated according to the IERP software model and its deployment information. The reference model customization component, combined with the software configuration management component, can provide a guarantee on the customization and configuration of target IERP models by reusing IERP reference models from a reusable component library. And new customized IERP components can also be added into the component library for future reuse.

Reusable component library and toolset layer: This layer consists of two parts: the reusable component library and the toolset for managing these components. The reusable component library is the depository of components of all kinds, including business components, industry reference models, industry-specific reference models, and ERP system reference modes. "Industry reference models" are general reference models that can be widely reused among different industries. "Industry-specific reference models" have an explicit industry orientation, which reflects the processing requirements and characteristics of target industries. Both industry and industry-specific reference models can be viewed as a subset of enterprise reference models, which usually provide multiple views such as the business process view, the function view, the resource view, and the organization view to express common knowledge about the enterprises of different industries and can be tailored by ERP vendors for use in building IERP software. One of the most well-known techniques is the ARIS reference model, which was adopted by SAP as an integral part of its ERP software. Baan, once a well-known ERP software vendor, has delivered a flexible and customizable ERP package in which the reference model library played a vital role. The reference model in a Baan ERP package has three parts: a general model, an industry-specific model, and an enterprise-specific model. Subindustry reference models can be built further if necessary. In this section, all reference models are treated as reusable software assets with low granularity for effectively constructing IERP systems. In an ERP system reference model, the reusable components and their relationships are described formally. Typically, reusable component management tools include acquisition, classification and description, retrieval and instantiation, and assembly and configuration. The integration of this layer with an IERP construction and customization layer and a teamwork supporting layer will play a key role in implementing the strategy of IERP software construction.

2.5.2.9 Summary

As ESs have become a basic business information processing requirement for many industries, the research on IERP for specific industries has been given attention. In this section, an IERP approach to tackling common ERP implementation issues such as excessive configurative parameters, low industry pertinence, and long implementation cycles has been introduced. The componentization of IERP systems and business software platforms has also been introduced. The three aspects regarding IERP have been introduced: (1) the concept of an IERP system and comparing IERP with general-purpose ERP, (2) an IERP-oriented software componentization approach, and (3) a business software platform framework.

Since an IERP system is large and complex, developing a good framework is not easy. It requires solid domain knowledge and experience in developing applications in the target industry. Implementing IERP may be influenced by many factors. Future research should study what specific factors may influence IERP implementation. How do these factors influence IERP implementation? Can the proposed framework be successfully applied in order to minimize the implementation risks involved? How can architectures, modeling methods, and tools be used to reduce the cost of software implementation and to increase user acceptance of IERP software solutions? In addition, research has been focused on the integration and coordination issues among IERP systems from different IERP vendors, system maintenance and upgrade issues, version-based IERP software customization, and several related management issues. More studies and best practice examples are needed to seek strategies that may be used to deal with the challenges and make

IERP implementation successful. The findings from research can be further used to improve the business software platform framework introduced in this section and to guide the implementation and maintenance of IERP systems.

2.6 Entire Resource Planning

In 2008, a new theory on MFs called Comprehensive Material Flow Theory (CMFT) revealed the essence of MF objects and phenomena. In this theory, MFs are specified in terms of the MF in the economic dimension, the social dimension, and the natural dimension, as well as their interrelationships. It was pointed out that the MF is not only an economic phenomenon but also a social and a natural one. In other words, there not only exists an economic MF, but also social and natural MFs. The economic MF is the core of the MF, whereas the social and natural MFs are the basis of the MF.

This theory provides extremely important insights into future EA and integration, especially when considering sustainable economic and societal development and growth. Based upon this, as well as upon the fact that considerations such as environmental factors and the scarcity of natural resources have been a concern, and the fact that the theory involves critical infrastructures, the concept of a next-generation ESs has been proposed. The next-generation ES is called ERP or CRP; in it, the concepts of ERP, ERP II, and ERP III have been integrated together, and the design has been extended to comprehensively encompass the resources used and produced by enterprises in different industrial sectors, within the context of economic and societal development. In ERP, not only is economic MF included, but social and natural MFs are included as well. ERP is considered a significant step forward in the evolution of ESs. Figure 2.13 shows the evolution of ESs in terms of its three stages of development efforts.

The major factors that have contributed to the birth of ERP are obvious. The past decade has brought fundamental changes both to the global economy and to

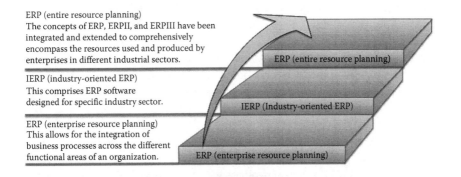

ERP (entire resource planning)
The concepts of ERP, ERPII, and ERPIII have been integrated and extended to comprehensively encompass the resources used and produced by enterprises in different industrial sectors.

IERP (industry-oriented ERP)
This comprises ERP software designed for specific industry sector.

ERP (enterprise resource planning)
This allows for the integration of business processes across the different functional areas of an organization.

ERP (entire resource planning)

IERP (Industry-oriented ERP)

ERP (enterprise resource planning)

Figure 2.13 The evolution of enterprise systems.

global business operations. The main challenges facing the enterprises are that (1) both the globalization of operations and the implementation of SCM are moving toward a deeper level and (2) the expectations for sustainable economic growth and global environmental protection are rising. The limitations of the existing systems restrain their flexibility to cope with the challenges described earlier, that is, global operations at a deeper level and the increasing requirements for sustainable economic growth and global environmental protection. The existing systems are just a part of the solution, and a mere modest modification of the existing systems will not meet the requirements. Only a new type of system will be capable of coping with the challenges. The key characteristic of the new system is the provision of comprehensive coverage of all relevant types of resource planning. Its main objective is to close the gap between the existing systems and the proposed systems.

2.6.1 Comprehensive MF Theory

Comprehensive MF theory reveals the essence of MF objects and phenomena. The MF in the natural world existed before the appearance of human society. Such materials existed in the natural world, and originally they were not economic commodities. The MF in the natural world can benefit human beings (e.g., energy generation by water, wind, and tide); on the other hand, it can also bring natural disasters to the human society (e.g., flood, windstorm, sand storm, debris flow, polluted atmospheric currents, and water currents).

The MF in the society is the MF phenomena peculiar to the human society. During the primitive society, agricultural production activities produce relevant primitive MFs. MF is of substantial importance for human survival, since human beings cannot survive without MF. Materials exist in the natural world that can also be economic commodities.

The MF in the economic sphere is an important MF phenomenon emerging at the most recent stage. Since commodity exchange and social labor division emerge from the production and development of agricultural society, the MF in the economic sphere is developing more and more quickly due to the demands on economic development. However, restricted by the MF impetus, it is small-scaled. By the beginning of the industrial society, the development of transportation greatly expedites the MF in the economic domain. The MF in the economic world includes the MFs of industries and the flow of various materials. As one important component of humans' economic behaviors, it is also a business behavior aiming at creating values and surplus values. Its characteristics include the following: materials are economic commodities, the impetus for flow originates from economic activities, and this kind of MF is a profitable economic behavior.

MF exists not only in the economic world but also in the social world and in the natural world. As such, it is not only an economic phenomenon but also a social and natural phenomenon. It is obvious that complex relationships exist. It can be said objectively that there exists a complex MF phenomenon rather than a simple

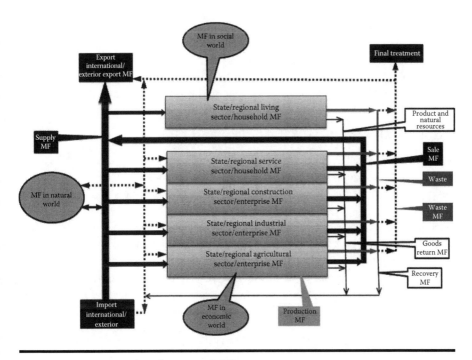

Figure 2.14   Comprehensive material flow.

MF phenomenon. Therefore, Xu (2008) called this "comprehensive MF phenomena." The theoretical framework for such comprehensive MF phenomena is called "comprehensive MF theory."

Figure 2.14 provides an introduction to comprehensive material flow, reflecting the MF in the economic world, the social world, and the natural world, as well as the interrelationships among them, which also present various MFs in the MF in the economic world (Xu, 2008). In overall MF, each of the MFs (such as MF in the economic world, the social world, and the natural world) is an important component of the overall MF rather than representing the overall MF. Among them, the MF in the economic world is the core of the MF, while the MF in the social world and the natural world are the foundations of the MF.

X party MF (XPMF) is an exemplary work that originated from the CMFT (Hou et al., 2007). Hou et al. have studied fundamental issues of XPMF, including the concept, characteristics, and operational framework of XPMF. In their study, both the theoretical and practical issues have been researched. According to Hou et al., XPMF is the collective term for MF service models. The first-party, second-party, third-party, fourth-party, and fifth-party logistics are all single-party MF model. According to the systems research, all single-party MF models decompose the supply chain MF into subunits based upon static points of view, although they do not include the interactions and structural relationships among subunits. XPMF is the collective term for MF service models, in which the value of X is determined through dynamic interactions. The synergistic effect of multiple-party MF can be considered. XPMF further possesses the properties for which MF resources can be reconfigured, reused, and expanded. Figure 2.15 shows the three-pyramid synergetic operational model of XPMF proposed by the authors (Hou et al., 2007).

2.6.2 Comprehensive MF Theory and ERP

As mentioned earlier, MF theory has directly contributed to the birth of ERP. The past decade has brought fundamental changes to the global economy and to global business operations. The main challenges facing the enterprises are that (1) both the globalization of operations and the implementation of SCM are moving toward a deeper level and that (2) the expectations for sustainable economic growth and global environmental protection are rising. These two factors involve the multiple MFs that retrain the existing ESs to cope with the challenges described earlier, especially with the increasing requirements of sustainable economic growth and global environmental protection. As such, the most recent development is that, since the existing systems are just a part of the solution, efforts to modify the existing ESs have been made. Shi et al. (2012) reported a study on developing a new type of ESs, which can consider both economic and natural MFs. The consumption of industrial or nonindustrial batteries has been increasing sharply. If the used batteries are not properly disposed of, the hazardous materials they contain may lead to environmental pollution. Research shows that one battery of the most commonly used battery category, if not treated and left rotten in soil, can render useless one square meter of land. Thus, the need to reverse the logistics of disposal of used batteries in order to lowering carbon emission becomes an important research topic in which information integration of reverse logistics is the key for the implementation of reverse logistics systems. In the framework of MF theory, ERP, and IIIE, Shi's study analyzes the processes and modeling of the reverse logistics of used batteries and investigates the integration of related information flows in preparation for developing ESs that are implementable and which comprise economic MF, social MF, and natural MF. As mentioned earlier, in general, only a new type of system will be able to cope with the challenges. The key characteristic of the new system is the provision of comprehensive coverage of all relevant types of resource planning.

2.7 Integrating ESs: Future Prospects

In today's highly competitive global economy, EI plays a major role in enterprise performance improvement. At present, for many enterprises, integration of ESs constitutes a rapidly growing need. EI has been applied to numerous types of enterprises, even to the integration of enterprises from different sectors. The drive for an integrated enterprise is mainly generated from the need of enterprises to provide

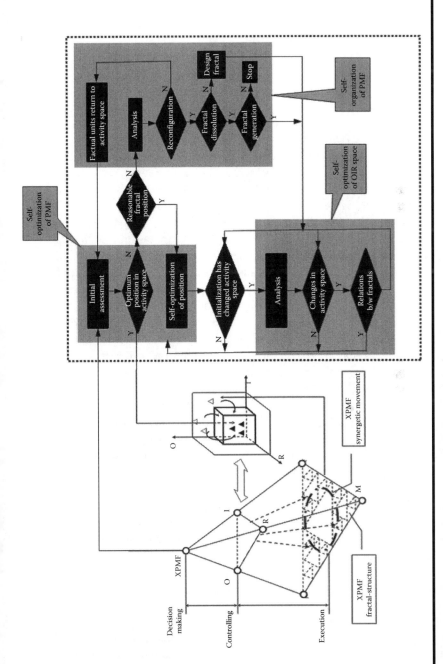

Figure 2.15 Three-pyramid synergetic operational model of XPMF.

timely availability of data and information, both intra- and interorganizationally. EI is an area of research developed in the 1990s as an extension of CIM. EI research has mainly been carried out within two distinct research subjects: enterprise modeling and information technology (in which EM refers to EI as a set of concepts and approaches that allow the defining of the structure of an enterprise, and the consistency of a system-wide function goes beyond the borders of functions). Enterprise modeling is a prerequisite to EI. In other words, the first task in EI is enterprise modeling. It is the process of generating abstractions about the business functions and business processes that are useful in describing the states of the enterprise. Enterprise modeling is an approach that consists of modeling enterprise organization, structure, and behavior. Lim et al. (1997) provided a general description of enterprise modeling and integration in terms of goals, domains, modeling, and other aspects. In EI, reference architectures such as ARIS, CIM-OSA, GIM, and PERA are frameworks that guide integrating systems through a structured methodology.

In the second subject, information technology, EI is carried out through the integration of ESs such as ERP, SCM, CRM, and Business Process Management Systems and also by developing applications such as CAD, CAM, and CAE. It is necessary to adopt different techniques to the concrete needs of each type of enterprise activity. Enterprises must adopt the form of organization and operations that allow them to obtain the maximum benefits from their resources.

EI can be approached in various ways. EI classification in terms of drivers, goals, domains, and types has been proposed by Lim et al. (1997). In general, three levels of integration are considered: (1) physical integration (interconnection of devices, NC machines, etc., via computer networks); (2) application integration (which deals with the interoperability of software applications and database systems in heterogeneous computing environments); and (3) business integration (coordination of the functions that manage, control, and monitor business processes). Researchers also consider that integration can be obtained in terms of (1) data (data modeling), (2) organization (modeling of processes and systems), and (3) communication (modeling of computer networks). Integration concerns the connections among devices, applications, and the exchanges of data among applications. It involves dealing with the heterogeneity that exists in systems that have not been integrated. Organizations have integrated applications in a "point-to-point" style; in such a style, interfaces must be developed directly between two applications. As the number of system applications increases, the cost of creating and maintaining a large number of interfaces becomes prohibitively expensive and results in spaghetti integration. Taking into consideration the new requirements for integration, organizations seek advanced EAI. The main components involve middleware infrastructure via the communications in information systems, adapters to connect applications with middleware infrastructure, and an integration broker that captures the business rules applied to workflow, distribution, and the processing of messages among applications.

Many vendors (including IBM) provide EAI tools. Modules in an EAI can be customized to specific needs by selecting and configuring adapters and defining special business rules for brokers. A case study provided by Lam (2005) showed a significant reduction in processing and decision-making time for business applications, seamless information integration and transformation, minimal disruptions to business and existing information applications, scalable and robust solutions, and an acceptable level of security. The networked enterprise, developed under a model of extended enterprise or virtual enterprise, is one of the latest emerging paradigms. This is an enterprise model that results from quick changes in the economic environment and forces suppliers and manufacturers to work with clients in a more direct way. With the globalization of manufacturing activities and the increasing requirements for interoperability, standardization plays an important role in R&D and in day-to-day activities related to EI and engineering. A number of research studies have been conducted on standardization in enterprise engineering and integration.

As mentioned earlier in this chapter, it is known that the scope and the complexity of enterprises have grown dramatically. Great challenges have arisen for EI, which requires the establishment of scientific underpinnings to face such challenges. Along with the evolution of ESs, especially from enterprise resource planning to entire resource planning, research is required to develop new frameworks and theories of EA, EI, and EAI. EI is a substantial challenge, a major systems challenge of these times. In particular, due to the emergence of the new MF theory, the concept of ERP as entire resource planning, the existing scope of enterprise modeling and EI, and the main reference architectures such as ARIS, CIM-OSA, GIM, and GIM need to be expanded to new and higher levels. All of them concern, in one way or another, the integration of data, processes, applications, and systems within an enterprise or an extended enterprise in order to achieve advancements at higher levels of industrial information integration. However, having a thorough understanding of the complexities involved and relationships between the theories, frameworks, techniques, and methodologies is a task of tremendous challenge. The objective of this chapter is to provide an introduction to this challenging task, which was achieved via the collection of relevant literature under key aspects, including an introduction to EI, manufacturing integration, engineering integration, customer integration, IERP, ERP, and CMFT.

References

Bartl, M., Füller, J., Mühlbacher, H., and Ernst, H. 2012. A manager's perspective on virtual customer integration for new product development. *Journal of Product Innovation Management*, 29(6), 1031–1046.

Bi, Z. 2002. On adaptive robot systems for manufacturing applications, PhD thesis, University of Saskatchewan, Saskatoon, Saskatchewan, Canada.

Edvardsson, B., Kristensson, P., Magnusson, P., and Sundstrom, E. 2012. Customer integration within service development—A review of methods and an analysis of insitu and exsitu contributions. *Technovation*, 32, 419–429.

Forsell, M. 2002. Improving component reuse in software development. *Jyväskylä Studies in Computing*, Vol. 16, University of Jyväskylä, Finland. ISBN 9513911616. http://books.google.com/books/about/Improving_Component_Reuse_in_Software_De.html?id=UOAJtwAACAAJ

Gao, X., Li, Z., and Li, L. 2008. A process model for concurrent design in manufacturing enterprise information systems. *Enterprise Information Systems*, 2(1), 33–46.

Hou, H., Xu, S., and Wang, H. 2007. A study on X party material flow: The theory and applications. *Enterprise Information Systems*, 1(3), 287–299.

Kulvatunyou, B. and Wysk, R. 2000. A functional approach to enterprise-based engineering integration. *Journal of Manufacturing Systems*, 19(3), 156–171.

Lam, W. 2005. An enterprise application integration (EAI) case-study: Seamless mortgage processing at Harmond Bank. *Journal of Computer Information System*, 46(1), 35–43.

Langerwalter, G. 2000. *Enterprise Resources Planning and Beyond: Integrating Your entire organization*. Boca Raton, FL: CRC Press.

Li, Z. and Xu, L. 2003. Polychromatic sets and its application in simulating complex objects and systems. *Computers and Operations Research*, 30(6), 851–860.

Lim, S., Juster, N., and de Pennington, A. 1997. Enterprise modelling and integration: A taxonomy of seven key aspects. *Computer in Industry*, 34(3), 339–359.

Lorincz, P. 2007. Evolution of enterprise systems. *Proceedings of the International Symposium on Logistics and Industrial Informatics*, Wildau, Germany, pp. 75–80.

Panetto, H. and Molina, A. 2008. Enterprise integration and interoperability in manufacturing systems: Trends and issues. *Computers in Industry*, 59(7), 641–646.

Shi, X., Li, L. X., Yang, L., Li, Z., Choi, J. Y. 2012. Information flow in reverse logistics: An industrial information integration study. *Information Technology and Management*, 13(4), 217–232.

Staley, S. and Warfield, J. 2007. Enterprise integration of product development data: Systems science in action. *Enterprise Information Systems*, 1(3), 269–285.

Su, Y. and Yang, C. 2010. Why are enterprise resource planning systems indispensable to supply chain management? *European Journal of Operational Research*, 203, 81–94.

Warfield, J. 2007. Systems science serves enterprise integration: A tutorial. *Enterprise Information Systems*, 1(2), 235–254.

Wu, S., Xu, L., and He, W. 2009. Industry-oriented enterprise resource planning. *Enterprise Information Systems*, 3(4), 409–424.

Xu, L. et al. 2006 Integrating knowledge management and ERP in enterprise information systems. *Systems Research and Behavioral Science*, 23(2), 147–156.

Xu, L. et al. 2009. Modeling and analysis techniques for cross-organizational workflow systems. *Systems Research and Behavioral Science*, 26(3), 367–389.

Xu, S. 2008. The concept and theory of material flow. *Information Systems Frontiers*, 10, 601–609.

Yin, Y. and Xie, J. 2011. Reconfigurable manufacturing execution system for pipe cutting. *Enterprise Information Systems*, 5(3), 287–299.

Zhang, Y., Li, Z., Xu, L., and Wang, J. 2011. A new method for automatic synthesis of tolerances for complex assemblies based on polychromatic sets. *Enterprise Information Systems*, 5(3), 337–358.

Chapter 3

Extended Enterprise Integration in Supply Chain

3.1 Interenterprise Collaboration

Enterprise collaboration is a practical approach used to enhance an enterprise's competitive advantage. Enterprise collaborations are not limited merely to the vertical collaboration between the upstream and downstream enterprises—they can also be horizontal collaborations. Enterprise collaborations can be multifaceted. Collaborations among enterprises may involve manufacturing, logistics, R&D, and many other aspects. Interenterprise collaborations are networks of enterprises that are collaborating with each other. The success of interenterprise collaborations relies on a solid establishment in a dynamically evolving extended enterprise environment. This, in turn, requires the systematic selection of collaboration partners beyond the enterprise's autonomous operations. Interenterprise collaboration in general will have a positive effect on an enterprise's increase in value.

With the rapid development of global economic cooperation, now more than ever, enterprises are collaborating with each other to increase opportunities. A major issue in enterprise collaboration and cooperation is the development of the capability of interoperation, that is, interoperability, as a component in an enterprise's collaboration infrastructure. Having enterprise interoperability can improve the efficiency of the cooperation, which is a key to enhancing the competitiveness of enterprises. The success of interenterprise collaborations relies on an infrastructure that takes care of issues such as collaboration coordination, trust management, and interoperability. Enterprise interoperation is one of the main characteristics of the collaborations among enterprises. Enterprise interoperation relates to both enterprise collaboration and enterprise interactions. Enterprise collaboration places emphasis on how enterprises cooperate to realize collaboration, and enterprise interaction focuses on how heterogeneous enterprise systems from different enterprises are capable of interacting with each other. "Enterprise interoperability" is defined as interoperation among two or more enterprises across organizational boundaries. "Interoperability" is the ability of two or more systems or components to exchange information and to use the information that has been exchanged (IEEE, 1990). Enterprise interoperability is also defined as the ability to communicate with peer systems and access the functionality of the peer systems (Vernadat, 1996). In addition to relating two or more systems together, the establishment of interoperability also implies the removal of incompatibilities between the systems. According to ISO 14258 and to some researchers' points of view, there are three ways to develop interoperability (Panetto and Cecil, 2013): (1) "integrated," if there is a standard format for all constituent systems. Diverse models will be interpreted according to the standard format; (2) "unified," if there is a common meta-level structure across constituent models, providing a means for establishing semantic equivalence; and (3) "federated," if models are dynamically accommodated rather than having a predetermined meta-model. The concept of mapping is applied at the ontological level, that is, the semantic level.

At present, enterprise interoperability is considered to be an area of research that is still in the embryonic stage (Jardim-Goncalves et al., 2013). Current research on enterprise interoperability focuses on important research problems such as the body of knowledge of interoperability. In spite of the research efforts that have been made so far, the theoretical foundations of enterprise interoperability have only recently begun to receive substantial attention. Jardim-Goncalves et al. (2013) studied this important issue recently by identifying the main characteristics related to the theoretical foundation of enterprise interoperability. In the proposed interoperability body of knowledge (IBoK), frameworks, theories, and models as three levels have been studied. The key research questions addressed include the following: How will the proposed IBoK contribute to establishing enterprise interoperability as a scientific subject? Two research questions have been put forward: (1) The scientific foundation of enterprise interoperability should provide IBoK with support at the level of frameworks, theories, and models and should provide a gap analysis on both where the existing research stands and where it should move toward; (2) analysis conducted at each IBoK level should provide problem-solving support regarding enterprise interoperability, to a certain extent. The efforts made on the IBoK demonstrate a growing interest in moving the subject toward becoming a scientific subject. The existing frameworks identified as related to enterprise interoperability include the Advanced Technologies for interoperability Heterogeneous Enterprise Networks and Applications (ATHENA), Business Interoperability Parameters, the CEN/ISSS eBusiness Roadmap, the C4 Interoperability Framework (C4IF), the

IDEAS Interoperability Framework, the European Interoperability Framework (EIF), Levels of Conceptual Interoperability, Levels of Information System Interoperability (LISI) C4ISR, NATO C3 Technical Architecture (NC3TA), and the Organizational Interoperability Maturity Model. Regarding the formal methods for enterprise interoperability, the disciplines that are considered relevant include mathematics, computer science, systems science, and the social sciences. The subjects that are considered relevant include calculus, graph theory, logic, set theory, Petri nets, and many others. The study completed by Jardim-Goncalves et al. (2013) on enterprise interoperability frameworks, theories, and models provide insights into the perspectives about the future direction of research.

Guedria et al. (2013) classify the major enterprise interoperability frameworks into AIF (the Interoperability Framework), ATHENA, EHIF (the E-health Interoperability Framework), EIF, and the Framework for Enterprise Interoperability (FEI). Reviewing the framework coverage with respect to enterprise interoperability, the main elements that have been identified are (1) three aspects for interoperability: organizational, conceptual, and technical; and (2) four tasks of enterprise interoperation: business, process, data, and service. Figure 3.1 provides a detailed summary of the review (Guedria et al., 2013).

There has been a good amount of research described in the literature that discusses ways to identify the relevant factors that influence interenterprise

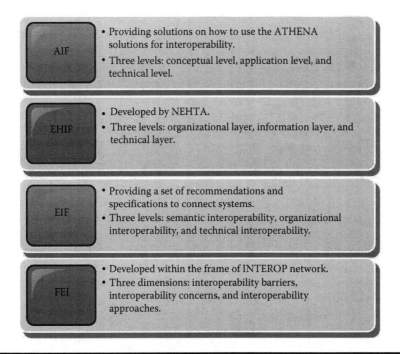

Figure 3.1　A comparison of AIF, EHIF, EIF, and FEI.

relationships. The relevant factors about collaborative relationships include strategic factors, organizational structure factors, business process and infrastructure factors, and cultural factors (Verdecho et al., 2012a,b). In order to achieve an effective collaboration, these factors are considered important; problems will occur due to a lack of attention on these factors. Since collaborative relationships do not always bring success to an enterprise, there is a need to study and measure the performance of interenterprise collaboration. There have been quite a few different frameworks introduced in the literature that can be used to measure the performance of interenterprise collaboration. One of the most widely accepted performance management or measurement frameworks is the balanced scorecard. To measure performance, a number of metrics are available. When a balanced scorecard is used, these metrics are generally grouped into financial perspective metrics, customer perspective metrics, internal business process perspective metrics, and learning and growth perspective metrics. These different metrics can be used to facilitate performance management at both the strategic and operational levels. From a methodological point of view, there is a need to further develop approaches that can manage the relationship between the earlier-mentioned factors and the interenterprise performance. Recently, a conceptual framework of collaboration for managing interenterprise collaborative relationships has been proposed (Verdecho et al., 2012a,b). This framework structures and classifies the relevant factors that influence collaborative relationships. Based on the existing models, Guedria et al. (2013) developed a maturity model for enterprise interoperability. It is structured on main enterprise interoperability frameworks and covers interoperability facets such as organizational, conceptual, and technical.

Different research methodologies have been applied to enterprise interoperation. Semantic frameworks are considered to have the potential to address interoperability issues through the use of ontology (Panetto and Cecil, 2013). Semantic web technology has been applied to an ontology-mediated function for interenterprises' manufacturing system engineering applications, and a manufacturing system engineering ontology model has been developed (Lin and Harding, 2007). This model includes manufacturing-relevant data that can enable interactions between data held by different enterprises in different formats. Through a common schema, it provides cooperating organizations with access, for the purpose of interenterprise collaboration. Through semantic schema matching, an enterprise can apply its own terminology to a mediated schema for exchanging and sharing information. This approach supports information autonomy for interenterprise collaboration by allowing individual enterprises to keep their own individual languages rather than requiring them to adopt standardized terminology.

The service-orientated computing paradigm is transforming traditional workflow management from one that is close and centralized to one that is more dynamic and more oriented toward interorganizational collaboration. Workflows serving interorganizational collaboration will take care of both intra- and interorganizational processes. Through integrating software agents, web services, and workflow ontology, an agent-based workflow model for interenterprise collaboration was developed, to support dynamic workflow definition and execution (Wang et al., 2006). The proposed approach is intended to integrate the heterogeneous software and hardware systems that exist within an enterprise or among collaborative enterprises.

3.2 Supply Chain Collaboration

A "supply chain" is defined as a set of activities that span enterprise functions from the ordering and receipt of raw materials, to the manufacturing of products, through to distribution and delivery to the customer. These activities are associated with the MF and information flow (Li, 2007). SCM is a set of synchronized activities for integrating suppliers, manufacturers, transporters, and customers efficiently so that the right product or service can be delivered at the right quantity, at the right time, to the right place. The ultimate objective of SCM is to achieve a sustainable competitive advantage (Li, 2007). In the context of economic globalization, enterprises no longer exist as independent entities, and the traditional competition between enterprises has gradually evolved into a competition between enterprise-centric supply chains. Enterprises within the same supply chain share resources and complement each other through cooperation. They form an integrated working unit, in the framework of interenterprise collaboration and operations, in order to promote overall competitiveness in the competitive supply chain environment. In the supply chain, partners' relationship becomes cooperative, as enterprises in the supply chain recognize that coordination among partners within the supply chain is a key factor in their success. In order to operate a supply chain efficiently in a cooperative manner, all of the related functions across the supply chain must operate in an integrated manner, in which the various partners within the supply chain must be efficient with respect to every aspect. This can cause increasing reliance on more collaboration, which, in turn, can be supported by better integration within the supply chain. Meanwhile, different supply chains are increasingly integrating with each other. More significantly, supply chains are becoming more interconnected with each other and are transforming into supply chain networks. Multiple supply chains, each of which may center on a dynamic center, may coexist in one or more industrial sector.

Enterprises within the supply chains or supply chain networks are required to share information in a timely fashion, to integrate resources effectively, and to adjust their business processes flexibly in order to adapt to the rapidly changing market. The increasing interactions between different supply chains call for an examination of supply chain networks as an integrated "whole" rather than as a collection of separate supply chains. Enterprise interoperability (as introduced earlier) holds critical importance in realizing such objectives.

Supply chain collaboration is defined as two or more enterprises working together to create a more competitive advantage than one that can be achieved by acting alone (Simatupang and Sridharan, 2005). These enterprises are required to work together actively toward common objectives; meanwhile, they share information, knowledge, profits, and even risks (Mentzer, 2001). The collaboration between partners in the supply chain may involve a two-level or three-level supply chain or a supply chain with more levels (Jonrinaldi and Zhang, 2013). In many cases, it involves more than three levels and multiple products. For example, a manufacturing supply chain may be a type of complex supply chain "consisting of tier-2 suppliers which produce raw materials to be supplied to tier-1 suppliers, tier-1 suppliers which produce parts for a manufacturer, the manufacturer which manufactures and assembles parts into finished products, distributors which deliver finished products to retailers, retailers selling products to end customers and a third party which collects used finished products and feeds reusable parts to the manufacturer" (Jonrinaldi and Zhang, 2013).

We described earlier that supply chain collaboration is defined as two or more enterprises working together to create a more competitive advantage than can be achieved by acting alone (Simatupang and Sridharan, 2005). Cao and Zhang (2011), however, define a supply chain collaboration as a partnership process in which two or more autonomous enterprises work closely to plan and execute supply chain operations toward common goals and mutual benefit. The collaboration between supply chain partners not only merely involves pure transactions but also leverages information sharing, market knowledge creation, and other important factors in an effort to achieve sustainable competitive advantage (Malhotra et al., 2005). Researchers have identified the problems of disconnected enterprise systems (often called "the information island") that can exist in enterprise collaborations. For example, the supply chain coordination system between Baosteel and Shanghai GM was not able to automatically exchange data, which made it impossible to successfully run vendor-managed inventory (VMI), JMI, and collaborative planning, forecasting, and replenishment (CPFR) modules. Research has revealed that information sharing can substantially improve overall supply chain performance. Researchers also agree about the importance of information sharing to enhancing coordination among the organizations in the supply chain, believing that it leads to better overall performance. Liu et al. indicate that the major success factors for supply chains include extensive data management capabilities and advanced interorganizational information systems that enable better information exchange (Liu et al., 2005). Researchers agree that supply chain collaboration involves information sharing (Manthou et al., 2004), knowledge sharing (Li, 2007), goal congruence (Angeles and Nath, 2001), decision synchronization (Stank et al., 2011), resource sharing (Sheu et al., 2006), and incentive alignment (Simatupang and Sridharan, 2005) among independent supply chain partners. Figure 3.2 shows a typical supply chain and the ways in which data, information, and knowledge are shared in an organization.

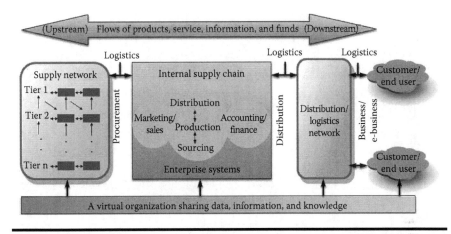

Figure 3.2 Typical supply chain and the way in which data, information, and knowledge are shared. (Adapted from Li, L., *Supply Chain Management: Concepts, Techniques and Practices,* **World Scientific, Singapore, 2007.)**

In general, collaborating enterprises hope to improve their performance in several different aspects, such as increasing profits, reducing costs, improving product availability, increasing economic value, improving inventory level, and others (Fawcett et al., 2008). Stank et al. (2011) found that collaboration with external supply chain entities will also enhance internal collaboration, which in turn will improve overall performance. In fact, supply chain collaboration helps enterprises reduce costs through process integration, thus increasing the possibility that the partners will behave in the best interest of the partnership. Supply chain collaboration also helps enterprises avoid internalizing an operation that may not be aligned with expected competencies. Through collaboration, supply chain partners can work as if they were part of a single enterprise (Lambert and Christopher, 2000). They can access and leverage each other's resources and enjoy the associated benefits. Such collaboration, of course, increases collaborative advantage and enhances enterprise performance.

To some extent, the meaning of "supply chain collaboration" deviates from that of "supply chain integration," although the terms are used interchangeably sometimes, since both of them refer to a coupling process between supply chain partners. "Collaboration" puts more emphasis on governance through relational means. "Integration" means the integration and control of those processes that were formerly carried on independently. Although there are some overlaps between supply chain collaboration and supply chain integration, "supply chain collaboration" is a construct to capture the joint relationship between autonomous supply chain partners, while "supply chain integration" captures the integrated relationships between formerly autonomous processes. Cao et al. (2011) create a model for supply chain coordination with seven components: information sharing, goal

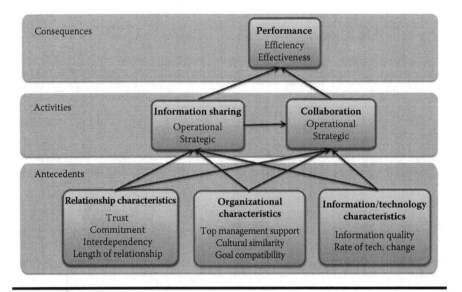

Figure 3.3 Integrative model for explaining how factors influence the performance of the supply chain.

congruence, decision synchronization, incentive alignment, resource sharing, collaborative communication, and joint knowledge creation. These seven components are mechanisms to reduce uncertainty, risks, and conflicts and to improve the relationships between partners. From a knowledge-and-learning perspective, joint knowledge creation enhances innovation and leverages resources.

An integrative model was proposed to explain how different factors influence the performance of the entire supply chain. Figure 3.3 shows the model (Lee et al., 2010).

In this model, "trust" is defined as the extent to which enterprises believe that supply chain partners will fulfill their responsibilities to each other in good faith. "Commitment" is defined as the extent to which supply chain partners will maintain and strengthen their business relationships. "Interdependency" is defined as the extent to which supply chain partners believe that their business relationships are necessary. "Length of relationship" is defined as the period for which supply chain partners have a business relationship. "Top management support" is defined as the extent to which top managers at both enterprises understand and support their business relationships. "Cultural similarity" is defined as the extent to which supply chain partners have similar values, beliefs, and management practices. "Goal compatibility" is defined as the extent to which supply chain partners have clear and agreed-upon transactional goals. Due to the increasing applications of information technology, supply chains have experienced many changes. "Information quality" is defined as the value of the information shared by supply chain partners. "Rate of technological change" is defined as the rate of technological change of the products of the supply chain partners. "Operational and strategic information sharing" is defined as the extent to which supply chain partners intend to share operational and strategic information, respectively. "Operational and strategic collaboration" refers to the extent to which supply chain partners intend to collaborate for operational and strategic purposes, respectively. "Efficiency" is defined as the extent of saving resources from the supply chain activities. "Effectiveness" refers to the extent of achieving goals from the supply chain activities.

It is obvious that many factors can restrict supply chain collaboration. Wang (2012) summarizes some of these obstructive factors in four aspects (see Figure 3.4). The strategies of constructing the collaboration system in the supply chain are positioning the role of the partners, constructing an incentive system, applying information technology, and creating a true collaborative system.

According to Cannella and Ciancimino (2010), one of the first frameworks of collaborative supply chains based on the degree of synchronization of partners' inventory control policies was developed by Holweg et al. (2005). Four supply chain archetypes were identified, and each is characterized by a combination of two dimensions: planning collaboration and inventory collaboration. "Planning collaboration" refers to the real-time sharing of market demand data for the generation of conjoint forecasting. "Inventory collaboration" refers to the real-time sharing of information on inventory levels and in-transit items for centralized

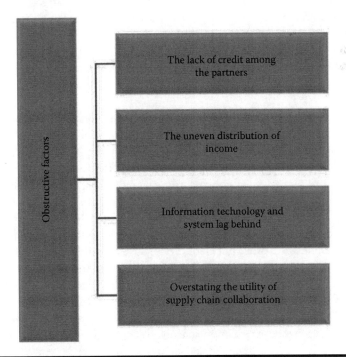

Figure 3.4 Obstructive factors in supply chain collaboration.

replenishment activities. The supply chain configurations are identified as follows (Holweg et al., 2005):

- Traditional supply chain (no collaboration): Each level in the supply chain issues production orders and replenishes stock without considering the situation at either the up- or the downstream tiers of the supply chain. It is a decentralized supply chain. Each member generates an independent production distribution plan on the basis of incoming orders from direct customers.
- Information exchange (planning collaboration): The retailer and supplier order independently, yet exchange demand information and action plans in order to align their forecasts for capacity and long-term planning. This is a decentralized supply chain. Each member generates an independent production distribution plan on the basis of incoming orders from direct customers and from market demand data.
- Vendor-managed replenishment (inventory collaboration): The task of generating a replenishment order is given to the supplier, who takes responsibility for maintaining the retailer's inventory (and subsequently the retailer's service levels). A centralized production distribution plan is generated by suppliers on the basis of the downstream partner's visibility of inventory levels and work in progress.
- Synchronized supply (inventory and planning collaboration): The supplier takes charge of the customer's inventory replenishment at the operational level and uses this visibility in planning its own supply operations. A centralized production distribution plan is jointly generated by supply chain members; it offers complete visibility of inventory levels, work-in-progress levels, and market demand.

Another framework for supply chain collaboration was proposed (Matopoulos et al., 2007). The proposed framework includes two main components. The first component is the design and government of supply chain activities. This component includes four elements. The first element concerns the selection of an appropriate partner. Enterprises make decisions regarding who they will choose as their partners in the supply chain environment. These decisions are typically made based on the expectations, perceived benefits and drawbacks, and the "business fit" of companies. The second element involves the selection of the activities in which collaboration will be established. Since not all of the business activities require collaboration, firms need to determine the specific activities upon which they will collaborate. The third element is the identification of the level at which firms will collaborate. The levels for collaboration include the strategic level, the tactical level, and the operational level. The fourth element is the selection of the appropriate technique and technology to facilitate information sharing. Technology investment and training are usually needed to help potential collaborators meet the requirements of collaboration in terms of technology and techniques.

The second component is the establishment and maintenance of supply chain relationships. A number of elements are important in establishing and maintaining a supply chain relationship. The crucial elements include trust, dependence, and a risk-and-reward-sharing balance. It is important to identify the ways in which these elements interact with each other and to note how they affect and determine the intensity of the collaboration, as well as the selection of appropriate information data sharing (Matopoulos et al., 2007).

The shift from the traditional supply chain toward integrated supply chain networks calls for innovative research methodologies. Analytical methods have been the mainstream methodology described in the SCM literature. Since many partners can participate in the chain, a coordination mechanism is needed for an effective collaboration between the partners in the supply chain. Much research has been conducted in this regard. A distributed constraints satisfaction problem algorithm that can be applied to manufacturing supply chain collaboration was proposed (Chang et al., 2009). Using a suitable algorithm, time and cost in the manufacturing supply chain can be reduced, and better effectiveness can be achieved for gaining competitiveness. For the supply chain composed of supplier and manufacturer, or of supplier and retailer, a revenue sharing contract has been proposed for realizing collaboration (Cachon and Lariviere, 2005). Wang and Wang (2008) have studied the collaboration between a single supplier and several customers through designing a wholesale discount strategy using the conditions of information symmetry and asymmetry. Chen and Xiao (2009) examined the collaboration problem between a single manufacturer and several retailers regarding demand disruption and found that discount strategy and price strategy can help solve the problem. Context-aware services can make an enterprise respond to its partners in the supply chain more quickly and timely, which, in turn, can facilitate collaboration among supply chain participants. Researchers analyzed the context-aware services in SCM and proposed a context-aware model for purchase order management between the buyer and the supplier, as well as a reasoning method that helps to analyze the change of context and provide proper services accordingly. Some supply chain problems can be attributed to increased inventory and poor customer service levels, which can increase the overall supply chain cost (Hudnurkar and Rathod, 2012). Two collaborative techniques, namely, VMI and CPFR, are considered to be effective in this regard. Experimental research using a beer game has been conducted to study the difference in the performance of a traditional supply chain, a VMI-based supply chain, and a CPFR-based supply chain in an industrial setting. Results affirm the ability of a collaborative supply chain to reduce inventory and the total cost of supply chain and to achieve high levels of service.

Systems science and the ideas behind it have penetrated many disciplines, and systems theory has been considered as the basis for many disciplines. The concept of a complex adaptive system (CAS) was introduced by John Holland in his book titled *Hidden Order: How Adaptation Builds Complexity*. Current knowledge and theory about CAS mainly originates from cybernetics. CAS is one of the focuses of complexity science in cybernetics, which provides a new perspective for recognizing, understanding, controlling, and managing complex systems. In a CAS, the members (or so-called agents) are adaptive; they communicate with each other. Recent application examples of CAS are limited not only to biology, sociology, and e-commerce, but also to supply chains. For example, Pathak and Dilts (2002) apply the concept of CAS to the investigation of the evolution and emergence of supply chain networks.

Information architecture for SCM has drawn a lot of attention in recent years. Recent research explores the role of service-oriented architecture (SOA), RFID, agent, workflow management, and the IoT as an enabler of real-time management and control in the supply chain. In the following sections, these cutting-edge technologies, which can be integrated as part of a supply chain system, are introduced (Xu, 2011).

3.2.1 Service-Oriented Architecture

"SOA" is a paradigm that helps to integrate or provide an architecture to group functionality in terms of interoperable services, based on business processes on top of the organization's existing enterprise information system. A "service" is considered to be an abstract business concept that represents the functionalities of business (Tao and Yang, 2008). As an architectural approach, SOA takes business applications and breaks them down into individual functions and processes, as services. This paradigm allows the recursive aggregation of services into new business processes and applications (Unger et al., 2009). Each of these existing services can be recomposited, reconstructed, and reused to create new applications. In other words, discrete components of software functionality can be recomposed and passed to other functions and systems as services that enable different applications to reuse common parts, and services can be pieces of application functionality that represent a reconstructed business task. In turn, alternative applications can be built by assembling these reconstructible components. Researchers point out that SOA enables organizations to create new applications dynamically to meet changing business needs, and its composability is considered to be an important benefit (Quartel et al., 2009).

SOA is gaining increasing importance in enterprise and SCM computing and is influencing corresponding developments in automating enterprises and supply chain operations. One of the key aspects of the significance of SOA to SCM is that services can be made available to others in the supply chain; thus, SOA has been considered to be one of the key technologies that enable globally integrated supply chains (Butner, 2009). SOA enables access to supply chain information for many different partners. With SOA, an organization can create a more flexible and responsive environment for its supply chain partners. Services can be built and maintained in one place but can be made available to applications in other places in the supply chain for incorporating them into their own functionalities.

For enterprise computing, Candido et al. (2011) recently proposed a service-oriented infrastructure to support the deployment of evolvable production systems.

Many enterprises are currently adopting a demand-driven model (Li, 2007). A successful demand-driven supply chain requires prompt responsiveness along the entire supply chain. Throughout such a supply chain, information processing continues to become more important as it requires a fully integrated enterprise system characterized by real-time data exchange, real-time responsiveness, real-time collaboration, real-time synchronization, real-time visibility, and a sophisticated level of information integration. SOA is considered suitable for such demand-driven supply chains. SOA can integrate reconstructable component business services in the process of managing supply chains in real time across organizational boundaries (Butner, 2009). In particular, SOA can integrate supply chain processes and information, and sharing such information can help create a better environment for real-time data exchange, real-time responsiveness, real-time collaboration, real-time synchronization, and real-time visibility across the entire supply chain.

"Service-oriented computing" (SOC) is a computing paradigm that utilizes services as fundamental elements for the development of applications/solutions. Such services represent individual functions and processes in business applications; they can be reused and recombined to create new business applications. As the business processes are changed, services can be reassembled.

SOC relies on SOA to build the service model, since SOA is an architecture in which discrete sets of software functionalities can be componentized to deliver services that enable different applications to use common or reusable components when facing changing operations and building new applications (Papazoglou and Georgakooulos, 2003), and other applications can access such services and incorporate them into their own functionalities (Butner, 2009). Since SOA publishes business functionality in the form of programming and accessible software services, other application programs can use these services through published and discoverable interfaces that provide a new blueprint to solve software reuse and enterprise information system integration (Zhang et al., 2008).

SOA represents the latest trend in the integration of heterogeneous systems and has also received much attention as an architecture for integrating platform, protocol, and legacy systems; it has been considered to be suitable for enterprise architecture as it is characterized by simplicity, flexibility, and adaptability (van Lessen et al., 2009; van Sinderen, 2008). A recent study addresses the modeling aspects of SOA that propose a blueprint for an integrated information system in SCM. An ADL named SO-ADL is put forward, and a service-oriented development process based on SO-ADL that guides the service-oriented system exploitation is defined. Furthermore, using web service technologies, the usage and practicability of the modeling approach can be demonstrated with examples (Zhang et al., 2008).

SOA represents an emerging paradigm for enterprises to coordinate seamlessly in the environment of heterogeneous information systems, enabling the timely sharing of information in the supply chain and developing flexible large-scale software systems (Rosenberg et al., 2008). Example applications include the InLife project (Ribeiro et al., 2009) and information integration based on an SOA in the agri-food industry (Wolfert et al., 2010).

3.2.2 RFID and IoT

SCM attempts to effectively obtain real-time information and to enhance dynamic management and control via information sharing by involved partners in the supply chain. RFID refers to identification and tracking by using radio-waves. In recent years, RFID has become very popular in logistics, MF systems, and SCM as a representative technology for automatic identification and data capture. RFID is an automatic identification solution that streamlines identification and data acquisition. Integrating promising technologies such as RFID can help improve the effectiveness of the information flow in a supply chain. Partners in the supply chain are able to access information and to practice quality control based on the data shared through RFID and other technologies. An industrial information integration model for SCM based on RFID has been proposed (Hsu and Wallace, 2007).

RFID technology has been adopted in SCM for the purpose of improving tracing capability, since traceability is inherently linked with supply chain integrity (Kumar and Budin, 2006). The application of RFID technology, one of the most cutting-edge technologies for supply chain traceability and integrity, has enabled enterprises to facilitate real-time traceability. In the construction industry, RFID can be assisted with personal digital assistants (PDAs) to enable on-site engineers to integrate work processes seamlessly at job sites. Wang et al. presented a web-based portal system that incorporates RFID to improve the efficiency and effectiveness of on-site data acquisition and information sharing in the supply chain. The system not only improves the efficient acquisition of data on-site using RFID but also provides a monitor to control construction progress. The study demonstrates the effectiveness of an RFID-based SCM application in the construction industry as it responds efficiently and enhances the information flow in the supply chain environment, since real-time monitoring and control is always desirable (Wang et al., 2007). Kumar and Budin gave two examples of massive recalls that could have been avoided if tracing capabilities had been incorporated, since tracing capabilities enable recalls of partial products instead of massive recall (Kumar and Budin, 2006). In the food industry, RFID can greatly reduce the number of recalls, as well as their negative impacts.

RFID systems are one of the most promising technologies to improve management across the supply chain. RFID shows great promise in SCM, since it allows nearly autonomous tracking of and passing of information. RFID technology can also be assisted by other technologies to enable the seamless integration of work processes. The IoT is a term that has been introduced in recent years to describe objects that are able to communicate via the Internet. The IoT is an emerging

Internet-based information architecture that can be employed to facilitate information flow in global supply chain networks. Haller et al. have provided the following definition: "A world where physical objects are seamlessly integrated into the information network, and where they, the physical objects, can become active participants in business processes. Services are available to interact with these 'smart objects' over the Internet, query their state and any information associated with them, taking into account security and privacy issues" (Haller et al., 2009).

With the assumption that objects have digital functionality and can be identified and tracked automatically, the IoT can dramatically streamline how the supply chain will be managed (Kranz et al., 2010). The significance of the IoT to SCM, especially the new supply chain information transmission approach based on IoT, has been emphasized and proposed (Yan and Huang, 2009). Since the IoT will have an impact on the global supply chain networks, many new opportunities in applying IoT to supply chain management can be foreseen in the near future.

3.2.3 Agent

An "agent" is defined as an autonomous software entity that is able to interact with its environment. An agent is able to respond to other agents and/or to its environment, and it has control over its internal state and actions. In a multiagent system (MAS), each individual agent is assigned to a problem or a subproblem, since a problem can be decomposed into subproblems. A MAS solves the entire problem through collaboration. A multiagent paradigm is useful for collecting data and for controlling activities in distributed environments (Kong et al., 2009); as such, MASs are considered to have the potential to improve SCM. Research on the application of MASs in industrial information integration presents a rich body of results. Distributed agents are increasingly used for the purposes of data collection, monitoring, and control.

A supply chain can be viewed as being managed by a set of intelligent agents; each is responsible for one or more activities in the supply chain, and each is interacting with others in executing their responsibilities (Fox et al., 1993). Agents can increase the level of flexibility in SCM and can enable enterprises in the supply chain to be more responsive. Agent-based systems can also provide opportunities to construct a large complex system out of relatively simple and autonomous components, since software agents are suitable for coordinating the complex interdependencies between the activities of independent organizations forming supply chains (Janssen, 2005). Nissen found that agents are of crucial importance for responsiveness (Nissen, 2001). Mangina considered that agents can enhance information visibility and quality (Mangina and Vlachos, 2005). Symeonidis et al. developed a multiagent architecture that combines multiagent and data mining technologies to provide intelligent decision support to SCM (Symeonidis et al., 2003). Janssen presented a MAS to improve SCM (Janssen, 2005). Li and Wang proposed a multiagent-based model in SCM with a particular emphasis on e-business (Li and Wang, 2007). MAS technology offers means and tools for supply chain collaboration. MAS has been introduced into supply chain coordination, and an agent-based framework for supply chain coordination in construction has been proposed (Xue et al., 2005). Cheng (2011) has adopted MAS for the purpose of realizing supply chain collaboration management as MAS shares many similarities with supply chain collaboration management. For example, a supply chain consists of many business entities that are geographically dispersed and have a variety of functions; similarly, an MAS consists of agents that are capable of various functions and playing various roles. Ezzeddine and Abdellatif (2012) developed a distributed architecture based on MAS and semantic web services (SWSs) for facilitating collaborative decision making in an extended enterprise environment. This architecture ensures cooperation and information exchange between different actors in the supply chain, while respecting their autonomy and distribution. This is an example of managing supply chain coordination through MAS. One method that has great potential in modeling complex systems is agent-based modeling (Zhang and Bhattacharyya, 2007). Agent-based modeling undertakes a bottom-up approach to the modeling of agents and the way they act and interact, and the overall dynamics emerge from the interactions of either heterogeneous or homogeneous agents. According to Zhang and Bhattacharyya (2007), agent-based modeling has been employed in supply chain research for the following reasons: (1) agent-based modeling is noted for its usefulness when attempting to solve problems that cannot be mathematically modeled, or for which analytical solutions are not readily obtainable; (2) supply chain networks, given their complexity, present an ideal application area for agent-based modeling. In the agent-based modeling process, agent behaviors are usually required to be calibrated. Various learning techniques are available, in which Q-learning, a type of reinforcement learning method, has been consistently performing well. In Zhang and Bhattacharyya's research, a supply chain network model was constructed on the basis of a general-purpose agent-based supply chain network framework. Q-learning was used; the multiple agents in the model simultaneously and independently learned to set values for a parameter.

The rapid development in the field of agent-based systems offers new opportunities to SCM, although the research of agent-based supply chain systems' use has still been limited so far. Supply chain performance can significantly benefit from the constant monitoring of supply chains using agent systems.

3.2.4 Workflow Management

Emerging e-business, e-commerce, virtual enterprises (VEs), and SCM are leading the trend of managing workflows across organizational boundaries, since today's enterprises must often operate across organizational boundaries, interacting with each other in order to face increasing competitive challenges. As Internet-based e-business is becoming the new frontier of business opportunities, partners in the supply chain are increasingly participating in dynamic global networks that involve complex business process. As a result, IT efforts have been focusing on cross-organizational business processes that link various organizations in the supply chain.

Crossorganizational workflows are composed of an intraorganizational workflow and an interorganizational workflow. A changing business environment requires an organization to adjust and integrate both the intra- and interorganizational processes frequently and dynamically. Wolfert et al. define inter- and intraintegration, and process and application integration in this way: "intraorganizational integration" is to overcome fragmentation between organizational units, while "interorganizational integration" is to integrate enterprises in the supply chain; "process integration" is to align tasks through coordination, and "application integration" is to align software systems to reach cross-system interoperability (Wolfert et al., 2010).

Since process integration is one of the main types of integrations and since such integration can be either intra- or interorganizational, intra- and interorganizational workflow monitoring and management capability can enhance the overall performance of supply chains. An intra- and interorganizational workflow monitoring and management capability can also enhance information sharing at both the intra- and interorganizational levels. Such information sharing in the supply chain will enable all of the partners in the extended supply chain system to improve product quality, optimizing operation performance, enhancing collaboration, and gaining competitive advantage.

With the emergence of e-business and e-commerce, another important trend appears: as workflow management is operated across multiple organizations, so that enterprises have the opportunity to reshape their business processes beyond their organizational boundaries. Business processes and their related workflow systems have gained greater interest since the early 1990s; research on enterprise business processes has become a prominent area that attracts attention from both academia and industry. Although workflow monitoring and management spans a broad continuum, the key idea of workflow management is to track process-related information and the status of each instance of the process as it moves through an organization (Oppong et al., 2005). As SCM has been rapidly developed worldwide, the scope of enterprise interoperation and integration is now extending from intra- to interorganizational; more efforts have been focused on the integration of interorganizational systems to form an interenterprise architecture. Consequently, it is necessary to study both intra- and interorganizational business processes with a scientific approach. One of the approaches to SCM is the managing and optimizing of operations based on the information provided by workflow systems of the enterprises in the supply chain. This requires not only intra- but also interorganizational process information.

Interenterprise workflow architecture can support SCM, since it focuses on the implementation of interoperability between independent enterprises and the deployment of business processes over multiple enterprises (Liu et al., 2005). In the information-based economy occasioned by the Internet and e-commerce, it is essential to consider workflow interoperability in a more complex and dynamic environment (Popova and Sharpanskykh, 2008; Prisecaru, 2008; Tan et al., 2008). Liu et al. provided an example of an interenterprise workflow architecture for SCM that is an integrated system which provides information exchange between customers and suppliers (Liu et al., 2005). The architecture consists of a workflow-supported inner supply chain system and an integrated interface. Bechini et al. described the role played by interorganizational information systems in developing an industrial traceability system (Bechini et al., 2008). An industrial traceability system for SCM requires tracing MFs as well as corresponding information flows. As a result, both the intraorganizational and interorganizational information systems are expected to play a major role.

The early idea of the automation of business processes dates back to the 1970s. Workflow Management Coalition (WMC) defines a "workflow" as a computerized facilitation for the automation of a business process, in whole or in part. The concept of workflow is closely related to a business process that consists of a number of tasks that need to be carried out and a set of conditions that determine the order of the tasks. Three types of workflows are generally recognized in literature (van der Aalst, 1998). "Production workflow" is associated with routine processes and is characterized by a fixed definition of tasks and an order of execution. "Ad hoc workflow" is associated with nonroutine processes, which could result in a novel situation. In an "administrative workflow," cases follow a well-defined procedure, but the alternative routing of a case is possible. Compared with the other two types, production workflows correspond to critical business processes and possess a high potential to add value to an organization. Hence, they are the focus of most of the studies on workflow modeling (Ha and Suh, 2008).

Workflow management plays an important role in cooperative business domains such as SCM. Workflow management has now been accepted as one of the most successful types of systems supporting cooperative enterprise operation (Dourish, 2001). Workflow management is considered to be an efficient way of monitoring, controlling, and optimizing business processes through IT support; it is playing an important role in improving organizational performance through the automation of business processes (van der Aalst, 1998). The crossorganizational workflow offers businesses the opportunity to reshape business processes beyond the boundaries of own organizations (van der Aalst, 2000). Interconnecting business processes across systems and organizations provides obvious benefits, such as greater process transparency, higher degrees of integration, facilitation of communication, and a higher throughput in a given time interval (Schulz and Orlowska, 2004). Workflow systems have been considered to be efficient tools that enable the automation of organizational business processes. Workflow management provides increased process efficiency through automation, process standardization, improved information availability, automated assignments of tasks, and process monitoring through specific management tools, that is, workflow management systems (WfMSs).

A WfMS defines, manages, and executes workflow through the execution of software. WfMS has become a standard solution for managing complicated processes in many business organizations since its appearance in the early 1990s. Despite a few failures associated with the introduction of WfMS, workflow technology has managed to become an indispensable part of enterprise systems. By having a dedicated automated system in place for managing business processes, such processes could theoretically be executed faster and more efficiently. Workflow technology can be used to improve the business process and to increase performance, since the improvement can be quantified with respect to lead time, wait time, service time, utilization of resources, etc. (Curtis et al., 1992). WfMS can also be used as a platform for knowledge sharing and learning in supply chain management. The knowledge workers in each enterprise can perform creative intellectual activities. WfMS can be employed as a repository of valuable process knowledge and can act as a vehicle for collecting and distributing knowledge across the entire supply chain. An additional workflow may need to be constructed in order to realize the benefits of knowledge sharing in the crossorganizational context.

Most traditional WfMSs assume one centralized enactment service, and most are able to support workflows inside one organization, but most have problems dealing with workflows crossing organizational boundaries (van der Aalst, 1999). It is critical to ensure that problems such as inconsistency and duplication of work do not arise due to the lack of transparency across different organizations. The inherently hybrid nature of business processes, particularly since such processes spread across multiple locations, resources, and organizational entities, presents challenges for researchers.

According to the framework proposed by Basu and Kumar, workflow research can be viewed in terms of three layers (Basu and Kumar, 2002). The first layer pertains to issues regarding "intraorganizational workflow," which links activities between different departments within one organization. The second layer corresponds to "interorganizational workflow," which covers distributed processes between different organizations; both of these comprise the crossorganizational workflow. The third layer concerns workflow technologies in e-commerce settings that have not yet been fully investigated.

Effective management of business processes relies largely on perfect workflow modeling and analysis. The following are the perspectives from which workflows are commonly modeled and represented (van der Aalst, 2000):

1. *Control-flow/process perspective*: Workflow processes are defined to specify which tasks need to be executed and in what order. This perspective also shows how a specific case is executed according to the specified routing.
2. *Resource/organization perspective*: Organizational structure and population (resources, ranging from humans to devices) are specified; this perspective focuses on problems such as defining which resources are involved in each task.

3. *Data/information perspective*: This perspective mainly deals with (1) the informational entities involved in the process and (2) the structure of these entities and their interrelationships.
4. *Task/function perspective*: This perspective describes the elementary operations performed by resources while executing a task for a specific case.
5. *Operational/application perspective*: In this perspective, the elementary actions are described; this perspective specifies which tools and applications are used to execute these actions.

Among the modeling techniques, most have shown capability in graphical representation and formal semantics in modeling workflows in an intraorganizational context. Currently, there is an urgent demand for translation between various models so that different WfMSs can interoperate. It is an interesting research topic that had not received much attention until few years ago. Further research could lead to methods for integrating heterogeneous models into a common framework.

Each of the earlier-mentioned techniques has its own advantages as well as disadvantages. Interorganizational workflow modeling moves deeper into studying the architectures for combining different organizational workflows while it continues to reconcile the relationships between each of the systems. Some approaches are specifically proposed for modeling interorganizational workflows; among these are the routing approach and the interaction model. Several cognitive approaches are also suggested to support the dynamic routing of information, and new languages (e.g., XRL and SGML) are being proposed to handle the routing of information among organizations; the latter deals with the patterns created for the coordinated partners to interact with each other. It also appears that a selective, yet partial, exposure of private workflow data to external partners will be possible in the future.

In terms of evaluation, qualitative analysis mainly focuses on checking the structural soundness, which can usually be done through validation and verification of workflows. Quantitative analysis requires the calculation of the performance indices related to workflows. The existing techniques include computational simulation, Markovian chain, and queuing theory.

Although analytical methods have been the mainstream methodology in the SCM literature, a major limitation of the analytical approach is its limits in modeling complex systems such as an integrated supply networks. This difficulty has been noticed in the literature. As a result, in addition to different types of models and algorithms, researchers have applied new technology in this regard. Supply chain integration is enterprise "collaboration" in nature, and general integration does not sufficiently support this new endeavor (Hsu et al., 2007). Using the techniques and technologies introduced earlier (and others, as well), one of the main technical issues facing collaboration in a supply chain is how to effectively integrate distributed computing resources to realize collaborative computing and to support the demand for new collaborative computing applications. Wu and Wang (2010) propose a service-oriented conceptual computing model to facilitate the integration of computing resources in a supply chain. In their model, each computing resource that participates in collaborative computing is encapsulated as a collaborative computing service and is coupled into a loosely collaborated computing alliance. A collaboration business-event-driven mechanism has also been designed for supporting dynamic and flexible collaborative computing among collaborative computing services.

Technologies based on web service composition (WSC) can provide an integrated environment for enterprise collaboration (Zhang and Xu, 2010). Supply chain collaboration can rely on the restructuring of the nodes in the supply chain, in which one node interacts with the other to complete the complex business collaboration. Each node of the supply chain may relate to different systems such as the procurement system, the inventory system, and the manufacturing system. A system in the supply chain can invoke the remote web services published by other nodes, through which business collaboration can be realized. The study shows that WSC is effective in integrating heterogeneous systems.

Similar to the work using SWS introduced earlier (Ezzeddine and Abdellatif, 2012), SWS has been proposed to be used in supply chain coordination (Liu and Nie, 2007). With the help of web service ontology in supply chains based on OWL-S, agents can analyze and interpret acquired information. Based on this, it is possible to facilitate communication. Service providers are able to publish semantic descriptions of their service capabilities and interaction protocols, which are useful for assisting requestors to understand and to select the appropriate services. Service requestors can formulate their objectives as well as form requests to service providers by using service providers' semantically described service interfaces as guides. Moreover, they can adjust current service flexibly, discover new ones automatically, and substitute the new ones for ones that are no longer available when the service interface, the form of service, or the interaction protocol change.

Supply chain collaboration requires communication among different platforms, which refers not only to data interchange but also to the integration of business processes among different partners. A collaborative configuration model in the supply chain based on the existing pattern of business operations and with interenterprise scenarios depending mainly on the partner interface process (PIP) of RosettaNet was proposed by Shi and Chen (2007). The model helps improve business operation and information sharing.

Figure 3.5 shows an integrated supply chain system that has been proposed (Chen et al., 2007). During the process of integration, distributed information interaction and collaboration are required for multienterprises based on web services by the core technology of UDDI. UDDI helps to store descriptions about enterprises and the services. UDDI supports service registration and service discovery to aid in the sharing of business information among enterprises.

A recent trend is to enable supply chain collaboration in a hybrid cloud, since many enterprises are adopting the infrastructure as a service (IaaS) model for developing purposes. Chebrolu (2012) discusses the successful enablement of

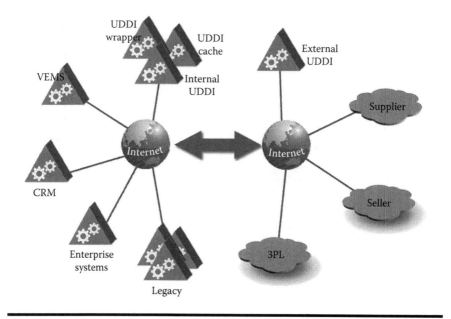

Figure 3.5 Integrated supply chain system.

collaboration among several supply chain partners using Cisco's enterprise collaboration platform, Quad, at an IaaS public cloud provider, Savvis. Several security mechanisms are implemented, including federated authentication and authorization, data level security, user to site SSL-VPN, web application vulnerability assessment, multiple-level firewalls, host hardening, and end-to-end transport layer security, in order to enhance the security posture of collaboration among supply chain communities in a hybrid cloud environment.

It is becoming a key requirement for enterprises in a supply chain to collaborate electronically in order to support more efficient business processes such as process innovation as well as product innovation (Cassivi, 2006). Another trend is related to the use of e-collaboration tools in a supply chain environment. Enterprises participating in the supply chain must develop the competencies needed for e-collaboration. With the continuous development of sophisticated e-collaboration tools, enterprises must learn to use these tools to collaborate with partners in order to manage and execute complex supply chain activities.

3.3 Integrating Supply Chain

Supply chain integration involves integration at both intra- and interorganizational levels. There are many different types of supply chain integration. "Supply chain integration" has been defined broadly as the extent to which supply chain partners work cooperatively in order to achieve mutually beneficial outcomes (Kannan and Tan, 2010). In supply chain integration, one of the main distinctions is between upstream and downstream integration. Another distinction is between intraorganizational (internal) and interorganizational (external) integration. New (2004) argued that integration can be understood in three ways: operational integration (which coordinates inventory, scheduling, transport, new product development), functional integration (which manages different managerial functions such as purchasing and inventory management), and relational integration (which improves boundary relations). In other research, distinctions have been made between technology initiatives (such as web-based integration), enterprise systems integration, and relational initiatives (such as crossfunctional involvement). More and more researchers agree that supply chain integration is multidimensional in nature.

Existing literature generally agrees that information sharing is one of the most important foundations for ensuring the success of supply chain integration (Lau et al., 2010). The literature also suggests that enterprises require a higher level of coordination for knowledge sharing, resource sharing, and information sharing. It would be very difficult to realize supply chain integration without the earlier-mentioned information sharing, knowledge sharing, and resource sharing. Kulp et al. (2004) indicate that for process collaboration, information sharing is considered as a basic dimension for a successful supply chain integration. Sahin and Robinson (2005) define information sharing and decision-making coordination as two major dimensions of supply chain integration at the operational level. As such, supply chain integration is a construct consisting of information sharing and operational coordination.

In the area of supply chain integration, a variety of models have been proposed. The role of collaboration between supply chain partners for improving the performance of inventory control is of interest. Tu and Yang (2010) developed a single-vendor multiple-buyer integrated inventory model with the objective of minimizing the total relevant annual cost for the assessment of collaboration. A procedure for finding the optimal solution has also been developed to illustrate the algorithmic process of the proposed approach. Coordination is a core problem in supply chain integration. Supply chain integrated planning is another important measure in realizing coordination between supply chain partners. Most existing research emphasizes two-level supply chain architecture rather than multilevel ones. Ji and Guo (2009) study the problem of the integrated planning of a three-level supply chain. A model of integrated supply chain planning has been proposed with the objective of minimizing the cost of production, inventory, and transportation. A hybrid genetic algorithm is used in problem solving. Hu and Hu (2010) analyzed the intelligent collaboration mode of supply chain. Shi and Bian (2010) developed a multiproduct, multistage integrated SCM model based on multiple-objective programming for the coordinative operation of the supply chain.

Technologies play a major role in integrating supply chains. Integrative studies on supply chain performance evaluation are scarce. Jia and Zhou (2007) proposed an integrated framework of system evaluation based on multiagents. The framework emphasizes five factors including the interactive agent, the management agent, the qualitative evaluation agent, the quantitative evaluation agent, and the comprehensive evaluation agent. As an implementation of SOA (introduced earlier in this chapter), web services ensure a loosely coupled integration model that allows for the flexible integration of heterogeneous systems in a variety of domains including business-to-consumer, business-to-business, and enterprise application integration (Chen et al., 2007). The agility of a supply chain requires a dynamic alliance of enterprises. For any enterprise to join the supply chain, it requires a process of information integration of the different enterprises and a dynamic integration of original information systems. One current research direction calls for the construction of enterprise application infrastructure for a supply chain by web services. The goal of web services effort is to achieve universal interoperability between applications in accordance with web standards (Chen et al., 2007).

Gu and Gao (2011) suggest that the IoT is the new environment for the application of remanufacturing/manufacturing (R/M) in an integrated supply chain. This includes the forward supply chain and the reverse supply chain. In this IoT environment, the R/M integrated supply chain based on IoT will have new characteristics. The structure of R/M integrated supply chain based on IoT has been proposed, with three stages: without RFID/EPC, with RFID, and with RFID/EPC, as the R/M integrated supply chain is introduced and compared.

3.4 Extended Enterprise Integration

Some researchers consider that there is no universally accepted definition for extended enterprises. The concept of extended enterprise emerged almost as the same time as those other concepts that emphasize the interorganizational collaboration, such as VE and enterprise networks (Lehtinen and Ahola, 2010). "Extended enterprise" refers to the collaborative relationship between supply chain partners from which they can obtain a competitive advantage and achieve higher customer satisfaction against other supply chains (Davis and Spekman, 2004). Extended enterprise is an abstract notion showing participating enterprises sharing core competencies (López-Ortega and Ramírez-Hernández, 2007). According to Camarinha-Matos et al. (2009), an extended enterprise represents a concept in which a dominant enterprise "extends" its boundaries. Mansouri and Mostashari (2010) define extended enterprise as "a wider system of enterprises that represents the holistic concept of enterprises alongside its internal components and external connectors including business partners, suppliers, and customers." From a holistic perspective, an extended enterprise integrates all of its processes, applications, people, and knowledge in pursuit of higher efficiency and effectiveness (Markus, 2000). Childe (1998) defines an extended enterprise as "a conceptual business unit or system that consists of a purchasing company and suppliers who collaborate closely in such a way as to maximize the returns to each partner." Bititci et al. (2005) define an extended enterprise as "a knowledge-based organization which uses the distributed capabilities, competencies, and intellectual strengths of its members to gain competitive advantage to maximize the performance of the overall extended enterprise." The extended enterprise is a philosophy in which member organizations strategically combine their core competencies and capabilities to create a unique competency (Bititci et al., 2005). Within the extended enterprise, core functionalities are provided separately by different enterprises, all of which come together for the purpose of cooperation. Individual enterprises must be able to provide their own core competences to a network of enterprises in the extended enterprise. Core capabilities must be shared by participating enterprises so that they can jointly exploit the best that each enterprise can possibly provide.

The extended enterprise is also defined as a long-term agreement among individual and complementary industrial units formed in a well-defined yet evolving process. Enterprises remain separate legal entities that keep control over their own systems and resources while providing their own core capabilities (Goethals et al., 2004). They not only extend their boundaries but also allow a substantial extension of the cooperation time frame.

The extended enterprise is an evolution of the integrated supply chain and can be considered as a complex set of collaborative enterprises, both upstream and downstream, from raw material to the final consumer, working together to bring value to each other. The advantages of the extended enterprise derive from the ability of enterprises to use their full networks of suppliers, vendors, customers, and clients (Lehtinen and Ahola, 2010). The extended enterprise is one of the main approaches to improving collaboration among individual enterprises. In order to manage complexity, enterprises aim at optimizing the processes of the entire extended enterprise for rapidly developing new products, processes, and services in a collaborative environment where all partners can share (Mengoni et al., 2011). During the past decade, more and more enterprises have changed from operating as stand-alone entities to producing goods and services through a network of independent or semi-independent organizations, in order to be able to operate in an extremely complex environment. An "extended enterprise system" (EES) is defined as "a complex network of distinctive yet distributed and interdependent organizational systems that are connected in an autonomic way to achieve objectives beyond reaching capacities of each" (Mansouri and Mostashari, 2010). An EES is composed of a myriad of components and subcomponents that function in combination with each other in order to achieve the overall objective (Ganguly and Mansouri, 2012). A completely developed EES will meet the following requirements (Mansouri and Mostashari, 2010): (1) required resources; (2) motivation, information, and knowledge (essentials for competition); (3) business-oriented characteristics (guaranteeing self-sustainability); (4) effective and efficient operations and communications (ingredients of growth); and (5) responsiveness to environmental circumstances (for strategic development purpose).

A simplified example of an extended enterprise is provided by Furst et al. (2001). In their example, the engineer uses the engineering component that is installed in enterprise *x*, the project manager (PM) controls the projects using the PM component in enterprise *y*, and the salesman works with his salesman component in enterprise *z*. Depending on the complexity of the interactions, the user interacts with an Internet browser or locally installed software. Needed data transmissions between components are automated and are made transparent to the user. The Internet functions as a low-cost means for data transmission. Furst et al. (2001) also came up with five basic requirements for extended enterprise information systems: (1) functionality: they are system supporting and/or fully automating business processes; (2) integration: they offer system-to-(internal or external) system communication; (3) usability: they offer effortless communication between the human user and the system; (4) security: they work to protect enterprise knowledge; and (5) flexibility: they are easily adjustable to a fast-changing business environment.

López-Ortega and Ramírez-Hernández (2007) indicate that an extended enterprise must be formed by the appropriate combination of the following: (1) product configuration, (2) key resources, (3) organizations, and (4) manufacturing processes. The mentioned categories of data are defined in a concept that is called the extended enterprise elements set, or E³S. Based on this concept, an extended enterprise can be seen as a family of sets, as described by the following equation:

$$E = \left\{ E^3 S_r, E^3 S_p, E^3 S_o, E^3 S_m \right\} \tag{3.1}$$

$$E^3 S_i = \left\{ e_i \big| P(e_i) \right\} \in E \tag{3.2}$$

In the second equation, the subindex *i* can be substituted by *r* (resources), *p* (product), *o* (organization), or *m* (manufacturing processes). López-Ortega and López de la Cruz (2009) indicate that in the current competitive and interconnected environment, manufacturing organizations are exploring a variety of forms of collaboration in order to sustain competitiveness. Extended enterprise is one of the important paradigms in facilitating collaboration among individual business units. This paradigm mainly claims that core capabilities are actually shared by participating enterprises; therefore, they can be exploited by the extended enterprise. Post et al. (2002) characterize the extended enterprise as the element within a network of interrelated stakeholders that creates, sustains, and enhances its value-creating capacity. They extend this concept by not only including the enterprise's local interactions with other businesses but also its relationships with other stakeholders, that is, internal as well as external stakeholders. They strongly claim that the long-term survival and success of organizations is determined by their ability to establish and maintain relationships within their entire network of stakeholders. The stakeholder view emphasizes that it is the stakeholder relationships (which are based on the involvement and commitment levels) that are the most critical, as compared to the transactions (which determine and contribute toward sustainable organizational wealth).

Zhang et al. (2012) point out that in extended enterprises, real-time manufacturing information tracking plays an important role, and the aim is to provide the right information to the right person at the right time in the right format in order to achieve optimal production management among the involved enterprises. However, many enterprises lack accurate, timely, and consistent manufacturing data. Such lagging information flow or unmatched information transfer method can bring extended enterprise with uncertainty. Thus, Zhang et al. (2012) propose an RFID-enabled real-time manufacturing information tracking infrastructure to address the real-time manufacturing data capturing and manufacturing information processing methods for extended enterprises. Following the proposed infrastructure, the traditional manufacturing resources such as materials and machines are equipped with RFID devices in order to build the real-time data capturing environment. In addition, a series of manufacturing information processing methods are developed in order to track real-time manufacturing information.

Hsu et al. (2007) suggest that new extended enterprise models such as supply chain integration and demand chain management require a new method of information exchange that extends traditional technology. The new requirements stem from, first, the fact that an information exchange involves large numbers of enterprise databases that belong to a large number of independent enterprises and, second, the fact that these databases overlap with real-time data sources (such as wireless sensor networks and RFID systems). One example is the industrial push to install RFID-augmented systems to integrate enterprise information along the life cycle of a product. This new effort demands openness and scalability and leads to a new paradigm of collaboration, using all these data sources. This collaboration requires a metadata technology for reconciling different data semantics that works on thin computing environments (e.g., emerging sensor nodes and RFID chips) as well as on traditional databases. It also requires a new extended global query model that supports partners to offer/publish information as they see fit, not just to request/subscribe to what they want.

Extended enterprises require novel modes member organizing and collaboration managing. In order to manage complexity, enterprises aim at optimizing the processes of the entire "extended enterprise" for the rapid development of new products, processes, and services by adopting collaborative environments wherein all actors can share product and process data, integrate software applications, and be supported by flexible tools for BPM (Mengoni et al., 2011).

Sun and Jiang (2007) point out that the structure of an extended enterprise is complex. Owners of manufacturing resources require the protection of their privacy. Many problems put a barrier across the road to manufacturing information sharing and tracking. To enable a controllable and dynamic manufacturing information sharing and tracking mechanism, a manufacturing information sharing and controlling method has been proposed that can balance the requirements of information sharing and privacy protecting. A time-dependent instantiated template net can be used to describe product information's complex structure and dynamic properties. On the basis of a searching algorithm, manufacturing information tracking is implemented via a history view and an in-time view.

Dong et al. (2005) point out that extended enterprise is a kind of organization comprising interconnected enterprises. Based on knowledge and manufacturing resource sharing, the enterprises team with each so that their competitive advantage can be improved. While jointly providing comprehensive products or services for customers, each enterprise can achieve its individual objective. Extended enterprise is a giant system with complexity in the aspects of both organization and operation. It is also a highly nonlinear system. Use of fractal theory is proposed to develop an approach to the manufacturing collaboration system called a fractal extended enterprise. Developing a sound performance measurement model is a critical task for a supply chain and its members, in order to examine their current status and identify improvement opportunities for steering their future direction. Lin and Li (2010) propose an integrated framework for supply chain performance measurement. It adopts six-sigma metrics and includes three components, that is, team structure measurement, supply chain process measurement, and output measurement, to provide complete coverage of performance requisites. Through the implementation of this framework, an organization will have the capability of monitoring its progress at a given point of time at each level within a supply chain.

Traditional performance measurement cannot be directly applied to organizations that operate as a part of an extended enterprise (Bititci et al., 2005). Thus, there is a need to develop models that can measure and manage the performance of extended enterprises. Bititci et al. (2005) developed an extended enterprise performance measurement model. In their model, the extended enterprise comprises a number of extended business processes. Each extended business process is the integration of the business processes of individual enterprises. The extended enterprise performance measurement model comprises scorecards such as (1) enterprise scorecards, which are specific to each enterprise collaborating in the extended enterprise. Essentially, these are conventional strategic scorecards; (2) business unit scorecards, which correspond to the collaborating business unit of an enterprise; (3) business process scorecards, which are the operational scorecards internal to each enterprise; (4) extended enterprise or metalevel scorecards, which include strategic interenterprise coordinating measures; and (5) extended business process scorecards, which include operational interenterprise coordinating measures.

There are a number of critical risks that might stand in the way of the proper functioning of extended enterprise integration and EESs (Ganguly and Mansouri, 2012). Through an extensive literature review, interview, and surveys, Ganguly and Mansouri (2012) identified a set of five major categories of possible risks (see Table 3.1).

Table 3.1 Risks Using Extended Enterprise Systems

Risk Categories	Operational Definition
Risk of financial collapse	Vulnerability against collective failure of financial stability of the network caused by systemic problems
Risk of disconnection	Vulnerability of the network's actors in regard to losing connectivity with their environment caused by ineffective communication
Risk of ineffective knowledge diffusion	Vulnerability of the network in making right decisions caused by the unavailability of appropriate knowledge or ineffectiveness in knowledge management and/or knowledge sharing
Risk of performance delay	Vulnerability of the network in timely response to its needs and demands caused by systemic delays of interaction among the network's actors
Risk of ineffective governance	Vulnerability of the network to productivity as well as to engaging in conflict resolution arguments and activities due to ineffective governance structures, regulation, and management of resources for the entire network

Subsequently, Ganguly and Mansouri (2012) ranked the risks through a survey analysis. As a result, they found that the most important category of risk facing EESs is the risk associated with ineffective governance. Since an EES is composed of a number of interdependent components, ineffective governance of the system might result in a breakdown of the overall structure of the EES, thereby resulting in huge financial losses. Overall, the risks are ranked in this order: ineffective governance, financial collapse, performance delay, the risk of ineffective knowledge diffusion, and finally, the risk of disconnection. Having a set of possible risks associated with an EES and understanding the relative importance of the risks would greatly help monitor and address the risks.

Another issue with EESs is related to real-time information capturing and tracking in extended enterprises. Extended enterprise integration must address the issues of performance delay, lagging information transfer flow, and the unmatched information transfer method, which can bring uncertainty to extended enterprises. Using real-time information to track extended enterprises is becoming more and more important. Zhang et al. (2012) propose an information infrastructure of RFID-enabled real-time manufacturing for extended enterprises. This infrastructure provides a model and methods for configuring RFID devices capturing real-time manufacturing information, and tracking and processing real-time manufacturing information throughout the extended enterprises.

3.5 Examples of Recent Research

3.5.1 Enterprise Collaboration: An Agent-Based Model

MASs can provide a cooperative environment for sharing data, information, and knowledge in a distributed environment. The concept of enterprise process framework has been studied by many researchers. In this section, MAS and web-based techniques are applied to large-scale complex enterprise dynamic modeling, in particular, to enterprise process collaborative modeling (Xu et al., 2008). Figure 3.6 shows the enterprise process collaborative modeling process (Xu et al., 2008). Figure 3.7 shows the framework of the enterprise modeling (Xu et al., 2008).

Figure 3.8 illustrates enterprise modeling process in two ways. Forward engineering is a process of model construction from planning, through to design and implementation and finally to maintenance. It requires enterprise dynamic modeling technologies to support it, such as process definition, process simulation (PS), process optimization (PO), and process enactment (PE). Reverse engineering modeling, unlike forward engineering, is a process that analyzes an enterprise system, abstracts its model, locates defects in present processes, and reengineers enterprise

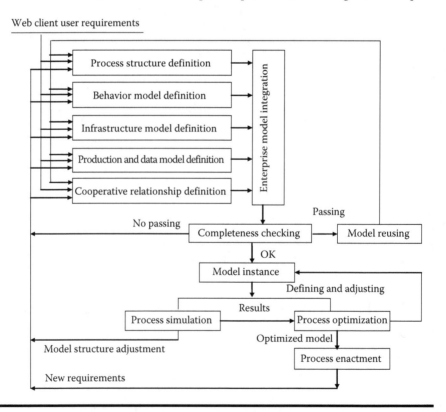

Figure 3.6 Enterprise collaborative modeling process.

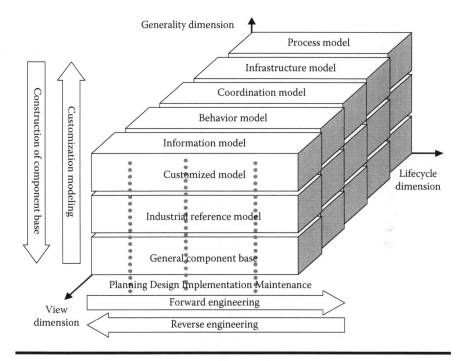

Figure 3.7   Framework for enterprise modeling.

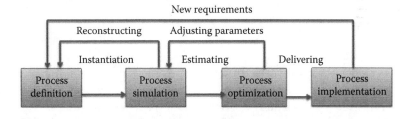

Figure 3.8   Enterprise business process reengineering.

processes. In reverse engineering, all applications should be mapped back to higher-level design abstractions. These design abstractions should be mapped back to the conceptual business models, in order to provide traceability. Reverse engineering is considered to be important for enterprise business process reengineering and evolution.

Complex enterprise modeling is an abstract description of enterprise requirements elicited from analysis. Figure 3.9 shows the layered complex enterprise modeling and mapping relationship, as large-scale complex enterprise processes can be analyzed well in a multilevel modeling framework. Lower-level modeling is a fraction of the combined activity in the upper-level model. As shown in Figure 3.9, the

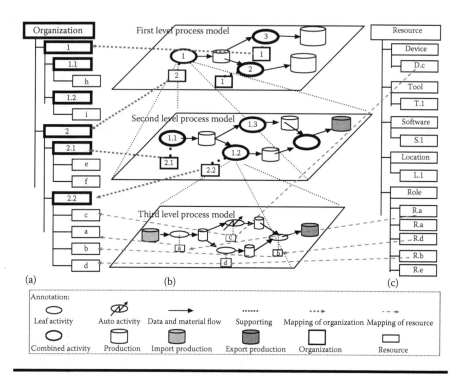

Figure 3.9   Layered complex enterprise modeling and mapping relationship. (a) Organization tree, (b) multilevel process model, (c) resource tree.

second-level process model is the interpretation of combined activity in the first-level process model, and the third-level model is the fraction of combined activity in the second-level model (Xu et al., 2008).

During enterprise modeling, only the organizations and resources defined in the infrastructure model can be used for process modeling. An infrastructure model includes an organization model and a resource model that are built in tree structures. They connect with process model via a relationship defined in the process model.

An agent-based enterprise collaborative modeling environment (AECME) using agent techniques has been developed. The architecture of the AECME is shown in Figure 3.10 (Xu et al., 2008). It includes engineering data management (EDM), engineering project management, job management, user management, and system integration, along with different kinds of enterprise modeling software tools (such as process definition, simulation, optimization, and enactment software tools).

An "e-engineering server agent" (ES agent) is the gateway for users to define, manage, and monitor information about enterprise modeling. Manipulations of engineering data related to the enterprise model, the modeling job, and the tasks are done by the EDM agent, which has two main functions: visual process modeling

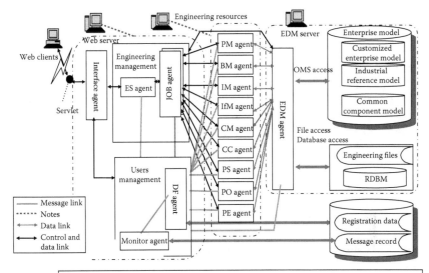

PM (process modeling), BM (behavior modeling), IM (information modeling), IfM (infrastructure modeling), CM (cooperative modeling), CC (completeness checking), PS (process simulation), PO (process optimization), PE (process enactment), DF (directory facilitator), ES (engineering server), EDM (engineering data management).

Figure 3.10   Architecture of AECME.

language based on an object management system (OMS), and the management of the engineering file and RDMS access. As a business job starts running, the ES agent generates a job agent (JA) dynamically. Then, the reverse process proceeds when the ES detects that a job is finished.

A "JA" communicates with an EDM agent to store and retrieve job data in an OMS model, engineering files, or databases; with the directory facilitator agent for searching matched engineering resource agents such as PM agents and PS agents; and with engineering resource agents for negotiation based on the modeling tasks' allocations. A JA is created and dissolved dynamically by the ES agent to control the modeling job it represents.

A "directory facilitator agent" (DF agent) has all the registration service functionalities for other agents and resides in this MAS; it keeps up-to-date agent registration and informs all registered agents with updated registry; and it provides lookup and match-making services to JAs.

A "monitoring agent" (MA) is specially designed to facilitate the monitoring of agents' behaviors in an MAS. In a distributed MAS, as information is distributed and controlled by each individual agent, it is necessary to have an agent that can accumulate information from various resources when required. Through this MA, dynamic condition of the system, the behavior of this multiagent environment, as well as all individual agents can be conveniently monitored or reviewed through a graphical tool provided to the user.

An "interface agent" (IA) is the web-based user interface for user requests to enter the agent system. Several Servlets are responsible for receiving requests to the web server. The IA then catches the user requests from the Servlets, translates the requests to messages, and initiates corresponding conversations to the related agents. On the other hand, when the agent system reaches a result, the IA creates updates to user interfaces based on the replies from the agents. The IA is a kind of application that functions mainly as a two-way bridge connecting the web-based user interfaces and the back-end agent systems.

"Engineering resource agents" such as PM, BM, IM, CM, IfM, PS, PO, and PE are the actual engineering problem-solving agents for enabling enterprise collaborative process modeling. They not only carry out the communication and negotiation functions on behalf of an engineering software but also implement the execution thread to perform the related model definition, analysis, simulation, optimization, or enactment, based on the EDM agent. In collaborative enterprise modeling, five models established through PM, BM, IM, CM, and IfM agents can be integrated and automatically complete syntax and semantic analysis checking by the PM agent.

A "process modeling agent" (PM agent) is a graphical editor with the capability of syntax and semantics checking for enterprise process structure definition. According to project management protocol and task assignment, it supports multilevel visual process modeling for collaborative operations and provides notations for activity, auto-activity, batch activity, combined activity (also call subprocess), production, messages, roles, machines, location, tools, and their connection relationships (data flow connection and related connection).

"Behavior modeling agent" (BM agent) is a window-based editor for activity behavior function. An "information modeling agent" (IM agent) is a window-based editor for production, message, and definition of various variables.

An "infrastructure modeling agent" (IfM agent) is a double graphical editor, one for the definition of organization and the other for the definition of the enterprise resource. Organization and resource models are in tree structures.

A "cooperation modeling agent" (CM agent) is for collaborative schedule/rules definition, which includes eight types of scheduling strategies such as highest priority first serve (HPFS), minimum slack time first serve (MSFS), first come first serve (FCFS), service in random order (SIRO), shortest operation time (SOT), longest operation time (LOT), longest remaining processing time (LRPT), and shortest remaining processing time (SRPT).

A "PS agent" is an actual engineering problem-solving agent for diagnosing business processes, analyzing system performance under different load conditions, predicting the impacts of organizational changes, and exploring new business opportunities. It can provide immediate feedback on how certain combinations of changes will affect process, and it can provide dynamic information about the activity flow, product flow, personnel flow, resource flow, etc. This information is then used to identify the key impacts of a BPR project upon the organization for reducing cycle time, increasing customer focus, increasing productivity, and improving quality.

A "PO agent" is a process model parameter optimization tool that uses an FR-TS algorithm consisting of the Fletcher–Reeves method and the Tabu search method, wherein the Fletcher–Reeves method is used to obtain a set of local optimal solutions and the Tabu search algorithm is used to discover feasible solutions in undiscovered areas. These two methods will be alternately iterated many times so as to get the set of global optimum solutions. The outputs of PO are a set of optimized models and a recommended model to assist in decision making.

A "PE agent" is an actual engineering problem-solving agent for enterprise process enactment. It executes the selected optimum enterprise process model for process monitoring and control and provides a task-table that orders storing tasks according to priority. During the process operation, some new requirements are fed back to PM if a change is detected, and the process needs to be improved.

"Engineering data management agent" (EDM agent) is a proactive engineering data service agent. EDM agent has the knowledge of database location, connection configuration, engineering file directories, and location and configuration of FTP server, among other things. It provides an interpretation function for enterprise process execution, as well as database and file system operation services, such as creating data sets for a new job, transferring engineering data, updating design data, retrieving design tasks, and helping to send data files proactively to the target agents before a job starts.

Figure 3.11 shows the metamodel of enterprise process modeling. An enterprise system model should contain a process model, a behavior model, an object model, a resource model, an organization model, and a cooperation model (Xu et al., 2008). An enterprise process model is usually composed of business processes, which consist of some sorted activities supported by specific resources with inputs and outputs. In PS and PE, activity scheduling and resource allocation are controlled by the specific cooperative rules defined by the cooperation model.

Umar (2005) proposed a conceptual framework for analyzing the IT infrastructure needed to support the next-generation enterprises (NGEs) (Umar, 2005). According to research, NGEs conduct business by relying on automation, mobility, real-time business activity monitoring, agility, and self-service over widely distributed operations. Enterprise applications that support computer-supported cooperative work have been discussed at three levels: (1) by definition of the cooperation model; (2) using the dynamic PERT/CPM technique to support flexible scheduling in PS according to the cooperative behavior rules defined in the cooperation model; and (3) via cooperation scheduling in enterprise process enactment, that is, task scheduling, which supports enterprise application integration and process integration.

In the proposed modeling environment, the cooperative behavior editor provides eight kinds of scheduling strategies. Process modeling engineers can define their own rules within the cooperative behavior editor. The scheduling strategies are shown in Table 3.2. Figure 3.12 shows the process-driven Enterprise Cooperative Scheduling Mechanism (Xu et al., 2008).

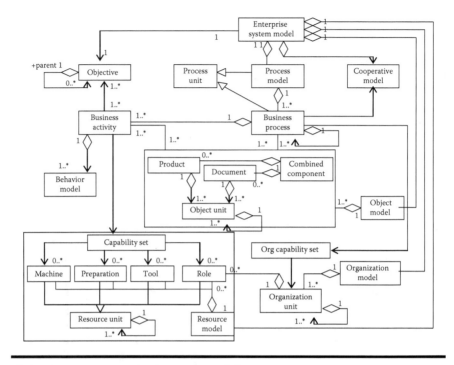

Figure 3.11 Framework of process model.

Table 3.2 Eight Strategies Used in AECME

Selectable Rules	Description of Rules
Rule 1: HPFS	Highest priority first serve
Rule 2: MS	Minimum slack time first serve
Rule 3: FCFS	First come first serve
Rule 4: SOT	Shortest operation time
Rule 5: LOT	Longest operation time
Rule 6: LRPT	Longest remaining processing time
Rule 7: SRPT	Shortest remaining processing time
Rule 8: SIRO	Service in random order

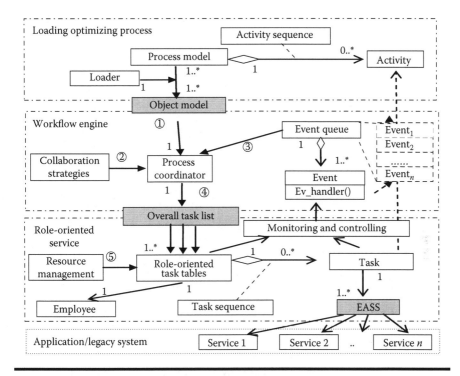

Figure 3.12 Process-driven enterprise cooperation scheduling mechanism.

3.5.2 VE Collaborative Operation: A Grid-Based Model

A "virtual enterprise" (VE) is characterized by a set of enterprises that form a temporary consortium to produce a product through sharing data and technology, since an individual enterprise cannot manufacture on its own. VE utilizes external resources to achieve agile manufacturing. The characteristics of a typical VE include, but are not limited to

- A temporary consortium of independent member companies and individuals, as they come together to seek a particular business opportunity
- A place where enterprises assemble themselves based on their agile manufacturing capabilities in terms of product customizability, cost-effectiveness, and product uniqueness, regardless of organization size, geographic location, computing environment, deployed technologies, or implemented processes
- A place where enterprises share core competencies, skills, and costs, which collectively enable them to access global markets with a worldwide competitive advantage that is not available individually
- Manufacturing customized products as the basis for the survival of VE

Efficiently and effectively managing and operating an enterprise entity (EE) call for the implementation of a performance management system (PMS). PMS is a dynamic system that supports decision-making processes through gathering and analyzing information.

The PMS framework is composed of quantifiable indicators that are able to measure and estimate the efficiency and effectiveness of an enterprise, not only partially, but also overall. In addition, there is a supporting infrastructure that enables data to be acquired, analyzed, interpreted, and disseminated. An internal/external monitoring system monitors the changes and developments that occur within/outside the enterprise. Warning messages will be sent when some indicators reach or exceed their limit. A review and decision system within the PMS checks the objectives and priorities according to the information from the monitoring system, which is a system at a higher level. Finally, an internal deployment system is used for system configuration in order to meet objectives and priorities. PMS is a dynamic system that integrates all of the resources within the enterprise. VE needs support from PMS. For a typical VE, performance management must be provided with a specific platform. Thus, the experiences acquired through the creation process of VE can be accumulated and utilized to support a new VE to join the consortium.

The original philosophy of grid technology in enterprise systems is adapted from the electric power grid. In the distribution of electric power, electricity producers are distributed across different geographic regions and feed electric power to the same public grid. The electricity consumers do not need to know the source of their electric power. Current research on manufacturing grid (M-Grid) systems has mainly concentrated on their concept, architecture, application prototype platform, and application. In order to improve the efficiency of VE, Deng et al. (2008) designed an M-Grid and multilevel manufacturing system of VE. In their design, as member enterprises are selected and tasks are assigned based on M-Grid, the key activities are assigned to the appropriate enterprise members by task decomposition, enterprise node searching, and the matching of characteristics of the manufacturing resources according to certain matching strategies. Manufacturing resource information management and VE management are the preconditions for the M-Grid, which provides the foundation for the information sharing of the distributed manufacturing resources. Research on the information service of the VE within an M-Grid environment will provide a solid theoretical and technological foundation for the implementation of the M-Grid platform. A well-designed functional M-Grid platform can be used to support EE evolution and will provide better service to the management of VE. In Tan et al. (2013), a platform called the M-Grid-based virtual enterprise operation platform (MGVEOP) is proposed. The main goal of establishing the MGVEOP aims at addressing the following requirements: (1) How to establish a VE? (2) How to collaborate among the members of VE? (3) How to realize the semiautomated business management of VE and its EEs?

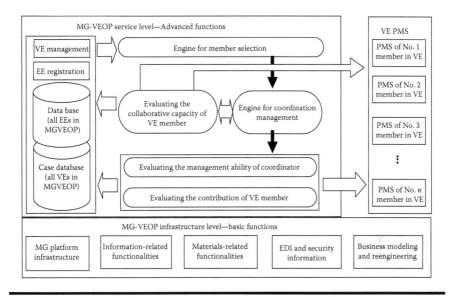

Figure 3.13 Architecture of MGVEOP.

(4) How to generate business opportunities with other EEs within MGVEOP? An architecture for MGVEOP is shown in Figure 3.13 (Tan et al., 2013).

MGVEOP can efficiently assist the member selection process: (1) In MGVEOP, EEs can communicate with each other within a common network; (2) MGVEOP is a platform that allows performance evaluation. All of the EEs participating in a VE will be evaluated. These evaluations will result in the criteria according to which the VE coordinator will determine the qualification of the EE as a candidate; and (3) MGVEOP is an intelligent and evolving platform. It is able to acquire experience from past VEs' management. These experiences can be utilized to facilitate the member selection process.

Figure 3.14 shows an activity diagram of MGVEOP, which displays the activities performed in three roles: (1) PRC or coordinator of VE, (2) PRPs, candidates of PRC, and (3) MGVEOP (Tan et al., 2010). If all of the functional positions have candidates, the VE is successfully generated, tasks are completed, and finally, the VE is dismissed (Tan et al., 2013). Until the VE is dismissed, MGVEOP records an evaluation of the leading ability of the coordinator and an evaluation on PRPs' contribution to VE. PRC or the coordinator and the PRPs do the performance evaluation with the VE, BPR, and performance management system. During this process, MGVEOP selects the candidate member meeting the requirements of the coordinator, choosing the optimal combination from candidates for generating a VE, and providing support to the management of the business process and performance evaluation for the VE and its members.

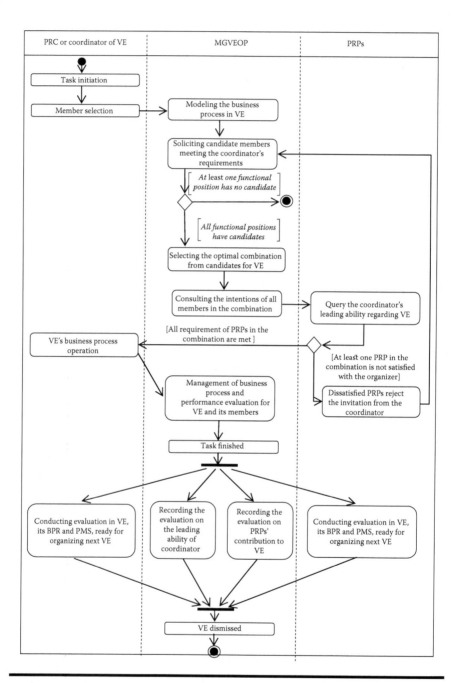

Figure 3.14 Activity diagram for MGVEOP.

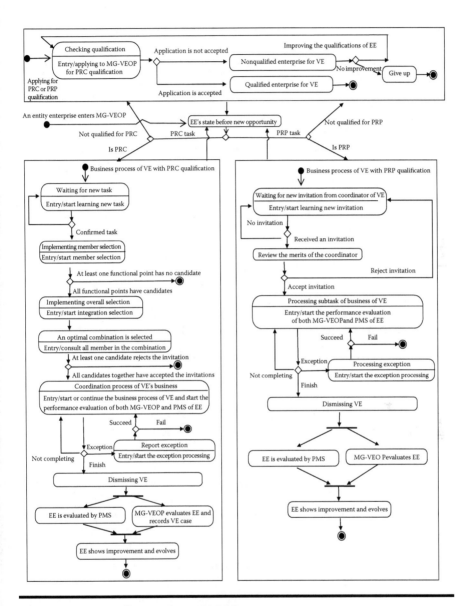

Figure 3.15   State diagram for MGVEOP.

The state diagram of a VE member within MGVEOP is illustrated in Figure 3.15 (Tan et al., 2013). The top part of Figure 3.15 illustrates the process of state changes of an EE applying for the qualification of VE membership (PRC or PRP). On the left part of the figure, a case of VE with PRC qualification is illustrated, which includes waiting for new tasks, implementing member selection, implementing overall selection, enabling optimal combination selection, coordinative processing of business of VE, dismissing VE, evaluating EE, and recording a VE case. The right part of the figure illustrates a business case of a VE member with PRP qualification that focuses on waiting and accepting invitations from the coordinator of VE, processing a subtask of business of VE.

Supply chains, extended enterprises, and VE have emerged in recent decades. For supporting consortium enterprises in organizing their manufacturing capabilities for producing desired products, MGVEOP is a well-designed platform. Through establishing VE within MGVEOP, the consortium enterprises or VE members are able to rely on the complementary core competences among members, to share production resources, to improve the ability of agile manufacturing, and to enhance the competitive advantage. The main contribution of this research is that it overcomes the limitation of the performance management of conventional enterprises and helps to realize whole value chain performance management by introducing a methodology for implementing VE management within an MGVEOP. This is a challenging research topic. In this research, the platform architecture and the fundamentals are introduced. The PMS within MGVEOP updates the balanced scorecard, which measures the performance of an organization based upon financial perspective, customer perspective, internal business perspective, and the learning and growth perspective. The system also has the ability of self-learning.

3.6 Summary

SCM has experienced many changes in the past decade as supply chains have become more interconnected and have transformed into supply networks (Li, 2007; Zhang and Bhattacharyya, 2007). The main reason is that, as the scope of enterprises has grown dramatically, different supply chains have become increasingly integrated with each other. It used to be that automated SCM was the purview of very large organizations such as General Motors and Boeing. The new trend is that today's businesses of all sizes need to share data with suppliers, distributors, and customers. Compressed product development cycles and lifetimes and just-in-time stocking imply that SCM systems must be interconnected, just as the applications that comprise the information systems of enterprises increasingly need to work together. The demand for supply chain integration has been increasing as a consequence of the increasingly global economy and has been made possible due to the advancement of information technologies.

Early enterprise systems did not support SCM. Early enterprise systems focused on intraorganizational integration as the benefits of intraorganizational information sharing have become obvious. An intraorganizational system (1) is aimed at providing a higher-level system related to activities that involve the coordination of business processes within the organization, and (2) is able to provide an integrated architecture to an organization for enhancing organizational performance. Since the last decade, as SCM has rapidly developed worldwide, more efforts have been focused on the integration of interenterprise systems, and more and more enterprises have moved toward interorganizational integration in order to support SCM through enterprise systems. Integrating both internal and external environment within and across organizations has been considered highly relevant to SCM (Liu et al., 2005). Interorganizational systems are able to allow communication between partners in the supply chain. Integrated enterprise systems can collect valuable SCM information for all related business processes across the supply chain. With ES integrated SCM, businesses can better predict market requirements, better perform in response to market conditions, and better align operations across global supply chain networks. The Demand Activated Manufacturing Architecture (DAMA) project is an example. DAMA has developed an interenterprise architecture and collaborative model for supply chains, which has enabled improved collaborative business across supply chains.

The emergence of interest in SCM has increased in recent years. However, most implementations have not yet been based on automated SCM. A new perspective based on blending the capabilities of SCM and current technologies is needed. This will produce the automated SCM systems required in a dynamic e-business environment. Businesses can harness the power of current technologies to dramatically improve their supply chain performance by adopting such new perspectives. Although the technologies and applications introduced earlier are, for the most part, not currently used in SCM, they are expected to have great potential to play a major role in the future. Technologies will prove that they will allow for real-time collaborative SCM. Since the basic enabler for effective supply chain quality management is information sharing, it has been and will be supported by advances in information technology. However, there are still many challenges and issues that need to be resolved for SCM systems to become more applicable.

We believe that successful SCM relies upon more sophisticated systems. With the coming of more advanced technologies, the overall quality of supply chain management will improve. One of the goals of SCM is to develop extended enterprise cooperative systems capable of promoting cooperation between enterprises in SCM. These systems should become virtual networks in which independent enterprises can be linked by information technologies. This evolution requires technologies that are able to facilitate an effective management in supply chains. To some extent, the development of such systems is a strategic move, which will allow the best SCM to be implemented.

Finally (but most importantly), this chapter addresses how SCM is more than what can be implemented with existing technologies; it shows how information technology can shape SCM. What we are addressing is an information technology effort to manage the supply chain issues from both the strategic and tactical approaches. We also believe that SCM can be accomplished with proper integration of existing and/or new technologies. This perspective will help researchers propose a research framework for a new generation of SCM system.

References

Angeles, R. and Nath, R. 2001. Partner congruence in electronic data interchange (EDI) enabled relationships. *Journal of Business Logistics*, 22(2), 109–127.

Basu, A. and Kumar, A. 2002. Research commentary: Workflow management issues in e-business. *Information Systems Research*, 13(1), 1–14.

Bechini, A., Cimino, M., Marcelloni, F., and Tomasi, A. 2008. Patterns and technologies for enabling supply chain traceability through collaborative e-business. *Information and Software Technology*, 59, 342–359.

Bititci, A., Mendibil, K., Martinez, V., and Albores, P. 2005. Measuring and managing performance in extended enterprises. *International Journal of Operations and Production Management*, 25(4), 333–353.

Butner, K. 2009. *Blueprint for Supply Chain Visibility*. New York: IBM.

Cachon, G. P. and Lariviere, M. A. 2005. Supply chain coordination with revenue sharing contracts: Strengths and limitations. *Management Science*, 51(1), 30–44.

Camarinha-Matos, L., Afsarmanesh, H., Galeano, N., and Molina, A. 2009. Collaborative networked organizations—Concepts and practice in manufacturing enterprises. *Computers and Industrial Engineering*, 57(1), 46–60.

Candido, G., Colombo, A., Barata, J., and Jammes, F. 2011. Service-oriented infrastructure to support the deployment of evolvable production systems. *IEEE Transactions on Industrial Informatics*, 7(4), 759–767.

Cannella, S. and Ciancimino, E. 2010. On the bullwhip avoidance phase: Supply chain collaboration and order smoothing. *International Journal of Production Research*, 48(22), 6739–6776.

Cao, M., Vonderembse, M., Zhang, Q., and Ragu-Nathan, T. 2011. Supply chain collaboration: Conceptualisation and instrument development. *International Journal of Production Research*, 48(22), 6613–6635.

Cao, M. and Zhang, Q. 2011. Supply chain collaboration: Impact on collaborative advantage and firm performance. *Journal of Operations Management*, 29(3), 163–180.

Cassivi, L. 2006. Collaboration planning in a supply chain. *Supply Chain Management: An International Journal*, 11(3), 249–258.

Chang, P., Juan, Y., and Liu, F. 2009. Incorporating DCSP algorithms into multi-agent system for supply chain collaboration—A case study for manufacturing supply chain. *Proceedings of 2009 World Congress on Computer Science and Information Engineering*, Los Angeles, CA, pp. 402–405.

Chebrolu, S. 2012. Enabling supply chain collaboration in a hybrid cloud. *Proceedings of 2012 IEEE Eighth World Congress on Services*, Honolulu, HI, pp. 309–312.

Chen, K. and Xiao, T. 2009. Demand disruption and coordination of the supply chain with a dominant retailer. *European Journal of Operational Research*, 197(1), 225–234.

Chen, T., Wang, L., Zhang, Y., and Sun, F. 2007. Research on methods of services-oriented integration for supply chain collaboration. *Proceedings of International Conference on Wireless Communications, Networking and Mobile Computing*, 2007 (*WiCom 2007*), Shanghai, China, pp. 4706–4709.

Cheng, R. 2011. A study on multi-agent based supply chain collaboration operation model facing time of delivery. *Proceedings of 2011 IEEE 8th International Conference on e-Business Engineering* (*ICEBE*), Beijing, China, pp. 102–104.

Childe, S. J. 1998. The extended enterprise: A concept for co-operation. *Production Planning and Control*, 9(4), 320–327.

Curtis, B., Kellner, M., and Over, J. 1992. Process modeling. *Communication of ACM*, 5(9), 75–90.

Davis, E. W. and Spekman, R. E. 2004. *The Extended Enterprise—Gaining Competitive Advantage through Collaborative Supply Chains*. Upper Saddle River, NJ: Prentice Hall Books.

Deng, H., Chen, L., Wang, C., and Deng, Q. 2008. Multilevel manufacturing system of virtual enterprise based on manufacturing grid and strategies for member enterprise selection and task assignment. *Journal of Shanghai University*, 12(4), 330–338.

Dong, H., Liu, D., Zhao, Y., and Chen, Y. 2005. A novel approach of networked manufacturing collaboration: Fractal web-based extended enterprise. *International Journal of Advanced Manufacturing Technology*, 26, 1436–1442.

Dourish, P. 2001. Process descriptions as organizational accounting devices: The dual use of workflow technologies. *Proceedings of the ACM International Conference Supporting Group Work*, C.A. Ellis and I. Zigurs, (Eds.), New York, pp. 52–66.

Ezzeddine, B. and Abdellatif, B. 2012. A multi-agent, semantic web service-based modelling for cooperation in the supply chain environment. *Proceedings of 2012 International Conference on Complex Systems*, Agadir, Morocco, pp. 1–6.

Fawcett, S. E., Magnan, G. M., and McCarter, M. W. 2008. Benefits, barriers, and bridges to effective supply chain management. *Supply Chain Management: An International Journal*, 13(1), 35–48.

Fox, M., Chionglo, J., and Barbuceanu, M. 1993. *The Integrated Supply Chain Management System*. Toronto, Ontario, Canada: University of Toronto.

Furst, K., Rodrigues, O., and OVE, G. 2001. Requirements and basic technologies for extended enterprises. *Elektrotechnik und Informationstechnik*, 118(11), 590–597.

Ganguly, A. and Mansouri, M. 2012. Evaluating risks associated with extended enterprise systems (EES). *IEEE Aerospace and Electronic Systems Magazine*, 27(5), 4–10.

Goethals, F., Vandenbulcke, J., and Lemahieu, W. 2004. Developing the extended enterprise with the FADEE. *Proceedings of the 2004 ACM Symposium on Applied Computing*, Nicosia, Cyprus, pp. 1372–1379.

Gu, Q. and Gao, T. 2011. R/M integrated supply chain based on IoT. *Proceedings of 2011 IEEE 14th International Conference on Computational Science and Engineering*, Dalian, China, pp. 290–294.

Guédria, W., Naudet, Y., and Chen, D. 2013. Maturity model for enterprise interoperability. *Enterprise Information Systems*, DOI: 10.1080/17517575.2013.805246.

Ha, S. and Suh, H. 2008. A timed colored Petri nets modeling for dynamic workflow in production development process. *Computers in Industry*, 59(2–3), 193–209.

Haller, S., Kanouskos, S., and Schroth, C. 2009. The Internet of Things in an enterprise context. In *Future Internet Systems* (FIS), Domingue, J., Fensel, D., Traverso, P. (Eds.), Berlin, LNCS, Vol. 5468, Springer, pp. 14–28.

Holweg, M. et al. 2005. Supply chain collaboration: Making sense of the strategy continuum. *European Management Journal*, 23(2), 170–181.

Hsu, C., Levermore, D., Carothers, C., and Babin, G. 2007. Enterprise collaboration: On-demand information exchange using enterprise databases, wireless sensor networks, and RFID systems. *IEEE Transactions on Systems, Man, and Cybernetics-Part A: Systems and Humans*, 37(4), 519–532.

Hsu, C. and Wallace, W. 2007. An industrial network flow information integration model for supply chain management and intelligent transportation. *Enterprise Information Systems*, 1(3), 327–351.

Hu, H. and Hu, D. 2010. Study on intelligent collaboration mode of supply chain. *Proceedings of 2010 IEEE International Conference on Service Operations and Logistics and Informatics (SOLI)*, Qingdao, China, pp. 205–207.

Hudnurkar, M. and Rathod, U. 2012. Collaborative supply chain: Insights from simulation. *International Journal of System Assurance Engineering and Management*, 3(2), 122–144.

IEEE Computer Society Standards Coordinating Committee. 1990. *IEEE Standard Computer Dictionary: A Compilation of IEEE Standard Computer Glossaries*. New York: IEEE.

Janssen, M. 2005. The architecture and business value of a semi-cooperative agent-based supply chain management system. *Electronic Commerce Research and Applications*, 4, 315–328.

Jardim-Goncalves, R., Grilo, A., Agostinho, C., Lampathaki, F., and Charalabidis, Y. 2013. Systematisation of interoperability body of knowledge: The foundation for enterprise interoperability as a science. *Enterprise Information Systems*, 7(1), 7–32.

Ji, X. and Guo, C. 2009. Supply chain integrated planning modeling and solution based on HGA. *Proceedings of the International Conference on Electronic Commerce and Business Intelligence*, Beijing, China, pp. 147–151.

Jia, X. and Zhou, X. 2007. Study on the integrated framework of supply chain performance evaluation system based on multi-agent. *Proceedings of International Conference on Wireless Communications, Networking and Mobile Computing (WiCom 2007)*, Shanghai, China, pp. 4898–4906.

Jonrinaldi, R. and Zhang, D. Z. 2013. An integrated production and inventory model for a whole manufacturing supply chain involving reverse logistics with finite horizon period. *Omega*, 41(3), 598–620.

Kannan, V. and Tan, K. 2010. Supply chain integration: Cluster analysis of the impact of span of integration. *Supply Chain Management-an International Journal*, 15(3), 207–215.

Kong, J., Jung, J., and Park, J. 2009. Event-driven service coordination for business process integration in ubiquitous enterprises. *Computers and Industrial Engineering*, 57, 14–26.

Kranz, M., Holleis, P., and Schmidt, A. 2010. Embedded interaction interacting with the internet of things. *IEEE Internet Computing*, March/April, 14(2), 46–53.

Kulp, S., Lee, H., and Ofek, E. 2004. Manufacturer benefits from information integration with retail customers. *Management Science*, 50(4), 431–444.

Kumar, S. and Budin, E. 2006. Prevention and management of product recalls in the processed food industry: A case study based on a porter's perspective. *Technovation*, 26, 739–750.

Lambert, D. M. and Christopher, M. G. 2000. From the editors. *International Journal of Logistics Management*, 11(2), pii–pii.

Lau, A., Yam, R., and Tang, E. 2010. Supply chain integration and product modularity: An empirical study of product performance for selected Hong Kong manufacturing industries. *International Journal of Operations and Production Management*, 30(1), 20–56.

Lee, B., Kim, P., Hong, K., and Lee, I. 2010. Evaluating antecedents and consequences of supply chain activities: An integrative perspective. *International Journal of Production Research*, 48(3), 657–682.

Lehtinen, J. and Ahola, T. 2010. Is performance measurement suitable for an extended enterprise? *International Journal of Operations and Production Management*, 30(2), 181–204.

Li, H. and Wang, H. 2007. A multi-agent-based model for a negotiation support system in electronic commerce. *Enterprise Information Systems*, 1(4), 457–472.

Li, L. 2007. *Supply Chain Management: Concepts, Techniques and Practices*. Singapore: World Scientific.

Lin, H. K. and Harding, J. A. 2007. A manufacturing system engineering ontology model on the semantic web for inter-enterprise collaboration. *Computers in Industry*, 58(5), 428–437.

Lin, L. and Li, T. 2010. An integrated framework for supply chain performance measurement using six-sigma metrics. *Software Quality Journal*, 18, 387–406.

Liu, J., Zhang, S., and Hu, J. 2005. A case study of an inter-enterprise workflow-supported supply chain management system. *Information and Management*, 42, 441–454.

Liu, Y. and Nie, G. 2007. Implementation of supply chains coordination using semantic web service composition. *Proceedings of IEEE International Conference on Service-Oriented Computing and Applications*, Newport Beach, CA, pp. 249–254.

López-Ortega, O. and López de la Cruz, K. 2009. Usage of agent technology to coordinate data exchange in the extended enterprise. In *Advanced Design and Manufacturing Based on STEP*, Xu, X. and Nee, A. Y. C. (Eds.), Springer Series in Advanced Manufacturing, pp. 399–418.

López-Ortega, O. and Ramírez-Hernández, R. 2007. A formal framework to integrate express data models in an extended enterprise context. *Journal of Intelligent Manufacturing*, 18(3), 371–381.

Malhotra, A., Gasain, S., and El Sawy, O. A. 2005. Absorptive capacity configurations in supply chains: Gearing for partner-enabled market knowledge creation. *MIS Quarterly*, 29(1), 145–187.

Mangina, E. and Vlachos, I. 2005. The changing role of information technology in food and beverage logistics management. *Journal of Food Engineering*, 70(3), 403–420.

Mansouri, M. and Mostashari, A. 2010. A systemic approach to governance in extended enterprise systems. *Proceedings of 2010 4th Annual IEEE Systems Conference*, San Diego, CA, pp. 311–316.

Manthou, V., Vlachopoulou, M., and Folinas, D. 2004. Virtual e-chain (VeC) model for supply chain collaboration. *International Journal of Production Economics*, 87(3), 241–250.

Markus, M. L. 2000. Paradigm shifts-e-business and business/systems integration. *Communications of the AIS*, 4. Article 10.

Matopoulos, A., Vlachopoulou, M., Manthou, V., and Manos, B. 2007. A conceptual framework for supply chain collaboration: Empirical evidence from the agri-food industry. *Supply Chain Management: An International Journal*, 12(3), 177–186.

Mengoni, M., Graziosi, S., Mandolini, M., and Peruzzini, M. 2011. A knowledge-based workflow to dynamically manage human interaction in extended enterprise. *International Journal on Interactive Design and Manufacturing*, 5(1), 1–15.

Mentzer, J. 2001. *Supply Chain Management*. Thousand Oaks, CA: Sage Publications, pp. 83–84.

New, S. 2004. Supply chains: Construction and legitimation. In New, S. and Westbrook, R. (Eds.), *Understanding Supply Chains: Concepts, Critiques and Futures*. Oxford, U.K.: Oxford University Press, pp. 69–108.

Nissen, M. 2001. Agent-based supply chain integration. *Information Technology and Management*, 2(3), 289–312.

Oppong, S. et al. 2005. A new strategy for harnessing knowledge management in e-commerce. *Technology in Society*, 27, 413–435.

Panetto, H. and Cecil, J. 2013. Editorial information systems for enterprise integration, interoperability and networking: Theory and applications. *Enterprise Information Systems*, 7(1), 1–6.

Papazouglou, M. and Georgakooulos, D. 2003. Service-oriented computing: Introduction. *Communications of ACM*, 46, 24–28.

Pathak, S. and Dilts, D. 2002. Simulation of supply chain networks using complex adaptive system theory. *Proceedings of 2002 IEEE International Engineering Management Conference*, Cambridge, UK, pp. 655–660.

Popova, V. and Sharpanskykh, A. 2008. Process-oriented organization modeling and analysis. *Enterprise Information Systems*, 2(2), 161–193.

Post, J., Preston, L., and Sachs, S. 2002. Managing the extended enterprise: A new stakeholder view. *California Management Review*, 45(1), 6–28.

Prisecaru, O. 2008. Resource workflow nets: An approach to workflow modeling and analysis. *Enterprise Information Systems*, 2(2), 101–124.

Quartel, D. et al. 2009. Model-driven development of mediation for business services using COSMO. *Enterprise Information Systems*, 3(3), 319–345.

Ribeiro, L., Barata, J., and Colombo, A. 2009. Supporting agile supply chains using a service-oriented shop floor. *Engineering Applications of Artificial Intelligence*, 22, 950–960.

Rosenberg, F., Michlmayr, A., and Dustdar, S. 2008. Top-down business process development and execution using quality of service aspects. *Enterprise Information Systems*, 2(4), 459–475.

Sahin, F. and Robinson, E. P. 2005. Information sharing and coordination in make-to-order supply chains. *Journal of Operations Management*, 23(6), 579–598.

Schulz, K. and Orlowska, M. 2004. Facilitating cross-organizational workflows with a workflow view approach. *Data and Knowledge Engineering*, 51(1), 109–147.

Sheu, C., Yen, H. R., and Chae, B. 2006. Determinants of supplier-retailer collaboration: Evidence from an international study. *International Journal of Operations and Production Management*, 26(1), 24–49.

Shi, C. and Bian, D. 2010. Integrated supply chain management by multiple objective programming. *Proceedings of 2nd IEEE International Conference on Information Management and Engineering (ICIME)*, Chengdu, China, pp. 531–534.

Shi, J. and Chen, X. 2007. The design of supply chain collaboration base on RosettaNet standards. *Proceedings of 2007 IEEE International Conference on Automation and Logistics*, Jinan, China, pp. 2212–2216.

Simatupang, T. M. and Sridharan, R. 2005. An integrative framework for supply chain collaboration. *International Journal of Logistics Management*, 16(2), 257–274.

Stank, T., Keller, S., and Daugherty, P. 2011. Supply chain collaboration and logistical service performance. *Journal of Business Logistics*, 22(1), 29–48.

Sun, H. and Jiang, P. 2007. Study on manufacturing information sharing and tracking for extended enterprises. *International Journal of Advanced Manufacturing Technology*, 34, 790–798.

Symeonidis, A., Kehagias, D., and Mitkas, P. 2003. Intelligent policy recommendations on enterprise resource planning by the use of agent technology and data mining techniques. *Expert Systems with Applications*, 25, 589–602.

Tan, W., Jiang, C., Li, L., and Lv, Z. 2008. Role-oriented process-driven enterprise cooperative work using the combined rule scheduling strategies. *Information Systems Frontiers*, 10(5), 519–529.

Tan, W., Xu, Y., Xu, W., Xu, L., Zaho, X., Wang, L., and Fu, L. 2010. A methodology toward manufacturing grid-based virtual enterprise operation platform. *Enterprise Information Systems*, 4(3), 283–309.

Tan, W., Xu, W., Yang, F., Xu, L., and Jiang, C. 2013. A framework for service enterprise workflow simulation with multi-agents cooperation. *Enterprise Information Systems*, 7(4), 523–542.

Tao, A. and Yang, J. 2008. Towards policy driven context aware differentiated services design and development. *Enterprise Information Systems*, 2(4), 367–384.

Tu, H. and Yang, M. 2010. An integrated model for the assessment of collaboration in the supply chain. *Proceedings of 2010 IEEE International Conference on Advanced Management Science (ICAMS)*, Chengdu, China, pp. 240–244.

Umar, A. 2005. IT infrastructure to enable next generation enterprises. *Information Systems Frontiers*, 7(3), 217–256.

Unger, T., Mietzner, R., and Leymann, F. 2009. Customer-defined service level agreements for composite applications. *Enterprise Information Systems*, 3(3), 369–391.

van der Aalst, W. 1998. The application of Petri nets to workflow management. *Journal of Circuits Systems and Computers*, 8(1), 21–66.

van der Aalst, W. 1999. Interorganizational workflows: An approach based on message sequence charts and Petri nets. *Systems Analysis Modeling Simulations*, 34(3), 335–367.

van der Aalst, W. 2000. Loosely coupled inter-organizational workflows: Modeling and analyzing workflows crossing organizational boundaries. *Information and Management*, 37, 67–75.

van Lessen, T., Nitzsche, J., and Leymann, F. 2009. Conversational web services: Leveraging BPEL for expressing WSDL 2.0 message exchange patterns. *Enterprise Information Systems*, 3(3), 347–367.

van Sinderen, M. 2008. Challenges and solutions in enterprise computing. *Enterprise Information Systems*, 2(4), 341–346.

Verdecho, M., Alfaro-Saiz, J., Rodríguez-Rodríguez, R., and Ortiz-Bas, A. 2012a. The analytic network process for managing inter-enterprise collaboration: A case study in a collaborative enterprise network. *Expert Systems with Applications*, 39(1), 626–637.

Verdecho, M., Alfaro-Saiz, J., Rodriguez-Rodriguez, R., and Ortiz-Bas, A. 2012b. A multi-criteria approach for managing inter-enterprise collaborative relationships. *Omega*, 40(3), 249–263.

Vernadat, F. 1996. *Enterprise Modelling and Integration: Principles and Applications*. London, U.K.: Chapman & Hall.

Wang, J. and Wang, S. 2008. Supply chain collaboration model centered on manufacturer in information asymmetry. *Journal of System Engineering*, 23(1), 60–66.

Wang, L., Lin, Y., and Lin, P. 2007. Dynamic mobile RFID-based supply chain control and management system in construction. *Advanced Engineering Informatics*, 21, 377–390.

Wang, S., Shen, W., and Hao, Q. 2006. An agent-based web service workflow model for inter-enterprise collaboration. *Expert Systems with Applications*, 31(4), 787–799.

Wang, W. 2012. A supply chain collaboration model based on system dynamics. *Advanced Materials Research*, 490–495, 1997–2001.

Wolfert, J. et al. 2010. Organizing information integration in agri-food-a method based on a service-oriented architecture and living lab approach. *Computers and Electronics in Agriculture*, 70, 389–405.

Wu, Y. and Wang, X. 2010. A Service-oriented computing model for supply chain collaboration. *Proceedings of 2010 International Conference on Computational Intelligence and Software Engineering (CiSE)*, Wuhan, China, pp. 1–3.

Xu, L. 2011. Information architecture for supply chain quality management. *International Journal of Production Research*, 49(1), 183–198.

Xu, L., Tan, W., Zhen, H., and Shen, W. 2008. An approach to enterprise process dynamic modeling supporting enterprise process evolution. *Information Systems Frontiers*, 10(5), 611–624.

Xue, X., Li, X., Shen, Q., and Wang, Y. 2005. An agent-based framework for supply chain coordination in construction. *Automation in Construction*, 14(3), 413–430.

Yan, B. and Huang, G. 2009. Supply chain information transmission based on RFID and Internet of Things. *Proceedings of 2009 ISECS International Colloquium on Computing, Communication, Control, and Management*, IEEE, Sanya, China, pp. 166–169.

Zhang, Q. and Xu, Q. 2010. Supply chain collaboration based on web services composition. *Proceedings of 2010 IEEE International Conference on Service Operations and Logistics and Informatics (SOLI)*, Qingdao, China, pp. 224–229.

Zhang, T., Ying, S., Cao, S., and Zhang, J. 2008. A modelling approach to service-oriented architecture. *Enterprise Information Systems*, 2(3), 239–257.

Zhang, Y. and Bhattacharyya, S. 2007. Effectiveness of Q-learning as a tool for calibrating agent-based supply network models. *Enterprise Information Systems*, 1(2), 217–233.

Zhang, Y., Jiang, P., Huang, G., Qu, T., Zhou, G., and Hong, J. 2012. RFID-enabled real-time manufacturing information tracking infrastructure for extended enterprises. *Journal of Intelligent Manufacturing*, 23(6), 2357–2366.

Chapter 4

Enterprise and Supply Chain Architecture

4.1 Enterprise Architecture

Enterprise architecture (EA) has been considered as an IT methodology to translate between business architecture and information architecture. EA is a term that has been broadly defined and used by both academics and practitioners. A number of EA efforts have been made, including the Department of Defense Architecture Framework (DoDAF), the Federal Enterprise Architecture Framework (FEAF), and the Open Group Architecture Framework (TOGAF). DoDAF is an architecture framework developed for the US Department of Defense (DoD) that provides structure for a specific stakeholder concern through viewpoints organized by various views. These views help users understand, assimilate, and visualize the scope and complexities of an architecture. It has been applied to large systems for realizing complex integration and has provided interoperability capability. The FEAF is an initiative of the US Office of Management and Budget and the Office of E-Government and IT for the purpose of representing the EA within the US Federal Government. EA became a recognized strategic and management tool in the US Federal Government with the passage of the Clinger–Cohen Act in 1996. TOGAF is an EA framework used for planning, designing, and implementing an enterprise information architecture. TOGAF® has been the registered trademark of the Open Group in the United States and in other countries since 2011. As a high-level modeling method that is typically modeled at four levels—business, application, data, and technology—TOGAF works to provide an overall as well as a holistic framework for information architecture tasks, which can then be further

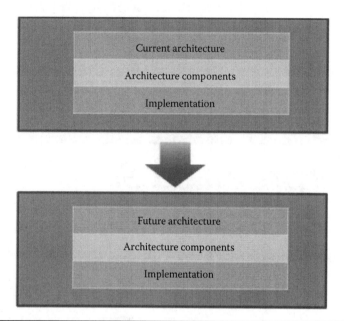

Figure 4.1   The EA effort is both continuous and iterative.

built upon. TOGAF is, in general, an approach that considers the entire life cycle of enterprise system architecture: from the planning and design to the implementation stage. It allows an enterprise to incrementally adapt to business changes through continuous iterations. The earlier-mentioned efforts not only pursue an in-depth understanding of business architecture but also support a technological line of interest. EA effort is characterized by both continuity and iteration. Figure 4.1 shows EA as a continuous activity for many organizations as well as the relationship between current and future architecture.

From a systems perspective, EA can be considered as the fundamental design of an organization as a whole, together with all of the relevant components, as well as the principles governing its design and evolution. The word "enterprise" here may refer to the context of extended enterprises. EA promises to help manage the ongoing enterprise processes in a consistent and systematic fashion, both intra- and interorganizationally, achieving and maintaining enterprise integration in the meantime. Some researchers consider that EA has already become an established subject; however, other researchers consider EA itself as a rather new field, since it still lacks a consistent and agreed-upon definition. As mentioned earlier, from a systems perspective, EA can be considered a type of holistic thinking about the architecture of an enterprise. EA is an architectural description of the enterprise, at both system and subsystem levels, to guide its implementation. The architecture consists of the structure of the system, subsystems, and their interrelationships. Meanwhile, EA strives to be a coherent whole of principles, methods, and techniques that can be used in the design and realization of various architectures. As such, EA's target is to represent an enterprise holistically at both the system and subsystem levels with sound methodologies. The EA's key subsystems include business architecture, information architecture, and other architectures. Business architecture addresses both business strategy and business processes. Information architecture addresses infrastructure, applications, and other aspects. Information architecture addresses the ontologies, taxonomies, data, and security associated with the enterprise. Some of these aspects interact across different architectures. EA aims to bring a coherent structure into key architectures for the system, subsystems, and then to align them systematically.

As EA can provide structure for a specific stakeholder concern as well as a larger and holistic perspective, EA can be applied to various specific tasks. In such specific tasks, EA can provide systematic support to map and trace identified tasks to enterprise artifacts modeled within the EA, supporting the overall strategy of an organization. The idea is to use EA descriptions to represent accurate information, allowing a better understanding of the components that can be affected from the manifestation of tasks.

EA can illustrate the systems relationships within an organization and the impact of the changes that an organization intends to make. EA can be described by its levels:

1. *System level*: This level deals with the design of a system and is mainly concerned with a system such as an enterprise, in terms of its structure and behavior. It deals with enterprise tasks such as enterprise development and enterprise integration and aims at structuring concepts and activities related to the system architecture.
2. *Subsystem level*: This level is mainly for sub-EA. Compared with the system architecture, this is more subsystem oriented.
3. *Meta-system level*: This level is mainly used for extended EA, as the design of an EA must be coherent with that of other enterprise systems in an extended enterprise context and must be aligned with SCM strategy.

For the three levels mentioned earlier, EA can be represented in terms of rules, text, graphics, etc., that relate to different levels. EA defines and organizes the generic concepts that are required to enable the creation of enterprise models for industrial organizations. Its main purpose is to provide an organizing mechanism so that concepts and knowledge about an enterprise and its interoperability, intra- and/or interorganizationally, can be represented in a structured way.

As mentioned earlier, EA is considered a framework that represents the interrelations of the whole system and its components. In EA, models are used to depict the overall infrastructure of the enterprise, as well as its main components and layers, thus aiding in the understanding of the enterprise. In this view, EA is able to define the business and technological components of an enterprise. Thus,

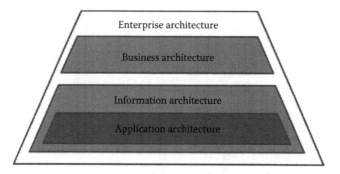

Figure 4.2   Relations between architecture layers and the integration through EA.

EA can be defined as a set of models that represent the structure of the whole enterprise. As such, EA can be employed as a base for designing and implementing enterprise systems. EAs are particularly useful in employing a consistent language to describe an enterprise. Furthermore, EA can be used as a reference model to note how an enterprise can achieve its objective by studying and comparing the architectures of different enterprises, relating to the architectures and their impact on the functionality of the enterprise. In Figure 4.2, each of the layers is a separate one, and the EA is the bridge that integrates each of them in a systematic manner.

EA is usually analyzed starting from the highest level, in order for it to adapt to the changes that are required. The business architecture reflects business strategies, processes, and functions of an enterprise. It is the base of the enterprise for which the business operations will be managed. Aligning the organization's business and information strategy is an important determinant for any enterprise's success. The importance of such alignment has been emphasized in the literature. For example, the measures of information management include focusing information on key business drivers and integrating data from across the business units to support business planning. Such measures highlight the roles of alignment, identifying key business drivers, and integration across business units to define information management. The application architecture facilitates the development and/or implementation of applications that fulfill the business requirements and assure that the required functions will be supported. Information architecture concerns the logical and physical aspects. In this architecture, computing services are identified and then planned to support the enterprise information infrastructure. Different architecture frameworks may share similar characteristics; it is beneficial to understand the similarities and differences of each in order to find a framework that fits the required organizational needs. The purpose of conducting an analysis of EA frameworks is to facilitate scientific analysis on the most suitable framework for creating an EA for a particular enterprise. Table 4.1 shows a comparison of different EA frameworks (from existing literature).

Table 4.1 Comparison of EA Frameworks Based on Certain Characteristics

Requirements	Structure	Top-Down	Abstraction	Artifact	Adoption	Research Focus
CIMOSA	High	Medium	Medium	Low	Low	Medium
FEAF	Medium	Medium	Low	Medium	Medium	Low
GERAM	High	Medium	Low	Low	Medium	Low
PERA	Medium	High	Medium	Medium	Low	Low
TOGAF	High	Medium	Medium	High	High	Medium
Zachman	High	Medium	High	High	High	Medium

In Table 4.1, certain characteristics are evaluated against different frameworks (Lapkin, 2004). Some researchers consider these characteristics to represent those attributes that an EA framework should have. At minimum, a framework should provide a logic structure, should be able to decompose the entire enterprise from its highest level, and should be able to abstract an enterprise business entity and provide artifacts to describe them. Some researchers consider that, relatively speaking, the Zachman framework is more suitable for the implementation purpose in terms of representing the context of EA. The reason is that the Zachman framework is extensive and flexible, in a relative sense, as it does not impose and/or restrict predefined artifact systems. In general, EA is recognized as a good tool for providing sound methods for analyzing the enterprise requirements through which an enterprise operates.

EA can have different layers, as follows (Winter and Fischer, 2006):

Business architecture layer: The business architecture layer represents the fundamental organization of the enterprise from a business strategy viewpoint. Design and evolution principles for the business architecture layer can be derived from approaches, based on different disciplines.

Process architecture layer: The process architecture layer represents the fundamental organization of processes in the relevant enterprise context. Design and evolution principles for this layer focus on effectiveness (creating specified outputs), efficiency (meeting specified performance goals), optimization, and other criteria.

Software architecture layer: The software architecture layer represents the fundamental organization of software artifacts (e.g., software services). A broad range of design and evolution principles is available for this layer. An example is introduced in Section 4.4.

Hardware architecture layer: The hardware architecture layer represents the fundamental organization of computing and telecommunications hardware and networks. A broad range of design and evolution principles is available for this layer.

Integration architecture layer: The integration architecture layer represents the integration aspects of information architecture components in the relevant enterprise context. The design and evolution principles for this layer focus on agility, cost efficiency, speed, and many other criteria. The main concern is achieving the goal of integration.

Figure 4.3 is an illustration of an EA that comprises the layers mentioned earlier. For each layer, integration is always a key concept and technology.

With the formation of architecture layers within the enterprise, methods and techniques are required for dealing with details and complexity. In a multilayer architecture, methods/techniques can be either layer specific or cross-layer. Layer-specific ones can be those business and/or process-oriented methods for the layers of the business architecture and process architecture. Cross-layer methods and techniques usually involve information integration. Based on the concepts of multilayer and cross-layer view and representation, EA represents all aggregate artifacts and their relationships across all of the layers of an enterprise. The way in which an enterprise system collaborates with other systems will inevitably depend upon its multilayer EA. A better EA is likely to result in a better-achieved goal.

As EA can be considered as a blueprint of the organization, the goals of EA include the following:

- Supporting organizational strategic goals through providing support for consistent and systematic design and evolution of artifacts on different layers

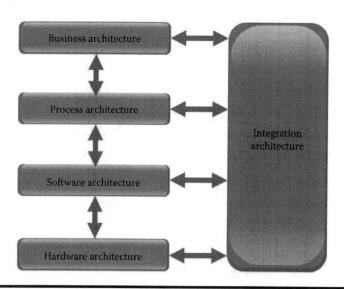

Figure 4.3 Multilayer EA architecture.

- Supporting business transformation, business process reengineering, new business development, etc., through providing a variety of analyses
- Supporting information integration efforts not only through documenting structures and relationships but also by allowing analyzing multilayer relationships

EA can bring many benefits to enterprises due to its impact on enterprise competitiveness, from both a business perspective and an IT perspective. Table 4.2 lists some of these benefits. EA can bring architectural alignment, coherence between strategy and operation, and collaboration among planning, operations, and infrastructure. Additional benefits include business IT alignment, consolidation, standardization, cost reduction, regulatory compliance, and agility, among others. EA enables enterprises to meet challenges such as integration, agility, and change. As the blueprint of an organization, an EA comprises most of the important assets of the organization, such as organizational structures and resources that are vital to the effective functioning of the organization. One EA can better satisfy some key goals than another EA. The concept of EA may emerge from the structuring of the organization's full vision, in all of its complexity and dimensions, in a theoretical fashion. EA can provide the framework to be used in the establishment of the business foundation for the enterprise; can effectively integrate processes, information, and diverse information technologies based upon the business' goals; and can lead to efficacious decisions pertaining to all of the relevant aspects of the business.

EA approaches have received considerable attention due to their ability to align business and IT in organizations, and they are making way for a business vision that can be reflected in both operations and supporting systems. However, due to the high-level nature of EA approaches, their ability to adapt to specific domains can be challenging and can require efforts both on the EA and on the application

Table 4.2   EA Benefits

Business-related benefits
• Knowledge management
• Adaption
• Improving operations
IT-related benefits
• Complexity management
• Resource management
• Visibility

domain. EA approaches still lack formal methods that can sufficiently represent the organizational context or goals in general and specifically require situation-specific assumptions, which can limit their interoperability with other components of the architecture.

In EA research, many existing frameworks, reference architectures, and methodologies are relevant and useful. In recent years, the ArchiMate language and framework has become available for EA design. Some researchers consider this situation similar to that of UML for software design—with its own international open standard. ArchiMate is an EA modeling language that supports the description, analysis, and visualization of architecture both within and across business domains. It is a technical standard from the Open Group and is based on the concepts of the IEEE 1471 standard. ArchiMate distinguishes itself from other languages such as UML by its enterprise modeling characteristics. ArchiMate divides the EA into three layers: the business, application, and technology layers. The "business layer" covers the business processes, services, functions, and events of business units. The "application layer" covers the software applications that support the components in the business with application services. The "technology layer" deals with the hardware and communication to support the "application layer." This layer offers infrastructural services needed to run applications. Each layer is self-contained, despite being a component of the integrated model.

Design and Engineering Methodology for Organizations (DEMO) is another major stream. In DEMO, an organization is viewed as consisting of three aspects: the B-organization (business) aspect, the I-organization (information) aspect, and the D-organization (data) aspect. A thorough understanding of the B-organization is a good starting point in designing and reengineering an organization, which can ultimately lead to developing software and systems for implementing business processes. DEMO takes a language-action perspective and looks at organizations at the ontological, info-logical, and data-logical levels. Central to DEMO are the basic patterns of a business transaction. DEMO further distinguishes the construction, process, state, and action aspects. Practical applications of DEMO have been reported for organizational composition and decomposition modeling.

4.2 Supply Chain Modeling and the Relationship with EM and EA Modeling

The interest in supply chain modeling has been steadily increasing ever since the topic of SCM began to evolve about two decades ago. During this period, many supply chain modeling methods have been proposed. A taxonomy for supply chain modeling has been proposed as shown in Figure 4.4 (Min and Zhou, 2002). Supply chain models are classified into categories such as (1) deterministic (nonprobabilistic), (2) stochastic (probabilistic), (3) hybrid, and (4) IT-driven. In particular,

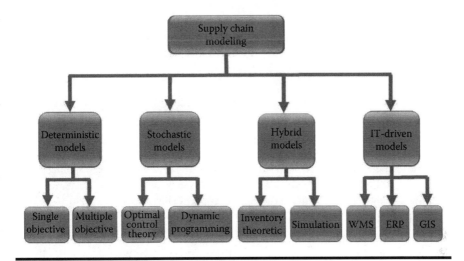

Figure 4.4 Taxonomy of supply chain models.

WMS, ERP, and GIS (under the IT-driven models) specifically relate SCM to enterprise systems, EM, and EA.

Supply chain models have also been classified into various frameworks with respect to the problem scope or application areas. The "problem scope" is viewed as a criterion for measuring the realistic dimensions of the model. The models that attempt to integrate different functions of the supply chain also appear. Considering that supply chain problems inherently cut across functional boundaries, supply chain models involve details from more than one business functions within the supply chain. Supply chain models can deal with the multifunctional problems of supplier selection/inventory control, production/inventory, location/inventory control, location/routing, and inventory control/transportation. See Figure 4.5.

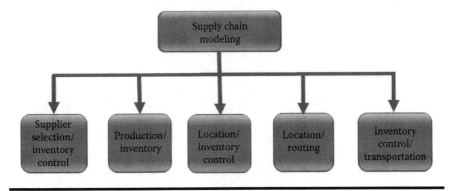

Figure 4.5 Types of integrated supply chain models.

The relationship between SCM, EM, and EA has been noted in the existing research (Min and Zhou, 2002). SCM, EM, and EA are intertwined with other and can be discussed in multiple views, including the orchestral view, the geographic view, and the institutional view:

- *Orchestral view*: SCM, EM, and EA represent the same line of the operations of the enterprise. SCM, EM, and EA have a close relationship due to their operational and governance mechanisms. SCM involves the controlling mechanisms of the physical artifacts of the enterprise, extended to the supply chain networks. More or less, the organization and governance of enterprises and SCM should be compliant with each other. Thus, SCM, EM, and EA orchestrate the same goals, that is, the goals of the enterprise and its extended enterprises.

- *Geographic view*: Most enterprises act in large supply networks. A supply chain may be implemented in an extended geography context presenting large differences in both space and time. The core of the architecture of an enterprise may include parts of the same supply chain network. For example, if the suppliers are from remote geographic areas, such geographic information need to be included in the EA. SCM is in many cases characterized by the significant geographical aspects that need to be represented in the physical artifacts of the enterprise. In supply chain networks, there is plenty of information about the artifacts. For example, multiple material flows as well their corresponding information flows (no matter whether they are synchronous or asynchronous, real-time or discrete) have to be encompassed within the EA.

- *Institutional view*: An enterprise may be subordinate to certain organizational entities in terms of the structure and operation of its systems and operational processes. Numerous business and regulatory requirements are out of the direct control of the enterprise but need to be considered in the business's operation. This forces the enterprise into a careful management of the degree of its compliance. International organizations and other institutions impose requirements on the enterprise and set up frameworks of compliance. In an extended enterprise context, for both enterprise and supply chain networks, there are numerous requirements for legality, security, and ethics. Many requisites are in a continuous process of change. An extended view about EA or an EA for extended enterprises is required. The issue here is for a business to use EA to comprehend and manage both enterprise and supply chains, leading to the shaping of the interorganizational infrastructure. This will continuously be influenced by the dynamic continuation and the reshaping of the enterprise and the extended enterprise's objectives, as well as the surrounding business and regulatory environments. EA must try its best to keep pace with such continuous changes.

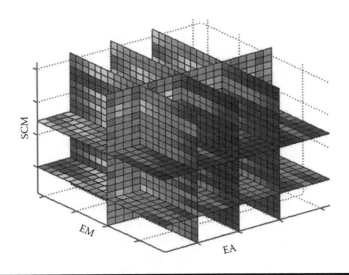

Figure 4.6   A new framework for SCM, EM, and EA modeling.

From the perspective of EM and EA, we can classify SCM models into five categories:

1. SCM models fit into the disciplines of operations research, operations management, and SCM (see Section 4.2.1).
2. SCM models involving intraorganizational interoperation (see Section 4.2.2).
3. SCM models involving interorganizational interoperation (see Section 4.2.3).
4. SCM models related EM and EA modeling methods (see Section 4.2.4).
5. SCM models representing a combination of (1) to (4) (see Section 4.3) that provides comprehensive modeling on SCM for EM and EA purposes and is suitable for extended enterprise environment (see Figure 4.6). The models in (5) are the most valuable ones for developing EAs. In the following, we briefly review the existing research and introduce selected models.

4.2.1 SCM Models in Operations Research, Operations Management, and SCM

Beamon (1998) divides supply chain modeling approaches into four categories: (1) deterministic analytical models, (2) stochastic analytical models, (3) economic models, and (4) simulation models. Biswas and Narahari (2004) classify supply chain models into three categories: (1) optimization models, (2) analytical performance models, and (3) simulation models. Huang et al. (2003) classify supply chain simulation models into three groups: (1) discrete-event simulation, (2) system dynamics, and (3) agent-based modeling and simulation. One of the major focuses of the optimization models is to find the location of different facilities (production, distribution, and warehousing) or paths through which different material goods will flow. These models correspond to the approaches proposed by Beamon (1998). The second category of approaches aims to address the stochastic and dynamic nature of supply chains by employing the methods that enable the modeling and representation of such systems. This category includes approaches such as Markov chains or Petri nets.

To be competitive, a supply chain must keep innovative through continuous improvement, which relies on an accurate and comprehensive evaluation of the supply chain. The Supply Chain Maturity Model is a technique that supports supply chain innovation by continuous improvement based on supply chain evaluation. In such models, supply chains are evaluated from a multidimensional perspective. Multilayer supply chains are modeled for minimizing the total cost of distribution related to distribution centers to be opened. The model can be solved by methods based on genetic algorithms. Then, the obtained results can be compared with other heuristic methods such as the Lagrangian method. The results indicate that with adequate input data, a genetic algorithm method can achieve good results. In addition, when the Lagrangian-based heuristic method cannot be applied because of changes in the model structure, a genetic-based method can be applied. Models have been developed for distribution networks in which products are delivered to customers in different ways. Pazoki et al. (2011) propose an approach for modeling supply chains. Instead of minimizing cost or maximizing benefit, they try to set the ratio of revenue-to-cost to a certain value in order to minimize the cost. Since this ratio is one of the decision criteria in an engineering economic context, it may be attractive to help industries decide whether or not to participate in the chain. Huang and Liu (2008) use the platform of the multiagent to model the supply chain and to verify the feasibility of supply chain simulation by use of multiagent supply chain modeling. There are many published papers in this category. The work introduced here shows just a few examples.

4.2.2 SCM Models Involving Intraorganizational Interoperation

Jureta and Faulkner (2005) propose an agent-based meta-model for enterprise modeling. Its aim is to develop an enterprise model that captures the knowledge of an organization and of its business processes so that an agent-oriented requirements specification of the system-to-be and its operational environment can be derived from it. To this end, the model identifies constructs that enable the capture of the intrinsic characteristics of an agent system. This approach allows a holistic perspective for integrating the human and organizational aspects in order to gain a better understanding of the business system. Sandkuhl and Kirikova (2011) focus on the use of the fractal paradigm in enterprise modeling. The model developed investigates whether the properties of fractal organizations can be applied in business analysis and whether this results in useful outcomes and/or new insights.

Based on an adaptation and operationalization of properties of fractal organizations, real-world cases are analyzed using the adapted properties. Just as in the example research introduced in the last section, there are many published papers in this category. The work introduced here shows just a few examples.

4.2.3 SCM Models Involving Interorganizational Interoperation

Supply chain integration involves market integration, organizational integration, resource integration, information integration, and other integrations beyond the intraorganizational context. Supply chain networks may be considered as a system of network systems. SCM models may consist of three layers including the basic SCM activity model, the SCM networking model, and the SCM system model. The models at higher levels are more interorganizational in nature (see Figure 4.7).

Researchers indicate that supply chains are typical complex systems that can be studied in the perspectives of complex networks. As such, the problems in supply chains can be analyzed based on complex networks. Supply chains can also be viewed as sociotechnical systems. Sociotechnical systems are systems that involve both complex societal and technical systems. The structure and behavior of societal and technical subsystems give rise to the overall behavior of a sociotechnical system. The sociotechnical systems theory can be of significance, since the system behavior of supply chains can be analyzed by considering both social and technical aspects and the interdependencies between them.

The models introduced in this section are generally related to interorganizational operations. As the enterprises in the supply chains are interdependent on each other and are supposed to work in coordination, any problem that occurs in any link will affect the entire supply chain. As such, supply chain risk can become a critical obstacle in SCM and can bring tremendous harm to the enterprises in the supply chain. It can eventually affect supply chain functioning. The model

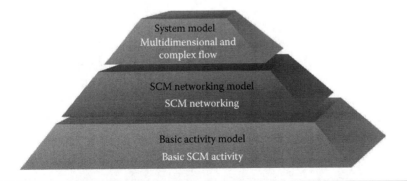

Figure 4.7 A system model of supply chain.

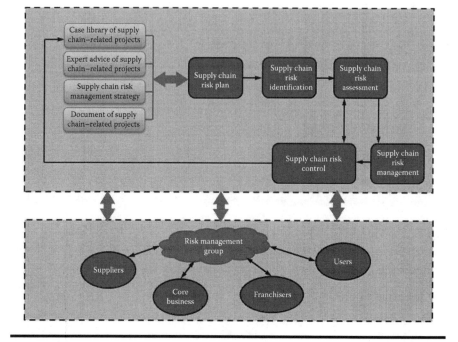

Figure 4.8   A model for supply chain risk management.

developed by Chen and Yuan (2009) indicates that great risks exist, due to the complexity of the supply chain. A supply chain risk management model is proposed, as Figure 4.8 shows. In the model, the members of the supply chain establish a risk management group. The model integrates a case library of supply chain–related projects, expert advice on supply chain–related projects, a supply chain risk management strategy, and the integration of qualitative and quantitative evaluation.

Uncertainty is the main attribute in managing supply chains (Mahnam et al., 2009). Managing a supply chain is difficult, since there are both various sources of uncertainty and complex interrelationships among the various entities in a supply chain. Uncertainty may result from customer's demand variability or unreliability in external suppliers.

Lu and Li (2006) apply the fuzzy sets theory to model the supply chain in an uncertain environment. They develop a fuzzy supply chain model based on possibility theory to evaluate entire supply chain's performance. The proposed model allows decision makers to express their opinions about the risk, to analyze the service level, and to select suitable supply chain strategies.

As Figure 4.4 shows, simulation is one of main research methodologies for studying supply chains. Long et al. (2011) point out that simulation, as a tool for studying complex systems, especially studying its dynamics and stochastic characteristics, has merits in the modeling and simulation of highly dynamic supply chains. Users can expect to understand the operation of the supply chain in a visual way and do a "what-if" analysis for different scenarios. In particular, multiagent-based distributed simulation is one of the most effective tools to model and analyze supply chains. Long et al. (2011) developed a multiagent model that is multilayered and can support the modeling and distributed simulation of complex supply chains. The model provides flexibility for simulation. Treating supply chains as a dynamic complex system, Liu et al. (2012) present a methodology for modeling both the structure and the dynamics of a complex supply chain, based on a process approach. Supply network modeling consists of a few segments in which each segment has certain characteristics, which are called submodels. The key submodels include the business information supply network chain structure submodel, the process submodel, the business environment submodel, and the constraints submodel. Relevant segments of the supply chain can be modeled while disregarding unessential details. The flexibility of the simulation model allows the researcher to study the behavior of a business information supply chain model in a real system environment. Simulation provides the ability to simulate the effect of particular events on a business information supply chain performance using different scenarios without a disruption in the process.

Other research methodologies have also been popular. Web services technology has been applied to implement a distributed simulation for supply chain modeling and analysis. Yoo et al. (2009) propose a supply chain simulation model that combines parallel and distributed simulation and web services technology. In the proposed model, parallel and distributed simulation provides the infrastructure for supply chain simulation while web services technology allows it to coordinate with the supply chain simulation model. The implementation demonstrates the viability of the proposed method. Shiau and Li (2009) model the interrelationships across a supply chain by using multiagent conflict detection and resolution methodology. The optimization and near-optimization solutions of supply chain networks can be determined. Ryu (2010) proposes a modeling methodology for supply chain operations that focuses on the relationships of supply chain entities in which supply chain operation problems are mathematically formulated into a multilevel programming problem, and a multiparametric programming-based computation methodology is proposed to compute the solution of the problems.

4.2.4 EM and EA Methods Related to SCM

As a result of economic globalization, we are witnessing an unprecedented amount of interorganizational cooperation. In contrast to the traditional intraorganizational and centralized business computing, contemporary interorganizational activities have been driven by highly adaptive collaborations and interactions. Consequently, enterprise modelers face the challenge of keeping pace with mirroring the interorganizational interoperations in global supply chain at multiple levels, including the conceptual level.

According to Vernadat (1996), there are five major motivations for EM: the management of system complexity, better management of all types of processes, capitalization of enterprise knowledge and know-how, business process reengineering, and enterprise integration. Charles and Lauras (2011) point out that EM could be defined as the art of externalizing knowledge that adds value to an organization or needs to be shared. The use of EM should enable the building of comprehensive pictures, which can explain complexity far better than a long and complicated explanation can. It also should enable the graphical formalization of a given knowledge and should be able to optimize its usage. The major advantage of EM is that it fosters the building of a common consensus about the way in which operations work or should work. The EM approach brings a number of methods and tools for representing the structure, behavior, components, and operations of a business entity in order to understand, (re)engineer, evaluate, and even control business operations and performance. It is also important to note that this approach can be implemented within a single organization as well as within an extended enterprise.

Enterprise models aim at representing the whole or part of an enterprise. They can be informal, semiformal, or formal. Enterprise modeling aims to construct a model of the whole or of part of the enterprise of any organization that can be considered as a system, in order to represent the structure of the organization and to analyze related characteristics and behavior. Languages of enterprise modeling allow the construction of models that are often characterized by a level of abstraction, such as conceptual abstraction. The formalization degree of the models varies; it can be informal (such as natural language), semiformal (such as language with graphic formalism), or formal (such as mathematical language). Most of the time, the models based on informal language are used to describe an existing situation while the models based on a formal language are used for scientific analysis. The enterprise modeling processes involve (1) obtainment of the necessary information to build the model, (2) construction of the model, (3) formalization of the model, and (4) explanation of the model.

Efforts have been made to study the required characteristics of EM methods. At least the following three aspects have been emphasized:

Multiview representation: Vernadat (2004) describes enterprise models as abstract representations of some enterprise perspectives. As such, an enterprise model may encompass various types of models, such as a process model, a resource model, an information model, and a decision model. EM should provide a structured representation, which is a selective representation of an enterprise. Model views are used to emphasize the aspects that are considered to be relevant and correspond to particular contexts.

System life cycle: EM should cover phases that are included in a system's life cycle, in general.

Genericity: EM should define the level of detail associated with the model.

EM is a subject in development; it has not yet reached maturity. As such, there is a great deal of latitude for future development and methodological improvement. Almeida et al. (2009) review a number of enterprise modeling approaches including ARIS, ArchiMate, DoDAF, Reference Model of Open Distributed Processing (RM-ODP), and BPMN. ARIS has been widely applied to EAs, but ARIS documentation has been hard to interpret. ARIS includes concepts such as the "organizational unit" and the "organizational unit type". Organizational units, organizational unit types, and other concepts can be related to a business process or its activities through an "executed-by" relation. According to their study, ArchiMate has a number of generic relations that can be applied between a number of modeling elements. However, the detailed semantics of such relations as they can be applied to particular kinds of concepts is not always clarified. DoDAF defines two products in the Operational View. These are the Operational Node Connectivity Description (OV-2) and the Organizational Relationship Chart (OV-4). DoDAF, though, has issues relating to a lack of consensus in the language representation of the concepts. RM-ODP, however, provides a rich conceptualization. BPMN focuses on business process modeling and does not provide constructs for organizational modeling. However, the activities in a business process may be related through using the "participant model" element. Almeida et al. have contributed a semantic foundation to the role-related concepts in enterprise modeling. Their efforts are positioned at conceptual and object-oriented modeling for a possible common foundation for these modeling domains.

Hoyland (2013) indicates that today's enterprise modeling practices are proprietary, time-consuming, and generally ineffective as tools for understanding details across all of the levels of an enterprise. The author combines the practice of EAs with systems perspectives and the semantic web to present a distinctive methodology for developing tools. A method called reusable quality technical architectures (RQ-Tech) has been developed. It demonstrates that a complex organization must creatively respond to a variety of events that can be holistically represented using a dynamic model. Specifically, the RQ-Tech technique provides ways to map and link multitudes of enterprise documents so that together they can effectively represent the nature and essence of the organization as one organic structure. The combination of enterprise documentation and the semantic web produces a holistic model of enterprise. Hoyland (2013) demonstrates that state-of-the-field research found that no product built by major vendors today claims to satisfy either Zachman's or PERA's criteria; none can be designated as capable of modeling the uppermost strategic layer of a large, complex enterprise. Hoyland's research covers three areas: EAs, systems theory, and semantic web, as shown in Figure 4.9.

According to Hoyland's study, a literature review of trends in these areas reveals that there is a specific set of information in each major area that supports the other two. Systems theory is shown to support semantic web methodologies in areas that are applicable to the practice of EA and form a foundation for understanding and

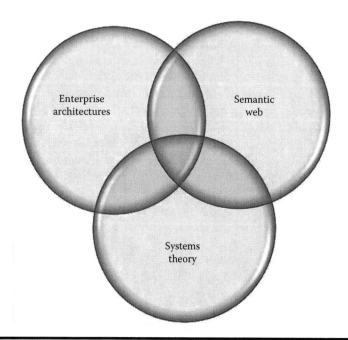

Figure 4.9   Three major research streams.

analyzing approaches to EA. Meanwhile, EA can be used to understand the complexity involved in an enterprise environment.

There is a great deal of latitude in applying the methodologies originated from other subjects to EM modeling. UML modeling language has been applied to modeling enterprise systems (Torchiano and Bruno, 2003). A modeling approach that is based on basic object-oriented concepts has been developed. In particular, the use of "instance models" as a key concept has been stressed. An object-oriented modeling approach for supply chain networks has been developed (Biswas and Narahari, 2004) in which objects belong to two categories: "structural objects" and "policy objects." UML is used to create a generic object model of supply chain elements. The example of a liquid petroleum gas supply chain has been used to illustrate the object-oriented modeling approach. An "ontology model," which explicitly defines the generic business processes relevant to supply chain operations, has been developed to support supply chain process modeling. Its ontology has been tested in the creation of the "information model" to support information exchange needs through industry case studies (Grubic and Fan, 2009).

Earlier, we briefly introduced EM modeling. In summary, EM mainly addresses the modeling of business goals, strategy, and processes. It is a set of modeling techniques for externalizing enterprise operations, which adds value to enterprises, extended enterprises, and supply chains. The scope of enterprise modeling thus corresponds to the enterprise and to supply chain operations by means of both intraorganizational and interorganizational integrations. On the other hand, EA is more concerned with aggregated models. It covers a broader scope, since an EA model can be an abstraction of all of the assets of an organization. Razak et al. (2011) define EA as a complete expression of the enterprise; it is a master plan that "acts as a collaboration force" between aspects of business planning such as visions, goals, strategies, and governance principles; aspects of business operations such as organization structures and processes; aspects of automation such as information systems and databases; and the enabling technologies such as computers, operating systems, and networks. EA is concerned with high-level logic for business processes and IT capabilities. EA is defined as a common business- and IT-related strategic approach that takes a systems perspective and views the entire enterprise as a holistic system encompassing multiple views such as the organization view, the process view, the knowledge view, and the enabling IT view in an integrated framework.

An EA framework provides (1) one or more meta-model(s) for EA description, (2) one or more method(s) for EA design and evolution, and (3) a common vocabulary for EA and maybe even reference models that can be used as templates or blueprints for EA design and evolution (Winter and Fischer, 2006). Goel et al. (2009) summarize several EA-related frameworks, models, and methods as follows:

- EA frameworks such as Zachman from Zachman Institute of Architecture, TOGAF from the Open Group, DoDAF from the US Department of Defense, MoDAF from the UK Ministry of Defence, the EA Cube from Scott Bernard, TEAF from the US Department of Commerce, FEAF from the US Federal Government, GERAM (ISO 15704: 2000), RM-ODP (ISO/IEC 10746), GRAI, ARIS, CIMOSA.
- EA methodologies such as TOGAF ADM, EAP, the EA Cube Method, and SEAM. Each of the US Government Frameworks includes a methodology (DODAF, FEAF, and TEAF).
- EA modeling techniques and notations such as ArchiMate, UEML, SysML, BPMN, ERD, and IDEF.
- EA tools such as Abacus from Avolution, Enterprise Architect from Sparx, System Architect from IBM Telelogic, BizzDesigner from Bizzdesign, ARIS Process from IDS Scheer, and Altova Enterprise from Altova.

Le and Wegmann (2013) suggest that modeling EA requires the representation of multiple views for an enterprise. This can be done by a team of stakeholders who have different backgrounds. Due to the multidisciplinary nature of EA, it is not suggested to choose a single modeling approach, even a widely recognized one, to build a truly comprehensive enterprise model. Developing a modeling framework that can be applied throughout the entire enterprise model and that can represent multiple views can be challenging.

In addition, EM and EA are subjects in development that have not reached their maturity stages yet. As such, there is a great deal of latitude for future development for methodological improvement. Lange and Mendling (2011) point out that although quite a few frameworks are called EA frameworks, they vary significantly, and the term is used very ambiguously. For example, four of the named EA frameworks are mostly combinations of enterprise ontologies and EA process methodologies that specify various tools, models, and artifacts. Zachman's is merely an enterprise ontology with no process or tooling implications. Additionally, the scope of these EA frameworks is about designing and building enterprise-wide information systems rather than managing and optimizing enterprises as a whole. Therefore, these frameworks, in practice, are often combined or adjusted to specific needs. Nevertheless, there are some elements that all of these frameworks have in common. Most EA frameworks provide architecture description techniques and the associated modeling techniques but do not specify a specific modeling language. Engelsman et al. (2011) consider that the current EA modeling techniques focus more on enterprise processes and applications and less on modeling the underlying motivation of EA in terms of stakeholder concerns and the high-level goals that address such concerns. They propose a language that supports the modeling of this motivation. The definition of this language is based on existing work on high-level goal and requirements modeling and is aligned with ArchiMate language.

An EA model is a representation of the as-is or to-be architecture of an organization. Aier and Gleichauf (2010) present a systematic approach to the capturing of dynamic transformation in enterprise models. This approach distinguishes a macro and a micro level, making the transformation process more transparent. On a macro level, a framework is proposed for temporal dimensions that are relevant in representing model transformation. From this framework, valuable propositions for adapting enterprise meta-models can be derived, in order to enable transformation. On a micro level, a procedure model is proposed for analyzing differences between as-is and to-be models with the objective of deriving effective transformation operations. The proposed procedure enables a consistency check between as-is and to-be models as well as the realization of an enterprise model transformation for planning purposes.

In EM, there is a great deal of latitude for applying the methodologies originated from other subjects to the EM modeling. Similarly, based on the RM-ODP, a standardization effort that defines the essential concepts for modeling distributed systems as well as ODP-related international standards/recommendations, a modeling framework called SeamCAD has been developed (Le and Wegmann, 2013). SeamCAD makes RM-ODP applicable in the context of multilevel EA and consolidates the SEAM, a family of methods for seamless integration purpose.

There is a growing need to develop effective EM and EA models that can integrate with existing supply chain models to help improve the designing and building of enterprise systems for the effective management and optimization of enterprises, extended enterprises, and supply chains as a whole. Development of these models becomes challenging when multiple flows need to be simultaneously modeled over a time horizon (see Figure 4.6). For instance, most research efforts on supply chains do not explicitly model the multiple entity flows of a manufacturing system that may be an explicit and important part of that supply chain. Researchers have studied the impact of flexibility on the lead time performance of the supply chain. Their study focused on flexibility at two levels: flexibility at the manufacturing system level and flexibility at the supply chain level. The studies indicated that both types of flexibility are equally important and that, to some extent, they can substitute for each other. However, the combined effect of these two flexibility types appears to be more beneficial than their individual effects.

4.3 Closing the Gaps between Existing SCM, EM, and EA Models

Currently, in the development of enterprise systems, the technology has not been developed in parallel with the advancement in related theories, techniques, and theoretical frameworks. Hoyland (2013) has illustrated the gaps that currently exist in the practice of EA in which systems theory has been ignored. Ma et al. (2011) indicate that the majority of current workflow systems have been developed with the perspective of information modeling without considering the actual requirements of SCM. Ma et al. (2011) developed a model for modeling and analyzing the interorganizational workflow systems in the context of lean supply chains (LSCs). An example of their research is introduced in Section 4.5. In order to gain real benefits from SCM, EM, and EA, relevant frameworks, methods, and techniques must span from its current scope and must truly reflect the details regarding the entire enterprise, the extended enterprise, and beyond. In order to deal with the complexity of such a requirement, methodology plays an important role. Currently, the urgent needs are to fill in the gap between the various types of SCM models, enterprise models, EA frameworks, and the next-generation type of EA, which will become the foundation of the new generation enterprise, as Figure 4.10 shows.

In other words, there is a great deal of latitude for future development in filling in the gap between the various types of SCM, EM, and EA models. This new challenge will require that new theoretical frameworks, models, techniques, and tools be developed. Enterprise Interoperability Science Base (EISB) is one of the recent efforts in the area of enterprise interoperability. The main models considered to be capable of supporting EISB include ARIS, DoDAF, OBASHI (OBASHI Business & IT methodology and framework), RM-ODP, TOGAF, and IBM's Zachman Framework (Jardim-Goncalves et al., 2013). EISB and the formal models that intend to be included will be useful in filling in the gap between the various types of SCM, EM, and EA models and the required next-generation new type of EA. For the list of formal models for the establishment of EISB, refer to Jardim-Goncalves et al. (2013).

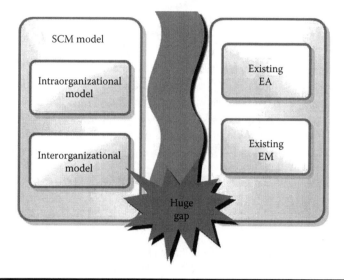

Figure 4.10 The huge gap between existing SCM, EM, and EA models.

4.4 Software Architecture: An Example of Recent Research

Enterprise systems are the key IT assets that enterprises use to plan, organize, schedule, and control their business processes. In particular, for SCM, enterprise systems have become critical enablers in streamlining processes and achieving effectiveness, efficiency, competency, and competitiveness. An essential component of an enterprise system is software architecture in EA, as Figure 4.3 shows. Software architecture describes a set of system components as well as their topological relations within an enterprise system.

Researchers have recently proposed many software architecture descriptions to accelerate industrial applications. These descriptions can be classified into domain-specific, distributed real-time control, embedded and dependable systems, agent platforms, and service-oriented architecture. In designing and implementing an enterprise system, software architecture must support the key business drivers; these drivers are also referred to as "quality attributes" or "nonfunctional requirements" (NFRs). Such support is aligned with the enterprise's mission and adds value at the system level (also see Figure 4.3).

When selecting software architecture for an enterprise system, multiple and often conflicting NFRs must be considered. For example, a system's flexibility and real-time performance can be in conflict with each other and must be balanced in software development. Despite the increasing number of proposals for software architecture options, few methods are currently available to evaluate software architecture against the requirements of a specific application. This has caused a hurdle in the development of an enterprise system, since the objectives of software architecture must be simultaneously considered to meet the requirements of today's IT-driven industrial automation.

In software development, functional requirements describe what the system can do and NFRs describe how well the system is able to fulfill required functions. NFRs, such as usability and reliability, represent the consideration of subjective performance at the system level. NFRs assist in decision making as software tools are selected in the hierarchical structure of an enterprise system.

The implementation of an enterprise system is usually focused on a few major NFRs. However, badly defining NFRs in selecting suitable software architecture could cause a serious error, which may be very hard to be fixed at the late stage of the system implementation. To uncover the key NFRs, a set of commercial software tools and the relevant publications on enterprise systems are selected and compared in Figure 4.11 (Niu and Xu, 2013). A diversified set of recurring NFRs have been explored by researchers, and various terminologies have been employed to describe similar meanings. Therefore, the concepts that are semantically close to the key NFRs are given in the second column. Each NFR's goal is expanded by the topics in the third column. A new perspective provided in Figure 4.11 is the distinction between business- and software-driven NFRs. The business drivers at one end of

NFR	Related Concepts	Topics	
Integration	Interoperability Coordination Synchronization	Linking and coordinating business processes over systems	 Business drive
Extensibility	Scalability	Adding new functionalities with ease	
Customer oriented	Customization Intelligence Flexibility	Aligning a company's businesses with customers' needs	
Performance	Efficiency Real-time Schedulability Memory usage	Optimizing system performance under multiple constraints	
Agility	Adaptability Flexibility Autonomy	Responding to change and uncertainties rapidly	
Reliability	Robustness Accountability Fault handling	Operating a system to resist system or product failure	Software drive
Security	Safety Info. protection	Being free from danger or threat	
Reusability	Reconstructability	Being reusable for new creations	
Testability	Reviewability	Verifying and validating a software artifact	
Usability	Simplicity	Being useful and usable	
Modularity	Recomposability Reconfigurability	Making systems decomposable and reusable	

Figure 4.11 Key NFR for ES.

the spectrum enable an enterprise to organize and promote its businesses; these capabilities are viewed as offering a superior advantage in contrast to the continuous evolution of traditional information systems. The NFRs at the other end of the range reflect the guiding principles that drive the software architecture design.

4.4.1 Types of Software Architecture

"Software architecture" describes system components as well as their external properties, and the internal relations of the components. This research field emerged in the 1990s when the major work was to establish the fundamentals of software architecture including description languages, formal logic, architectural styles, design patterns, and the like. In the context of industrial information integration, software architecture represents the business structures and processes of an enterprise system; it is a vital tool to support the assessment of design operations at an early stage.

A significant contribution to the development of software architecture is pattern codification, which can be used as the blueprint for components, constraints, and their relations. Patterns define the general solutions that can be reused to accelerate the software development process. Some methods used to apply the operational patterns for enterprise systems design are as follows:

- *General-purpose software packages* encapsulate data structures and algorithms to implement a generic but customizable solution of business problems, based on the best practices. Market-leading providers include SAP, Oracle, and Baan. In these packages, many operational patterns are exploited: database-centered data sharing, pipeline-based data processing, and event-driven message invocation, to name a few. The packaged software tools have been adopted by a variety of enterprises to optimize their business processes.
- *Domain-specific software architecture* is tailored to enterprise systems in a specified domain. This architecture includes some special components that differ from the common components of generic software architecture. For example, an industry-oriented ERP is capable of accommodating the requirements especially for a certain industry domain; some of the insignificant software elements and tools included in generic software packages can be removed in order to reduce the complexity. Major enabling technologies for DSSA domain-specific software architecture include Enterprise Java Beans (EJB), and Microsoft's Component Object Model (COM+).
- *Distributed computing* involves several interacting elements coordinated to achieve a system-level goal. Distributed programming typically falls into one of the following architectural options: client-server, n-tier architecture, and peer to peer. For the application of distributed computing, distributed architecture develops a flexible manufacturing cell using the Ethernet network. In addition, programming languages with parallel and concurrency supports

(e.g., C++) and middleware technologies (e.g., Common Object Request Broker Architecture or CORBA) are among the key enablers of distributed computing.

■ *Agent and multiagent systems* (*MAS*) have received much attention recently and have been deployed widely. "An agent" is an autonomous entity situated in the environment, whereas an MAS is composed of a group of agents; the agents within an MAS can cooperate or compete with each other to achieve goals at the system level. MASs have been successfully applied in manufacturing, behavior scheduling, workflow management, and business rules integration.

■ *SOA* can be viewed as a recent advance in integrating heterogeneous platforms including legacy software tools. An SOA allows an enterprise system to extend its capabilities by applying reusable software modules so that the development cost can be reduced without "reinventing the wheel." Equipped with methods such as Simple Object Access Protocol (SOAP), Web Services Description Language (WSDL), and UDDI, SOA has been introduced as a critical enabling technology to enterprise systems.

Efforts have begun to leverage the central ideas of software architecture, abstraction and separation of functions, to tackle the complexity of enterprise systems development. Some combined approaches are proposed; for example, software agents are incorporated into the SOA for the coordination of interactions in complex systems. As the number of enterprise systems solutions increases, it becomes important to systematically evaluate software architecture in designing and implementing an enterprise system.

4.4.2 Scenario-Based Software Architecture Analysis: A New Method

The benefits of fulfilling the enterprise systems NFRs are inarguable; however, there are many unsolved practical issues when NFRs are considered in the implementation of enterprise systems. For example, NFRs are usually subjective and can be hard to quantify. This calls for qualitative methods to reason how well the enterprise systems software can meet the NFRs. Moreover, users express their missions with different terminologies, even though there are several standards related to NFRs, for example, the International Organization for Standardization and International Electrotechnical Commission (ISO/IEC) 25030. Another challenge is that an enterprise system has to balance a set of conflicting objectives to determine its software architecture. Some examples of the conflicting objectives are flexibility versus productivity and scalability versus reliability; moreover, all of the objectives contribute to the cost factor. For success, it is necessary to consider all of the objectives simultaneously at the system level.

To meet the practical challenges in evaluating software architecture based on the given NFRs, a scenario-based method (as shown in Figure 4.12) has been

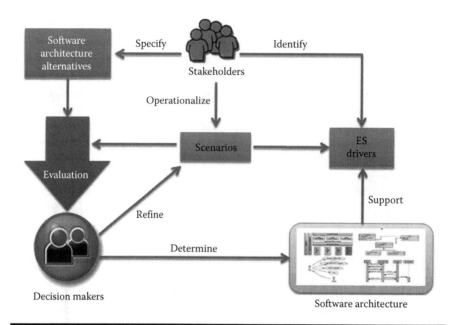

Figure 4.12   Framework for scenario-based ES software architecture analysis.

proposed, in which boxes and arrows represent entities and activities, respectively. The core component, "evaluation", is highlighted. As depicted in Figure 4.12, the scenarios play two important roles in the evaluation: First, they allow the abstract NFRs to be concretely defined, operationally measured, and meaningfully communicated among the stakeholders; second, they link architecture choices to the satisfaction of the enterprise systems drivers, which helps management to make an informed decision about the system that is best suited to their needs.

A large group of researchers has been focusing on developing software tools for enterprise systems, since factory automations are now driven by information technologies. Despite the fact that more and more informatics-related tasks are being carried out at the early stage of product development, the evolution of enterprise systems has not caught up with this trend. Specifically, the research on software architecture for enterprise systems remains inadequate to deal with today's IT-driven industrial automation. Thus, there is a critical need to evaluate software architecture and to select the most appropriate one to fulfill the specific business requirements, in particular, NFR.

In this section, we have identified the challenges confronting enterprise systems software architecture development and introduced a scenario-based method for the trade-off analysis of software architecture (Niu and Xu, 2013). There are several dimensions in which this work can be expanded in the future to contribute to enterprise systems software architecture as a component of IIIE. First, the level of the details in the reported empirical study can be increased, in order to enhance the strengths of the developed method. Second, identifying and codifying industrial-oriented NFR catalogs and analysis patterns should be in order. Third, when evaluating the enterprise systems architecture alternatives, it is worth comparing a proposed scenario-based method with other holistic approaches.

4.5 Modeling and Analysis of Workflow for LSCs: An Example of Integrative Modeling of SCM and EM (Ma et al., 2011)

A cross-organizational workflow generally comprises intraorganizational work-flows and interorganizational workflows. The modeling and analysis techniques for the intraorganizational workflows are relatively mature. Zisman (1977) pioneered the initial research on workflow management. Subsequently, over the last two decades, numerous researchers have applied Petri nets to workflow research. However, the majority of traditional workflow systems seem to assume one centralized enactment service and are not able to support interorganizational workflows; meanwhile, problems such as inconsistency due to the lack of transparency in cross-organization workflow systems exist. Developing modeling and analysis techniques of cross-organizational workflow systems are considered challenges for researchers.

With respect to modeling and analysis of workflows, researchers have defined a loosely coupled interorganizational workflow net based on Petri nets and message sequence charts (MSCs), discussed its verification and consistency, and presented a method based on Petri nets, which exploit the structure of the Petri nets to find potential errors in designing workflows and allow for the compositional verification of workflows. Modeling and analysis of interorganizational workflows in terms of Petri nets was also proposed; this idea focuses on the techniques for verifying the correctness of interorganizational workflows. One study focused on a loosely coupled workflow process in which two communication mechanisms interact between organizations via asynchronous and synchronous communication. An interorganizational workflow (IOWF) based on local workflow nets was also introduced. In addition, van der Aalst and Kumar (2003) proposed the techniques of an eXchangeable Routing Language (XRL) using XML syntax and the formalism of Petri nets, which make it possible to analyze the correctness and the performance of integration as described in XRL. Furthermore, Verbeek and van der Aalst (2004) presented XRL/Woflan, which is a software tool using state-of-the-art Petri nets analysis technique for verifying XRL workflows. XRL/Woflan uses eXtensible Stylesheet Language Transformations (XSLT) to transform XRL specifications to a specific class of Petri nets and to allow users to design new routing constructs. In XRL/Woflan, the Petri net representation is used to determine whether the workflow is correct.

In recent years, researchers and practitioners have paid close attention to the issue of cross-organizational workflow cooperation. Based on the workflow-view approach, Schulz and Orlowska (2004) discussed the entities of an architecture that provides execution support for mediated and unmediated cross-organizational workflows and argued that this architecture can support a workflow-view-based cross-organizational workflow model. This architecture uses a Petri net–based representation as the basis of the consideration of state dependencies between the tasks in a workflow and the adjacent task in a workflow view. This representation focuses on the structural aspects of a workflow, which include the internal states of tasks, the view task, and their state dependencies. In order to allow virtual enterprises to collaboratively manage business processes, Liu and Shen (2003) describe a novel process view that is derived from a base process to provide process abstraction for modeling a virtual workflow process. This process view model enhances the conventional activity-based process models by providing different participants with the various views of a process. In the context of supply chains, in the process of cooperation, enterprises must closely monitor internal processes and those of partners in order to streamline business-to-business (B2B) workflows. As such, a process view model has been proposed to extend beyond conventional activity-based process models for designing workflows across multiple enterprises. A process-view-based coordination model for B2B workflow management has also been developed. The proposed approach mitigates the shortcomings of interenterprise workflow collaboration. Chebbi and Tata (2005) presented a framework that enables the cooperation of interorganizational workflows. This framework allows the plug-in of any existing WfMS enabling external applications to provide a high degree of flexibility. It also addresses workflows cooperation requirements. Furthermore, Chebbi et al. (2006) presented a view-based approach to dynamic interorganizational workflow cooperation. The relevance of interorganizational workflows is best seen when considering emerging virtual organizational forms. This approach allows for partial visibility of workflows and their resources, thus providing powerful methods for interorganizational workflow configuration. Tata et al. (2008) presented a novel bottom-up approach and platform for workflow interconnection and cooperation in the context of short-term virtual enterprises. To take into consideration the privacy of the participants, the preservation of their preestablished workflows, and the integration of existing WfMS, Tata et al. proposed CoopFlow—a new bottom-up approach for the abstraction, matching and interconnection, and cooperation of workflows.

In the earlier-mentioned literature, important considerations such as time factor, material flow, and information flow for LSC collaboration have not been thoroughly considered. In order to create an efficient and lean supply chain, collaboration of business processes among its members via cross-organizational workflow management system in the light of required information, material flow, and capital flow during a stipulated time interval must be facilitated. In this example, the occurrence time, material flow, and information flow are taken into account when building model of cross-organizational workflow net for LSCs.

Du et al. (2007) proposed an interorganizational logical workflow net (ILWN) for modeling and analyzing real-time cooperative systems based on time

Petri nets (TPNs), workflow techniques, and temporal logic. Through attaching logical expressions to some actions of an ILWN model, the size of the model can be reduced. Thus, ILWNs can efficiently mitigate the state explosion problem to some extent. Moreover, with regard to issues such as the fact that the existing formal methods for analyzing CSs cannot properly deal with accountability and obligations, Du et al. (2009a) presented a new class of labeled Petri net (LPN) models. In this class, the behavior of each partner is represented by an LPN, while a CS is modeled by the combination of all partners' LPN models. The obligations were verified based on LPN languages and the nonblocking properties of action sequences, while accountability could be proved by the network conditions and local action sequences on each partner's side. With regard to e-commerce workflows (ECWs), Du et al. (2009b) introduced labeled workflow nets (LWN) and their semantics on the basis of LPNs, and then presented an interorganizational labeled workflow net (ILWN). Furthermore, they analyzed the soundness of ILWNs and the properties of the interactive actions. The proposed framework for ECWs is able to record the history of interactive events and to monitor the execution of interactive activities in order to achieve a common goal via recording messages among organizations.

In this example, the characteristics of LPNs and those of TPNs are integrated into labeled TPNs (LTPNs), which define labeled time workflow nets (LTWNs) and cross-organizational LTWNs (CLTWNs) in order to create an LSC, and their properties are analyzed by using Petri nets theory.

4.5.1 Lean Supply Chain

In the last two decades, lean manufacturing that originated from JIT production of the Toyota Motor Corporation, extended from the internal operation of the company to the supply chain as a whole, which has generated the concept of LSC. The definition of LSC has been discussed in the literature.

Reeve (2002) describes lean SCM as "planning, executing, and designing across multiple supply chain partners to deliver products of the right design, in the right quantity, at the right place, at the right time." Vitasek et al. (2005) have offered a more detailed definition of LSC: "a lean supply chain as a set of organizations directly linked by upstream and downstream flows of products, services, finances, and information that collaboratively work to reduce cost and waste by efficiently pulling what is needed to meet the needs of the individual customer." Rivera et al. (2007) summarized the definition of LSC: "a lean supply chain is a network of integrated organizations in which the capabilities of all entities are aligned with customer demand." To achieve this status, an LSC must possess the following characteristics:

First, close relationships must be established among all of the supply chain members who are sharing gains and responsibilities. A collaboration based on trust is the foundation for all of the activities that integrate the supply chain.

Second, information needs to be transparent throughout the supply chain, including the end customer's demand, opportunities, and responsibilities. Therefore, all supply chain members should be able to align themselves with customer demand and aim for overall benefits.

Third, lean logistics approaches should be implemented in order to physically carry out and benefit from lean thinking.

Finally, to sustain an LSC, performance must be monitored, maintained, and improved. Metrics that reflect the overall supply chain performance should be adopted, and the results should be visible to all members.

According to Vitasek et al. (2005), supply chain networks are coordinated efforts among partners to eliminate waste across the supply chain as a whole. They can be accomplished only through successful collaboration across common processes. Collaboration among enterprises in supply chains is the foundation that supports other building blocks, that is, lean logistics, information systems, performance measurement, and continuous improvement. Collaboration facilitates the implementation of lean logistics, information system integration among organizations, and interorganizational performance measurement systems.

In this example, a model of workflow is introduced that can reflect the characteristics of an LSC mentioned earlier, that is, modeling of workflow that can portray the material flow, the information flow, and the capital flow using formal methods.

4.5.2 Assumptions

To meet the requirements of the users for products and services, collaborative management of an LSC allows material flow, information flow, and capital flow to move among organizations during the assigned time intervals and at the minimal cost. According to the characteristics of LSC (and for the convenience of modeling and analysis of workflows), the assumptions are as follows:

1. The requirements of the user are relatively predictable.
2. Upstream enterprises regard meeting production requirement of downstream enterprises as their goal. Only upstream enterprises meet their targets first, and then downstream enterprises meet theirs.
3. The information flow is transmitted downstream to upstream or upstream to downstream; materials are delivered from upstream to downstream, and capital is transferred from downstream to upstream.
4. Business processes between upstream enterprises and downstream enterprises are collaboratively executed via material flow, that is, the execution time of business processes between organizations is coordinated in such a way that upstream enterprises can deliver materials for downstream enterprises in accordance with time and quantity of delivering materials required by the downstream enterprises.

5. Business processes between upstream enterprises and downstream enterprises are collaboratively executed via information flow and capital flow in accordance with the time of collaboration between enterprises.

4.5.3 Standardization of Collaborating Business Process between Organizations

In order to generate a stable LSC system, the processes of collaborating business process between organizations ought to be standardized. Bala and Venkatesh (2007) defined interorganizational business process standards (IBPS) as "technical specifications for interrelated, sequential tasks and business documents that are agreed upon and shared by trading entities to achieve a defined and common business objective. IBPS are designed and developed to automate, integrate, and facilitate value chain activities such as supply chain management, collaborative forecasting, new product development, and inventory management."

Therefore, the processes of collaborating business process between organizations are a series of sequence processes that are arranged by means of contracts between organizations. In general, the process of collaborating business process can be described as follows:

Step 1: The downstream enterprise sends requirement information to its upstream partners (suppliers or third logistics).

Step 2: After receiving the demanded information from the downstream enterprise, the supplier sends information of confirmation to that downstream enterprise.

Step 3: After manufacturing purchasing materials, the supplier delivers materials to the downstream enterprise on time.

Step 4: After receiving materials from the upstream supplier, the downstream enterprise (e.g., the manufacturer or customer) pays for the materials.

4.5.4 Modeling and Analysis of Cross-Organizational Workflow

Here, LPNs are combined with TPNs, and LTPNs are defined. Based on LTPNs, LTWNs and CLTWNs are defined. As deficiencies in a workflow definition may lead to a longer throughput time, low service levels, and a need for excess capacity (Basu and Kumar, 2002), it is important to well define a workflow process. The soundness of the LTWNs and the CLTWNs should be verified via theoretical analysis.

A cross-organizational workflow system is the component of enterprise systems that supports collaborative business processes among organizations in supply chains. Currently, the majority of workflow systems have been developed from the perspective of information modeling without the consideration of the actual requirements of SCM. In this example, the modeling and analysis of the cross-organizational workflow systems is focused in the context of an LSC using Petri nets. The details include (1) the assumed conditions of the cross-organization workflow net according to the idea of LSC, and the standardization of collaborating business process between organizations in the context of LSC; (2) the concept of LTPNs defined through combining LPNs with TPNs, and the concept of LTWNs defined based on LTPNS. CLTWNs defined based on LTWNs; and (3) the notion of OR-silent CLTWNS and a verifying approach to the soundness of LTWNs and CLTWNs.

4.5.5 Application Example

The proposed method is illustrated by an example. An example of collaboration among organizations in a manufacturing supply chain is shown in Figure 4.13. In this example, a retailer submits an order to a manufacturer, and the manufacturer negotiates with its upstream plants. From the most upstream plants to the

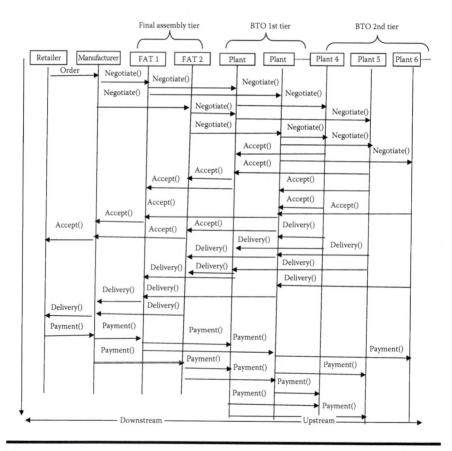

Figure 4.13 The sequence of cross-organizational collaborative process in a lean supply chain.

manufacturer, acceptance/refusal information is sent to the downstream plants according to the choices of the upstream plants as well as their production capacity (i.e., two-way information flows are transmitted in the LSC). If an upstream plant accepts its downstream plant's order, then it will start to produce the components, and it must deliver components to downstream plants on time (i.e., material flows are flowed from upstream plants to downstream plants). When the retailer receives the products, it pays the manufacturer for the products. In turn, the manufacturer pays its upstream partners, and they pay their upstream partners (i.e., capital flow moves from the most downstream ones to the upstream supplier, step by step). In Figure 4.13, we see that if each member in the supply chain can accomplish its own business process at the minimum cost, then this supply chain is an LSC. Figure 4.14 shows

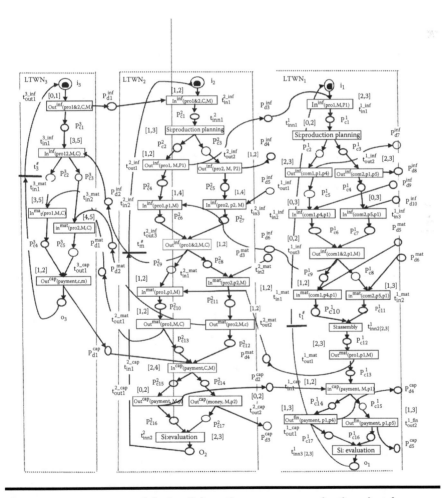

Figure 4.14   CLTWN model of collaboration among organizations in a lean supply chain.

the segment of the CLTWN model that comprises three local LTWN: the retailer's LTWN, the manufacturer's LTWN, and FAT1's LTWN. If there is an adequate space, the whole CLTWN can be drawn, and each LTWN can be numbered from the enterprise situated in the upstream supply chain to the enterprise situated in the downstream supply chain. In this example, $LTWN_3$ represents the retailer's LTWN (which is the most downstream in supply chain), $LTWN_2$ represents the manufacturer's LTWN (which is in the middle in supply chain, and $LTWN_1$ represent FAT1's LTWN. $LTWN_2$ is upstream of $LTWN_3$ and is downstream of $LTWN_1$. Three extra transitions, $t_1^\#, t_2^\#, t_3^\#$, are added in $LTWN_1$, $LTWN_2$, and $LTWN_3$, respectively. Each local workflow is modeled by an LTWN.

References

Aier, S. and Gleichauf, B. 2010. Towards a systematic approach for capturing dynamic transformation in enterprise models. *Proceedings of the 43rd Hawaii International Conference on System Sciences*, Honolulu, HI, p. 10.

Almeida, J., Guizzardi, G., and Santos, P. 2009. Applying and extending a semantic foundation for role-related concepts in enterprise modeling. *Enterprise Information Systems*, 3(3), 253–277.

Bala, H. and Venkatesh, V. 2007. Assimilation of interorganizational business process standards. *Information Systems Research*, 18(3), 340–362.

Basu, A. and Kumar, A. 2002. Research commentary: Workflow management issues in e-business. *Information Systems Research*, 13(1), 1–14.

Beamon, B. M. 1998. Supply chain design and analysis: Models and methods. *International Journal of Production Economics*, 55(3), 281–294.

Biswas, S. and Narahari, Y. 2004. Object oriented modeling and decision support for supply chains. *European Journal of Operational Research*, 153(3), 704–726.

Charles, A. and Lauras, M. 2011. An enterprise modelling approach for better optimisation modelling: Application to the humanitarian relief chain coordination problem. *OR Spectrum*, 33, 815–841.

Chebbi, I., Dustdar, S., and Tata, S. 2006. The view-based approach to dynamic interorganizational workflow cooperation. *Data & Knowledge Engineering*, 56, 139–173.

Chebbi, I. and Tata, S. 2005. CoopFlow: A framework for inter-organizational workflow cooperation. In Meersman, R. and Tari, Z. (Eds.), *Lecture Notes in Computer Science*, Springer, Heidelberg, Vol. 3760, pp. 112–129.

Chen, J. and Yuan, P. 2009. Research on supply chain and logistics risk management model. *Proceedings of the First International Conference on Information Science and Engineering (ICISE)*, Nanjing, China, pp. 4467–4470.

Du, Y., Jiang, C., and Zhou, M. 2007. Modeling and analysis of real-time cooperative systems using Petri nets. *IEEE Transactions on Systems, Man, and Cybernetics Part A: Systems and Human*, 37(5), 643–654.

Du, Y., Jiang, C., and Zhou, M. 2009a. A Petri net-based model for verification of obligations and accountability in cooperative systems. *IEEE Transactions on Systems, Man, and Cybernetics Part A: Systems and Human*, 39(2), 299–308.

Du, Y., Jiang, C., Zhou, M., and Fu, Y. 2009b. Modeling and monitoring of e-commerce workflows. *Information Sciences*, 179(7), 995–1006.

Engelsman, W., Quartel, D., Jonkers, H., and van Sinderen, M. 2011. Extending enterprise modeling with business goals and requirements. *Enterprise Information Systems*, 5(1), 9–36.

Goel, A., Schmidt, H., and Gilbert, D. 2009. Towards formalizing virtual enterprise architecture. *Proceedings of 13th Enterprise Distributed Object Computing Conference Workshops*, Auckland, New Zealand, pp. 238–242.

Grubic, T. and Fan, I. 2009. Integrating process and ontology for supply chain modelling. *Proceedings of International Conference on Interoperability for Enterprise Software and Applications China*, Beijing, China, pp. 228–235.

Hoyland, C. 2013. The RQ-tech methodology: A new paradigm for conceptualizing strategic enterprise architectures, PhD dissertation, Old Dominion University, Norfolk, VA.

Huang, G. Q., Lau, J. S. K., and Mak, K. L. 2003. The impacts of sharing production information on supply chain dynamics: A review of the literature. *International Journal of Production Research*, 41(7), 1483–1517.

Huang, W. and Liu, Z. 2008. Research on supply chain model and simulation based on multi-agent. *Proceedings of 2008 International Symposium on Knowledge Acquisition and Modeling*, Wuhan, China, pp. 384–387.

Jardim-Goncalves, R., Grilo, A., Agostinho, C., Lampathaki, F., and Charalabidis, Y. 2013. Systematisation of interoperability body of knowledge: The foundation for enterprise interoperability as a science. *Enterprise Information Systems*, 7(1), 7–32.

Jureta, I. and Faulkner, S. 2005. An agent-oriented meta-model for enterprise modelling. *Lecture Notes in Computer Science*, Vol. 3770, pp. 151–161, Springer, Heidelberg.

Lange, M. and Mendling, J. 2011. An experts' perspective on enterprise architecture goals, framework adoption and benefit assessment. *Proceedings of 2011 15th IEEE International Enterprise Distributed Object Computing Conference Workshops*, Helsinki, Finland, pp. 304–313.

Lapkin, A. 2004. Architecture frameworks: How to choose. Gartner Report G00124230, November 19, 2004, Gartner Publisher, Stainford, CT.

Le, L. S. and Wegmann, A. 2013. Hierarchy-oriented modeling of enterprise architecture using reference-model of open distributed processing. *Computer Standards & Interfaces*, 35(3), 277–293.

Liu, D. and Shen, M. 2003. Workflow modeling for virtual processes: An order-preserving process-view approach. *Information Systems*, 28, 505–532.

Liu, V., Zhang, J., Sun, X., and Huang, H. 2012. Business information supply chain modeling and simulation methodology. *Proceedings of International Conference on Modelling, Identification & Control*, Wuhan, China, pp. 339–344.

Long, Q., Lin, J., and Sun, Z. 2011. Modeling and distributed simulation of supply chain with a multi-agent platform. *The International Journal of Advanced Manufacturing Technology*, 55, 1241–1252.

Lu, C. and Li, X. 2006. Supply chain modeling using fuzzy sets and possibility theory in an uncertain environment. *Proceedings of the Sixth World Congress on Intelligent Control and Automation*, Dalian, China, pp. 3608–3612.

Ma, J., Wang, K., and Xu, L. 2011. Modelling and analysis of workflow for lean supply chains. *Enterprise Information Systems*, 5(4), 423–447.

Mahnam, M., Yadollahpour, M., Famil-Dardashti, V., and Hejazi, S. 2009. Supply chain modeling in uncertain environment with bi-objective approach. *Computers and Industrial Engineering*, 56(4), 1535–1544.

Min, H. and Zhou, G. 2002. Supply chain modeling: Past, present and future. *Computers and Industrial Engineering*, 43(1–2), 231–249.

Niu, N. and Xu, L. 2013. Enterprise information systems architecture-analysis and evaluation. *IEEE Transactions on Industrial Informatics*, DOI: 10.1109/TII.2013.2238948.

Pazoki, M., Fatemi Ghomi, S., and Jolai, F. 2011. A new approach in supply chain modeling. *Proceedings of 2011 IEEE International Conference on Industrial Engineering and Engineering Management (IEEM)*, Singapore, pp. 955–958.

Razak, R., Dahalin, Z., Ibrahim, D., and Yusop, N. 2011. Investigation on the importance of enterprise architecture in addressing business issues. *Proceedings of 2011 International Conference on Research and Innovation in Information Systems*, Kuala Lumpur, Malaysia, pp. 1–4.

Reeve, J. 2002. The financial advantages of the lean supply chain. *Supply Chain Management Review*, March/April, 6, 42–49.

Rivera, L. et al. 2007. Beyond partnerships: The power of lean supply chains. In Jung, H., Chen, F., and Jeong, B. (Eds.), *Trends in Supply Chain Design and Management*. Springer, Heidelberg, pp. 241–268.

Ryu, J. 2010. Modeling supply chain operations as multi-level programming problems and their parametric programming based computation methodology. *Korean Journal of Chemical Engineering*, 27(6), 1681–1688.

Sandkuhl, K. and Kirikova, M. 2011. Analysing enterprise models from a fractal organisation perspective—Potentials and limitations. *Lecture Notes in Business Information Processing*, 92, 193–207.

Schulz, K. and Orlowska, M. 2004. Facilitating cross-organizational workflows with a workflow view approach. *Data & Knowledge Engineering*, 51, 109–147.

Shiau, J. and Li, X. 2009. Modeling the supply chain based on multi-agent conflicts. *Proceedings of IEEE/INFORMS International Conference on Service Operations, Logistics and Informatics*, Chicago, IL, pp. 394–399.

Tata, S., Klai, K., and M'bareck, N. 2008. CoopFlow: A bottom-up approach to workflow cooperation for short-term virtual enterprises. *IEEE Transactions on Services Computing*, 1(4), 214–228.

Torchiano, M. and Bruno, G. 2003. Enterprise modeling by means of UML instance models. *ACM SIGSOFT Software Engineering Notes*, 28(2), 12.

Van der Aalst, W. and Kumar, A. 2003. XML-based schema definition for support of inter-organizational workflow. *Information Systems Research*, 14(1), 23–46.

Verbeek, H. and van der Aalst, W. 2004. XRL/Woflan: Verification and extensibility of an XML/Petri-net-based language for inter-organizational workflows. *Information Technology and Management*, 5, 65–110.

Vernadat, F. 1996. *Enterprise Modeling and Integration: Principles and Applications*. London, U.K.: Chapman & Hall.

Vernadat, F. 2004. Enterprise modeling: Objectives, constructs & ontologies. *Tutorial 1st EMOI Workshop at CAiSE*, Riga, Latvia.

Vitasek, K., Manrodt, K., and Abbott, J. 2005. What makes a lean supply chain. *Supply Chain Management Review*, 9(6), 39–45.

Winter, R. and Fischer, R. 2006. Essential layers, artifacts, and dependencies of enterprise architecture. *Proceedings of the 10th IEEE International Enterprise Distributed Object Computing Conference Workshops (EDOCW'06)*, Hong Kong, China, IEEE Computer Society.

Yoo, T., Kim, K., Song, S., and Cho, H. 2009. Applying web services technology to implement distributed simulation for supply chain modeling and analysis. *Proceedings of the 2009 Winter Simulation Conference*, Austin, TX, pp. 863–873.

Zisman, M. D. 1977. Representation, specification and automation of office procedures. PhD thesis, University of Pennsylvania, Wharton School of Business, Philadelphia, PA.

Chapter 5

Information Architecture for Enterprise and Supply Chain
A New Discipline of Industrial Information Integration

Today's business takes place in a collaborative environment. Collaboration can take a variety of forms, including intra- and interorganizational collaborations. To realize the collaboration, enterprises recognize the necessity to implement integrated enterprise systems, which can integrate their intra- and interorganizational business processes. During the past decade, enterprise systems have emerged as a promising tool for integrating and extending business functions and processes across the boundaries of organizations. In general, enterprise processes that were not designed as interoperable need to be integrated on an intra- and interorganizational basis. Enterprise systems (1) provide a single system that is central to the organizations and (2) ensure that information can be shared among all partners in the supply chain. An enterprise system is composed of a suite of modules and applications. An enterprise can build an enterprise system platform through integrating a number of modules and applications. In general, enterprise systems provide a platform that enables industrial organizations to integrate and coordinate their business processes. Enterprise systems are considered to be a revolutionary advance in the continuous evolution of computer applications in the business and industry.

5.1 Intraorganizational Systems

Enterprise systems started in the late 1960s with MRP. MRP was later superseded by MRP II in the 1980s. MRP II integrates material planning, shop floor operations, and more. In the 1990s, enterprise systems began to be called ERP, which integrates many more functions than its predecessors MRP II and MRP. As a consequence of the developments in this evolutionary process (especially since the 1990s), enterprise systems have effectively achieved intraorganizational integration, and the benefits of intraorganizational information sharing have become recognized. An intraorganizational system is aimed at providing (1) a system for the coordination of business processes within the organization and (2) an integrated architecture to an organization for enhancing organizational performance. A defining characteristic of the early systems was ensuring the integration of the internal business processes for an enterprise's internal coherent operation. The tools for bridging the various isolated systems include focusing on common databases and encouraging intraorganizational coordination. In the intraorganizational context, enterprises use enterprise systems to integrate all of their intraorganizational applications in order to achieve a high degree of integration. Intraorganizational systems enable the connectivity of disparate systems that exist within the organization. In fact, intraorganizational integration has successfully overcome the fragmentation between organizational units. Such systems do not specifically support SCM, which is more interorganizational in nature.

"CAD" is the use of computer systems to assist in the optimization of an engineering design. CAD can be used to dramatically improve the quality of design and has been proven to be an indispensable tool in many industrial sectors. In general, a specific type of CAD system is developed for a specific industry sector. For example, its use in electronic design is known as electronic design automation, while in mechanical design it is known as mechanical design automation. Early enterprise systems (such as MRP II systems) traditionally interface with engineering design systems to receive BOM and routing information (Langenwalter, 2000). Such integrative systems can serve intraorganizational purposes by integrating the design process, product data management, and project management with the rest of the enterprise, or even for interorganizational purposes, to integrate with supply chain partners for collaborative engineering design after a substantial extension.

The following describes a CAD system for a ceramic kiln with an object-oriented database (OODB) and a focus on an intraorganizational purpose. With the rapid development of the construction industry, the requirements for the quality, types, and sizes of the construction ceramics are varied. There can be many independent engineering systems for product design. These independent systems can be self-contained and can operate independently. In order to build an intraorganizational CAD system, the key of the integrated system is to integrate different systems in order to ensure data consistency. An "OODB" is a combination of object-oriented technology and database technology. The basic structure in OODB is an object. An object includes not only metadata, but also methods. OODB has the capacity to contain data objects of different types and of managing objects of different types. In the object-oriented data model, objects and their relations can be directly represented by the concepts of classes, subclasses, abstract classes, and super classes. A CAD system for integrating design, graph drawing, and computation is built according to the design characteristics of ceramic kilns (Chen and Xu, 2001). In the system, an OODB is developed, and design data, knowledge, and processes are highly integrated. The objective is to combine an engineering database with a knowledge base, as well as to develop an integrated intraorganizational CAD for engineering design. As OODB is suitable for application in an engineering environment, OODB is taken as the kernel of the system. The design system, drawing system, and knowledge-based system are operated through directly interfacing the engineering database; thus, the problem of data sharing is resolved. Figure 5.1 shows the overall system architecture.

The manufacturing environment has dramatically changed in the past few decades due to worldwide competition among manufacturers. The development of new manufacturing technologies has contributed to today's competitive situation in the manufacturing industry. Meanwhile, the demand for products of higher quality with a lower price and better performance in an ever-shorter delivery time is higher and higher. This environment has stimulated rapid changes in the manufacturing industry, causing a significant shift in how products are designed, manufactured,

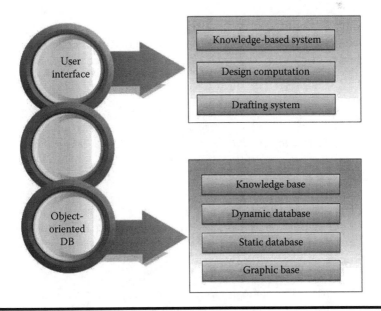

Figure 5.1   CAD system for ceramic kiln for intraorganizational application.

and delivered. In other words, competition has been forcing changes in the way product designers and manufacturing engineers develop products. In conventional product development, conceptual design, detailed design, process planning, prototype development, and testing are considered as sequential processes. Compared with the traditional sequential method, "concurrent engineering" is a systems approach to integrate concurrent design of products and their related processes. Concurrent engineering is intended to stimulate the product designers/developers to consider all of the elements of the product life cycle in the early phase of product development. The following describes an intraorganizational system for product design in concurrent engineering (Xu et al., 2007).

In order to improve product quality, lower cost, shorten the product development cycle, and fulfill customer requirements, engineering design requires that product designers take all of the factors involved in the life cycle of a product into consideration. Manufacturing, assembling, maintenance, and environment protection are typical stages of the life cycle of a product. As a result, quite a few related concepts have been proposed, such as design for assembly (DFA), design for manufacturability (DFM), design for serviceability (DFS), and design for environment (DFE). DFA, DFM, DFS, and DFE reflect different aspects of product design stages. It is obvious that overemphasizing one stage over another may not be a good choice; therefore, it is suggested that designers take all stages (such as DFA, DFM, DFS, and DFE, as well as related methods) into consideration.

Not only can concurrent product design stages be classified into stages such as initiation, DFA, DFM, DFS, and DFE but also careful evaluations and appropriate decisions regarding design alternatives must be made at each stage. From a systems point of view, product design is considered as an iterative process characterized by "design–evaluation–redesign." Such an iterative process is complicated for a number of reasons: (1) it is necessary to take all design objectives into account. However, some objectives conflict with each other such as precision versus manufacturing cost, material performance versus material cost, and so forth; (2) in the design stages, especially the early stages, it is difficult to quantify and weigh design objectives precisely due to lack of information or to the fact that objectives are still vague; (3) designers' subjective preferences could make the evaluation even more complicated. However, proper decisions still need to be made for product design in concurrent engineering.

With a comprehensive evaluation model based on mathematical approaches, a system is developed to provide support for multistage decision making (selecting best design alternatives) in concurrent engineering design. The overall objective is to develop a system that will help design engineers within an organization in a concurrent engineering environment. There can be different kinds of systems for concurrent engineering design, such as stand-alone systems or distributed systems in networking environments. An intraorganizational system for concurrent engineering is shown in Figure 5.2. It consists of conceptual design, assembly design, manufacturing design, and so forth. The example introduced here is an

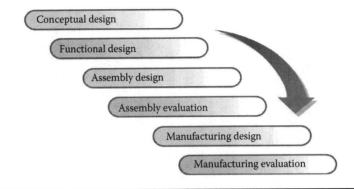

Figure 5.2   An intraorganizational system for product design.

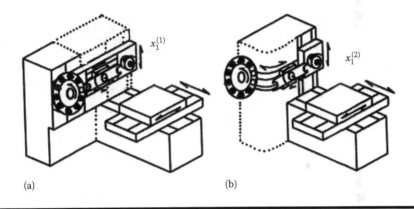

(a)   (b)

Figure 5.3   Design stage. (a) Alternative 1, (b) alternative 2.

intraorganizational subsystem for concurrent engineering that was developed and implemented in the real-world environment.

Using this system, conceptual design alternatives can be analyzed. Examples of machining center design alternatives are shown in Figures 5.3 through 5.5, which show the alternatives available in stages such as (1) the initial design (Figure 5.3), (2) the DFA (Figure 5.4), and (3) the DFM (Figure 5.5).

In this system, a multistage model for concurrent engineering product design is applied to improve the concurrent engineering process or practices through the improvement of the related decision-making process. The system is able to evaluate the alternatives comprehensively using mathematical methods and can then rank the alternatives. The system is beneficial in improving design capability since it enables engineers to evaluate design alternatives with interrelated criteria such as functionality, reliability, and manufacturability to achieve Design for x (DFx), with x as one of the criteria. The system provides decision support aids not only to capture the features of different concurrent design stages but also to perform

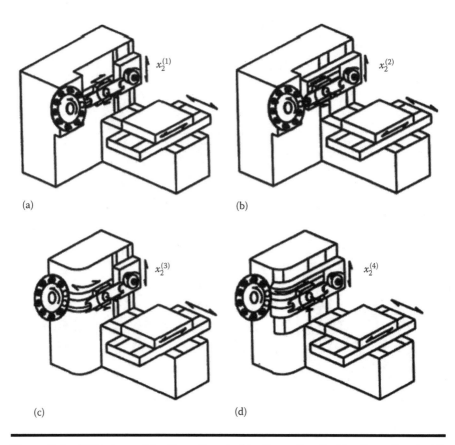

Figure 5.4 DFA stage. (a) alternative 1, (b) alternative 2, (c) alternative 3, (d) alternative 4.

automated decision support for DFx. The what-if-analysis (i.e., what would happen if a particular decision is taken?) is one of the useful functionalities provided by the system. The implementation results show that the system is practical and useful for concurrent engineering product design.

5.2 Interorganizational Systems

Advances in computing technology provide us with an enormous amount of quality information for industrial operations. One of the challenges we face is to make use of enterprise systems to collect, analyze, utilize, and distribute information. Integrative systems such as enterprise systems have broad applications in a variety of industry sectors. However, previously built CAD, CAE, CAM, and PDM systems were separately developed within an organization; also they can be heterogeneous to each other. First, for data collection, analysis, and utilization purposes, such

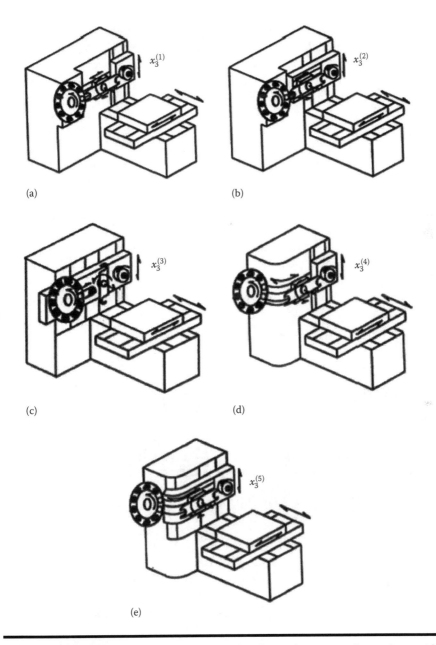

Figure 5.5 DFM stage. (a) Alternative 1, (b) alternative 2, (c) alternative 3, (d) alternative 4, (e) alternative 5.

systems should not be considered to be isolated. They should be integrated as intra-organizational systems to realize data, function, and process integration and to support the production life cycle. Second, since the last decade, the problem of disconnected information systems, or information islands that exist between business enterprises, has been identified. A typical example is that the supply chain coordination system between Baosteel and Shanghai GM was not able to exchange data automatically, which made it impossible to successfully run VMI, JMI, and CPFR modules. Thus, integrating enterprise systems across organizations has received a lot of attention.

In the aspect of developing integrative systems support, over the last decade, manufacturing systems integration has been paid a lot of attention. Since manufacturing enterprises may own many heterogeneous applications, these heterogeneous applications need to be integrated in order to enable interaction with one another within the organization (and possibly externally) to implement a collaborative manufacturing environment. In the collaborative manufacturing environment shown in Figure 5.6, we can see that many processes such as the forecasting, production planning, procurement, and manufacturing processes may not be able to exist by themselves and may need to interact with other partners' systems and applications for running businesses. At a higher conceptual level, systems, processes, and applications need to be integrated to a certain level in order to drive collaborative manufacturing initiatives.

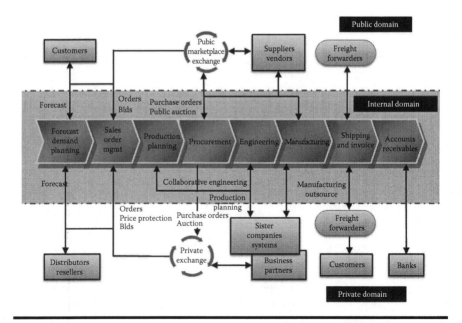

Figure 5.6 Collaborative manufacturing environment. (Adapted from Ho, L. and Lin, G., *Int. J. Prod. Res.*, 42, 3731, 2004.)

As the SCM initiative has been rapidly developing worldwide, more efforts have been shifted to integrating intraorganizational systems into interorganizational architecture in order to support SCM through enterprise systems. The scope of enterprise integration and interoperation has begun to extend from intra- to interorganizational.

In the global supply chain environment, interorganizational enterprise systems have been playing an important role in dealing with the challenge of competition. Interorganizational systems allow communication between the partners in the supply chain. Interorganizational systems have been defined as systems that are shared by two or more collaborative enterprises. In some sense, developing (and implementing) interorganizational systems is becoming a necessity. This is especially true for the enterprises within the SCM network or within the same industrial sector, since enterprises can, in general, greatly benefit from interorganizational interoperation, as interorganizational enterprise systems provide enterprises with infrastructures to effectively support their supply chain activities through facilitating the sharing of information and the flow of materials. The impact that interorganizational systems have on SCM thus has been recognized and emphasized. Research reveals that information sharing can substantially improve overall supply chain performance. Information sharing can help achieve supply chain integration, can enhance coordination among organizations in the supply chain, and can lead to a better overall performance. Liu et al. (2005) indicate that the major success factors for a supply chain include advanced interorganizational systems that enable better information exchange. Such systems can collect valuable SCM information for all of the related business processes across the supply chain. With integrated SCM, businesses can predict market fluctuations, can better innovate in response to market changes, and can align operations across supply chain networks. Studies suggest that (1) the use of interorganizational systems can reduce the interorganizational coordination cost and (2) the related business processes will, therefore, become much more efficient.

Humphreys et al. (2001) analyze the SCM from the interorganizational relationship research paradigm and from the interorganizational enterprise system service provider's perspective. A framework is proposed, which deploys interorganizational enterprise systems from an interorganizational system provider's perspective. The framework was applied to light industries in the context of SCM. The results show that a competitive advantage can extend beyond the focal enterprise and that the upstream and downstream enterprises can also benefit from such systems. The benefits include supply stability, reciprocity, efficiency, and legitimacy. Figure 5.7 shows a model of interorganizational enterprise system in SCM. It represents an interorganizational enterprise system consisting of a focal organization, its upstream organizational set, and its downstream organizational set. The area outside the interorganizational enterprise system boundary includes the possible environmental influence on the interorganizational systems.

Supply chain

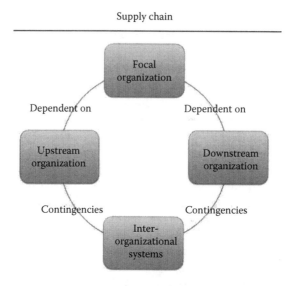

Figure 5.7 Interorganizational systems in supply chain management. (Adapted from Humphreys, P. et al., *Int. J. Prod. Econ.*, 70, 245, 2001.)

Although it has been showing promising future success for implementing interorganizational systems, coordination among systems is a challenging task. For this purpose, it is necessary to study interorganizational business processes from a systems approach, especially since today's enterprise systems are required to address complex processes taking place within and beyond the walls of enterprises, as changing business environments require an organization to adapt both the intra- and interorganizational processes more dynamically and frequently. Interorganizational integration occurs mainly through coordination. Process integration is one of the main types of integration. Process integration can be intra- or interorganizational, or both. Due to the closed connections and transformation mechanisms between process management and workflow management, intra- and interorganizational workflow monitoring and management capacity will be helpful for process integration and for information sharing at both intra- and interorganizational levels. Such information sharing in the context of the supply chain enables all partners in the supply chain to optimize operation performance, to enhance collaboration, and to gain a competitive advantage.

Interorganizational workflow architecture supports SCM, since it focuses on both intra- and interorganizational levels, facilitating (1) the deployment of business processes over supply chains and (2) the interoperations between enterprise processes. In the information-based economy featured by e-business, it is essential to realize workflow interoperability in a complex and dynamic environment. As e-business continues to prevail, the embracing of interorganizational

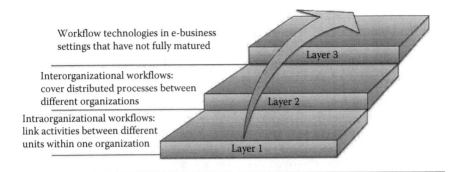

Workflow technologies in e-business
settings that have not fully matured

Interorganizational workflows:
cover distributed processes between
different organizations

Intraorganizational workflows:
link activities between different
units within one organization

Figure 5.8   Workflows in three layers.

BPM and the related workflow management has received a lot of attention. The workflow research can be viewed in terms of three layers (Figure 5.8). Figure 5.8 shows the position of interorganizational workflows as it comprises both intra- and interorganizational workflows. In Figure 5.8, the first layer pertains to the intraorganizational workflow, which links activities between different departments within one organization. The second layer corresponds to the interorganizational workflow, which covers the distributed processes between different organizations. Both layers 1 and 2 together comprise the interorganizational workflow. The third layer concerns the workflow in e-business settings that have now been intensely researched.

The development of information architecture and software architecture for interorganizational enterprise systems has garnered much attention in recent years. Examples of interorganizational workflow architecture for SCM have been proposed by researchers. Architecture can have many interpretations; some researchers consider an architecture to be a unique design method, one used to construct a group out of interrelated and interdependent constituents that are functionally related. Appropriate information architecture helps make enterprise systems as well as supply chain systems more integrated, effective, and responsive in facing complex and fast-changing market conditions. Software architecture is an essential component of an enterprise system. Software architecture describes a set of system components as well as their topological relations in an enterprise system. An advantage of architecture analysis lies in the early information sharing about a software system's high-level design. Building an integrated enterprise system requires appropriate architectural solutions.

Integrating the enterprise systems in different organizations requires coping with technical challenges. With fast-growing information technologies as their backbone, service-based interactions have become common. SOA and web services are playing a major role in enabling service-oriented applications to provide more than simple resource access and file sharing, but rather ubiquitous service provisions. In complex situations, a service workflow composed of several subservices is widely used for performing sophisticated functions. Precisely completing a complicated task usually requires integrated results from several services. WfMSs are notable for managing this complexity, since they offer service composition during workflow formation, service execution during workflow enactment, and workflow interoperation. Many current WfMSs have been extended to support more dynamic service workflows through coordinating disparate autonomous services that may be made available at distributed sites. However, the obstacle is that many WfMSs face the technical issues of scalability, dynamicity, and inconsistency in collaborations, since each participating service might possess disparate requirements and its own format that may preclude the simplification of global requirement integration and enforcement. WfMS may manage services without this awareness, resulting in reluctance of services to participate in it. The consequence is that the burden is pushed to the service side, as witnessed by each attempting to locally create its own protections. Due to the lack of generic patterns or general approaches in specifying requirements that fit into this context, difficulties can exist in service workflow interoperation. To alleviate these, a mechanism that is adaptive in dynamic open environment is required, in which requirements are locally defined and universally integrated. The lack of a formal and uniformed requirements specification (1) impedes the scalability that already tends to be limited within a certain set of similar domains, (2) fails to address the high level of dynamicity when each service acts in an unpredictable manner, and (3) results in inconsistency when requirements from each autonomous service are expressed differently. Moreover, automating time-consuming processes such as workflow creation, evaluation, and reasoning about services requires a strong foundation in terms of the syntax and semantics of a formal specification. Finding a solution would reduce the inefficiency in workflow interoperation and therefore would leverage the level of overall scalability and security.

SWSpec, a Service Workflow Specification language, which allows arbitrary services in a workflow to formally and uniformly impose their requirements, has been proposed (Viriyasitavat et al., 2012). The merit of this method is that it allows requirements to be independently and uniformly specified by each service, subject to the willingness of participation in a workflow. In terms of scalability, SWSpec is a key component for automatic compliance checking in a large-scale interoperation. SWSpec allows each service to modify (relax or restrict) requirements even during workflow executions. As a result, the solution enriches the proliferation of service provisions and consumptions over the Internet. In addition, the requirements specification in three modes, with respect to service workflows, is proposed. The algebraic operators and their syntax and semantics to form formal expressions are proposed, and their applications are demonstrated with business transaction examples. Now, we will use an example to illustrate. Suppose that there is an airline that provides flights plus booking services. Figure 5.9 shows the travel planning

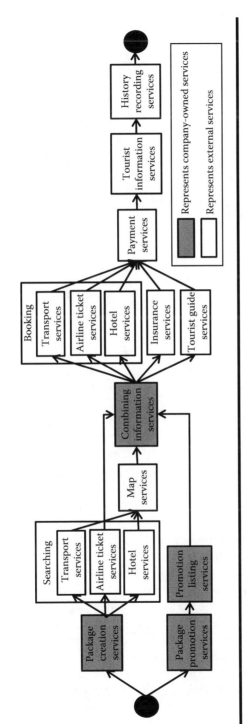

Figure 5.9 Travel planning process.

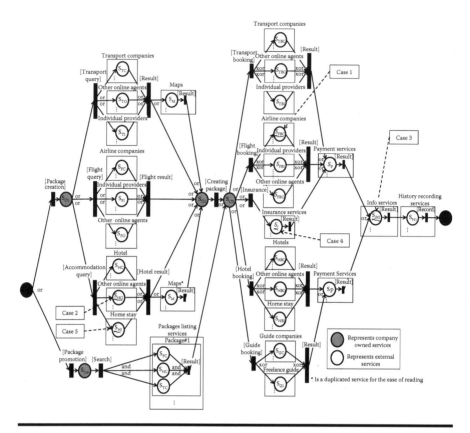

Figure 5.10   Petri-net-based workflow for the planning process.

process (Viriyasitavat et al., 2012). The sequences of the tasks are expressed by a service workflow, each of which is executed by a set of agreed services. Figure 5.10 shows the Petri-net-based workflow for the planning process (Viriyasitavat et al., 2012). In this context, rather than specifically describing every detail of workflow operations, a service workflow represents the high-level workflow describing the association between tasks and services. In other words, a service workflow is used to represent how and which services are connected to (responsible for) a particular task. The proposed solution is the development of a formal specification language of requirements, which serves as a basic element for automatic compliance checking about service workflows. It allows each service to uniformly express its own requirements, such that the impracticality and problems introduced can be alleviated. The formal specification allows the requirement of the service to be expressed in a formal way. Figure 5.11 shows the graphical illustration of the examples of algebra operators used in the formal methods (Viriyasitavat et al., 2012).

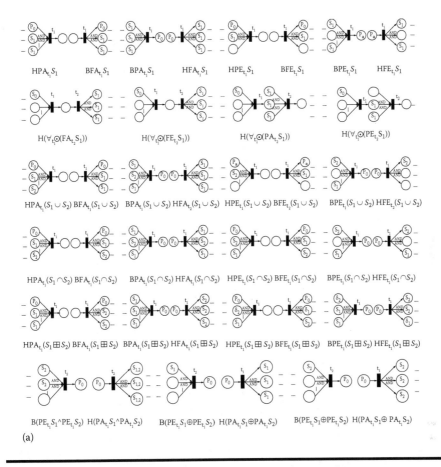

Figure 5.11 Graphical illustration examples of algebra operators. (a) AND connective.

(*continued*)

5.3 Model-Driven Architecture

"Model-driven architecture" (MDA) is a framework based on the Unified Modeling Language (UML) and other industry standards for visualizing and transforming software design models. MDA was originally proposed by Object Management Group (OMG), mainly for use in models and transformations between them in the software development process, in order to achieve an effective interoperability of enterprise models and software applications and to strengthen reusability. MDA moves the focus of software development more toward model building orientation, as the name reveals. As Debnath et al. (2008) describe it,

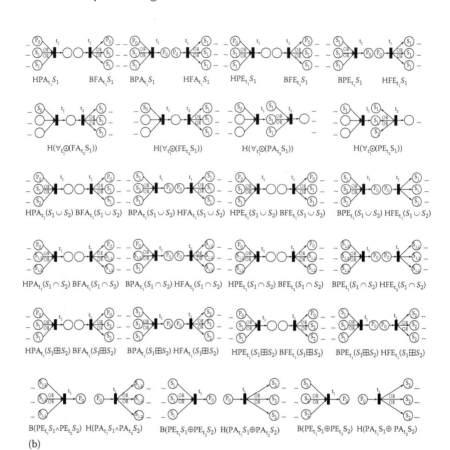

Figure 5.11 (continued) Graphical illustration examples of algebra operators. (b) OR connective.

MDA is model-driven because it provides a means for using models to direct the course of understanding, design, construction, deployment, operation, maintenance and modification. All artifacts such as requirements specification, architecture descriptions, design descriptions, and codes are regarded as models. One of the key features of this framework is the notion of automatic transformations that describe how a model in a source language (source model) can be transformed into one or more models in a target language (target model).

MDA is able to separate business and application logic from underlying platform logic; in other words, enabling conceptual models exist independently of platforms and thus make heterogeneous systems' interoperation easier. In addition, MDA enables the generalization of the existing component architectures to include the

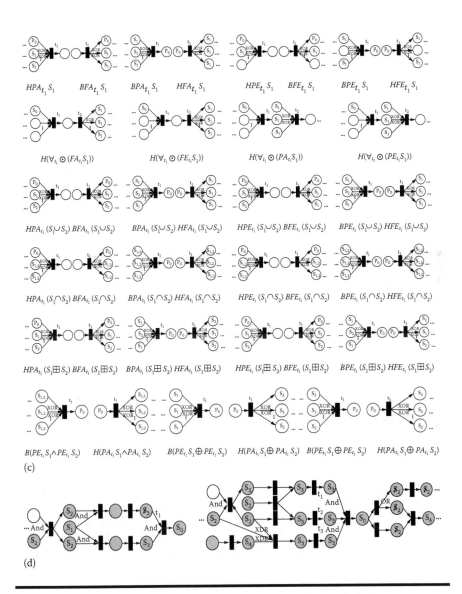

Figure 5.11 (continued) Graphical illustration examples of algebra operators. (c) XOR connective, (d) combinations.

entire system life cycle within a unified, platform-independent framework (Singh and Huhns, 2005). The life cycle includes acquiring business requirements, analysis, design, testing, implementation, management, and evolution in an iterative cycle to meet business requirements and technology change.

MDA is based on three modeling techniques. The first one is UML (as mentioned earlier), which includes sublanguages for a variety of software modeling needs.

However, unlike UML, MDA promotes the creation of machine-readable highly abstract models that are developed independently of the implementation technology and are stored in standardized repositories. Therefore, they can be accessed repeatedly and can be automatically transformed into schemas, code skeletons, integration codes, and deployment scripts for various platforms (Kleppe et al., 2003). The second is Meta Object Facility (MOF), which is built as a subset of the constructs of UML. Having a modeling capability such as MOF enables the exchange of models among development tools and middleware. Model exchange is standardized through XML Metadata Interchange (XML), which provides the data type definitions (DTDs) for UML, MOF, and Common Warehouse Metamodel (CWM) (Singh and Huhns, 2005). The third is the CWM, which standardizes the data warehouse application life cycle.

Through shifting the focus of software development strategy, the aim of MDA is to separate the specifications of system functionality and implementation. Since platform dependencies are added later, the same analysis model can be used in many different settings. Instead of writing a code on a manual basis, MDA promises to create application code from requirements models automatically, thus avoiding common sources of errors and consequently improving the resulting application quality.

Two essential design aspects promoted by MDA are the use of different models at different abstraction levels: from (1) the conceptual computation-independent model (CIM) to the logical platform-independent model (PIM) and then (2) from the logical to the physical platform-specific model (PSM). The idea is to facilitate the evolution from one CIM, describing the business system independently of the software system, to possibly *n* number of PSMs and eventually to their corresponding codes in a specific language. The models are in closed connections and transformation mechanisms, which facilitate passage from one stage to another and separate concerns by segregating implementation choices from business needs specifications. In this process, CIM captures the business requirements without reference to particular system implementations. A CIM is not intended to show details about the structure of software systems. PIM provides a logical view of the system from a platform-independent perspective; meanwhile, it is open to a number of possible platforms. PSM provides a view of the system that integrates PIM with a particular platform. Technology is defined by the choice of the implementation platform in a generic way. The method plays an important role in bridging the gap between business requirements and software implementations. As such, the ultimate solution is a mix of information coming from both. The Y symbol is frequently used to summarize these principles, as shown in Figure 5.12.

The MDA process can also be described through stages and their transformation mechanisms, as Figure 5.13 shows. A conceptual model is constructed at the stage of CIM. Based on the CIM, a PIM is created at the logical level. With a sufficiently complete and precise PIM, a PSM can be generated by using

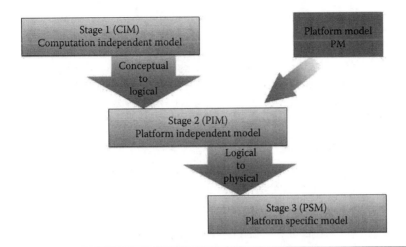

Figure 5.12   Model-driven architecture.

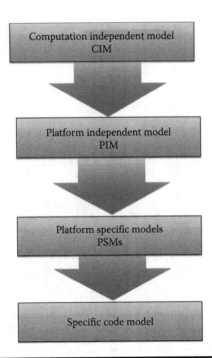

Figure 5.13   MDA process.

model-to-model transformation mechanisms at the physical level, and the specific code model (which can be viewed as implementation) can be automatically transformed from the PSM (Lin et al., 2012).

Figure 5.13 shows that MDA distinguishes between PIM and PSM. A PIM captures the essence of the structure and the function of the modeled system, whereas a PSM captures details specific to implementations. Figures 5.12 and 5.13 provide the detailed process of applying MDA. Specifically, the first step is to describe business requirements through CIM. The second step is to develop a logical model to satisfy the requirements described in the first step but without committing to a specific technology platform. The third step is to transform the PIM to a platform-specific PSM. The general systems idea here is to make modeling the primary focus of the development process, eventually reaching generating codes through modeling and model transformation, that is, the generation of a target model from an initial model. MDA-based methods have many applications in industrial information integration. An MDA-based method for the automatic generation of programs has been proposed (in XML-based industrial control system description language) to consolidate the modeling methodologies used to support the development phases. In this method, domain-specific views are identified, and the mapping between the software components is developed through an XML style sheet (XSL) that facilitates the transformation between domains and can further facilitate the portability of codes (Vyatkin, 2013). Quartel et al. (2009) propose model-driven development of mediation for business services using COSMO. The proposed framework is based on a service-oriented MDA technique. Figure 5.14 illustrates the steps, which can be iterative, that constitute the proposed method for service mediation. It is a process of transforming PSM to the respective PIM. Subsequently, the PIMs are semantically enriched by adding information that could not be derived automatically from PSM.

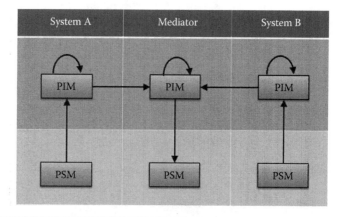

Figure 5.14 PIM and PSM in a model-driven development of mediation.

5.4 Service-Oriented Architecture

SOAs are emerging as new trends for building agile and interoperable enterprise systems, especially since SOA represents the latest trend in integrating heterogeneous systems and has thus received much attention as a new architecture for integrating enterprise systems in industrial information integration. SOA is a suitable paradigm that could aid enterprise systems integration including integrating platforms, protocols, and legacy systems, as it is characterized by flexibility, adaptability, and simplicity. From the perspective of SCM, SOA represents an emerging paradigm for enterprise coordination in a seamless manner in the environment of heterogeneous enterprise systems, enabling the development of flexible large-scale supply chain systems and the timely sharing of information in the supply chain.

As an architectural approach, SOA takes business applications and breaks them down into individual functions and processes as services. In SOA, a "service" can be considered as an abstract business concept that represents the functionalities of business (Iacob and Jonkers, 2009) as a business process on the top of the organization. SOA can help improve functionalities in terms of interoperable services. The systems characteristics (such as the iterative and recursive properties of SOA) also allow the recursive aggregation of services into new business processes and applications (Unger et al., 2009). In this systematic process, (1) each of existing services can be recomposited, reconstructed, and reused to create new applications; and (2) discrete sets of software functionalities can be recomposed and passed to other functions and systems as services that enable different applications to reuse common parts. In turn, services can be pieces of application functionality that represent a reconstructed business task. SOA makes the development of alternative applications possible through assembling reconstructible components; thus, SOA enables organizations to create new applications dynamically to meet changing business needs (Quartel et al., 2009). One of the key benefits of SOA in SCM is that services can be made available to others in the supply chain; SOA enables access to supply chain information for many different partners. Thus, SOA has been considered as one of the most promising technologies to enable globally integrated supply chains (Butner, 2009). By using SOA, an organization can create a more flexible and responsive environment for its supply chain partners. Services can be built and maintained by one partner but made available to other partners in the supply chain (for incorporating purposes). SOA has been successfully applied in industrial systems such as manufacturing systems, logistics systems, train car management systems, control systems for semiconductor processing equipment, order entry systems, electronics production systems, etc. For example, a factory's online business system can have a purchase-and-ordering application, which could communicate (1) to an inventory application on another web server to specify the items that need to be reordered or (2) to a web service from a credit bureau to request the credit history from loan services for prospective borrowers (He and Xu, 2013).

SOA builds and integrates applications using the concept of software services. As a method for building enterprise applications that promotes a loose coupling between components, SOA has the following characteristics: (1) services can be software components that have published contracts/interfaces; these contracts are platform, language, and operating system independent. XML and SOAP are the enabling technologies for SOA as they are platform independent, (2) users can dynamically discover services, and (3) services are interoperable (Wang et al., 2006).

Many enterprises are currently adopting a demand-driven model (Li, 2007). A successful demand-driven supply chain requires prompt responsiveness along the entire supply chain. Throughout such supply chains, information sharing continues to be important, as it requires a fully integrated enterprise system characterized by real-time data exchange, real-time responsiveness, real-time collaboration, real-time synchronization, real-time visibility, and a sophisticated level of information integration and exchange. SOA is considered suitable for such demand-driven supply chains. SOA can integrate supply chain information, thus helping to create a better environment for real-time data exchange, real-time responsiveness, real-time collaboration, real-time synchronization, and real-time visibility across the entire supply chain. In addition, in order to be responsive, SOA can integrate reconstructible "component business services" in the process of managing supply chains in real time across organizational boundaries (Butner, 2009).

SOC is a computing paradigm that supports enterprise system integration for the following reasons: (1) it utilizes services as fundamental elements for developing distributed applications/solutions in heterogeneous environments. Such services represent individual functions and processes in business applications; they can be reused and recombined to create new business applications. As the business processes change, services can be reassembled; and (2) it promises cooperating services that are loosely connected and creates agile applications that span organizations and platforms. SOC relies on SOA to build the service model, as SOA is the architecture in which discrete sets of software functionalities can be componentized to deliver services that enable different applications to use common or reusable components when facing changing operations and building new applications (Papazouglou and Georgakooulos, 2003), and other applications can access such services and incorporate them into their own functionality (Butner, 2009). Since SOA publishes business functionality in the form of programming and accessible software services, other application programs can use these services through published and discoverable interfaces that provide a new blueprint for software reuse and enterprise systems integration (van Lessen et al., 2009).

Not only an architecture for enabling a globally integrated supply chain in general, but SOA is also becoming a solid foundation for business process coordination and orchestration, and for assembling process services into larger end-to-end processes. In conjunction with BPM, SOA is very promising for developing interorganizational enterprise systems. The integration and coevolution of SOA and BPM are on the horizon (Lorincz, 2007). As BPM on SOA enables combining business requirements and software capability for an organization, BPM and SOA technologies have been integrated in end-to-end resource planning as a core technology (Li and Li, 2009). As an emerging architecture, the advantages of using SOA for developing interorganizational enterprise systems include cross-heterogeneous platforms, flexible functional integration based on loosely coupled services, the low cost of system establishment, standard protocols, reusability, and especially the ability to bridge the gap between business and information technology through a set of business-aligned services using a set of design principles, patterns, and techniques (Lorincz, 2007; Wang et al., 2008).

SOA can be used to build web services. A web service is a method of communication that uses standard Internet protocols and has a collection of functions that are packaged as a single entity and published to the network for use by other programs (Chaudhary et al., 2002). Web services can be accessed in many languages, using many component models and running on many operating systems. They utilize the HTTP protocol that allows function requests to pass through firewalls. XML is used to format the input and output parameters of the request; consequently, the request is not tied to any particular component technology or object-calling convention. Main web service protocols include SOAP, the protocol used to interact with a web service; WSDL, the language for specifying the interface to a web service; and UDDI, the repository for storing references to web services so that clients can find them (Newcomer, 2002).

Web services consist of three components, as Figure 5.15 shows (Gunzer, 2002)

1. A service broker that acts as a lookup service between a service provider and a service requestor
2. A service requester that asks the service broker where to find a suitable service provider and that binds itself to the provider
3. A service provider that publishes its services to the service broker

SOA, web services, and the enterprise service bus (ESB) provide a promising framework for the development, integration, and interoperation of distributed enterprise applications, and they work especially well for interorganizational communication. SOA, web services, and ESB are playing an ever-increasing role in enabling the interaction of software components across organizational boundaries and are being adopted by a number of enterprises across the globe. Service-oriented integration is an evolution of enterprise application integration (EAI) in which proprietary connections are replaced with standards-based connections over an ESB that is location transparent and provides a flexible set of routing, mediation, and transformation capabilities. SOA-based ESB is designed to work across different middleware products and standards to implement enterprise-wide SOA. It can shield the application of various types of heterogeneous systems to realize the smooth flow of

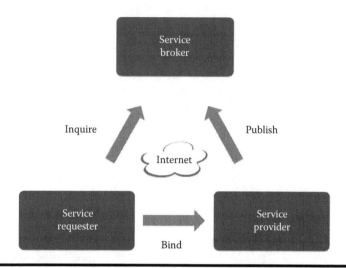

Figure 5.15 A service-oriented architecture using web services. (Adapted from Gunzer, H., Introduction to web services, 2002, available at http://archive.devx. com/javasr/whitepapers/borland/12728jb6webservwp.pdf.)

data between application systems, and it can also improve interoperability capacity between systems (Xu et al., 2010). Thus, SOA-based ESB is also viewed as a new type of middleware technology. A recent example of its implementation is an electronic power application system that was integrated by using ESB (Xu et al., 2010). Some examples of SOA-based ESB products include BEA AquaLogic Service Bus, IONA XMLBus, and IBM WebSphere Service Bus.

SOA is playing an increasingly important role in enterprise computing. High-level complexity characterizes the design processes of complex products. As products are becoming more complex, the design processes and the management of those design processes are also more complicated. PLM plays an essential role in managing product data throughout all phases of a product life cycle. In recent years, attention has been directed toward the application of SOA in PLM. Some standardization organizations have been involved in developing standards for PLM with SOA. OMG PLM Services is an OMG standard specification. OASIS PLCS PLM Web Services (ISO10303-239:2005) is an ISO Standard for the Exchange of Product Model Data. Hachani et al. (2013) analyze these two standards and conclude that the main research trend of SOA for PLM is to enable the integration of heterogeneous PLM systems in order to enhance collaboration between/among partners and to enable data exchange. Hachani et al. (2013) propose a flexible process support approach to the PLM system. The objective is to specify, design, and implement business processes in a flexible way so that the new business requirements can be quickly included in PLM solutions. Unlike existing approaches, the proposed approach introduces a service-oriented perspective, rather than an activity-oriented approach.

5.5 Interoperability Models

Interoperability is "achieved only if the interaction between two systems can at least, take place at the three levels: data, resource and business process with the semantics defined in a business context" (Chen and Doumeingts, 2003). Interoperability also refers to the ability of a system (or process) to use the information and/or the functionality of another system (or process) by adhering to common standards (Chen and Doumeingts, 2003). Broadly speaking, interoperability is the ability of two or more different entities (e.g., business units, processes, software, and systems) to perform interoperation. Interoperability helps to realize integration. It promotes the idea that integration has to be achieved using relevant frameworks, architectures, and standards (Touzi et al., 2009).

Different frameworks that characterize interoperability in different perspectives are available. These frameworks include the ATHENA Interoperability Framework (AIF) (Athena, 2003); the Command, Control, Communications, Computers, Intelligence, Surveillance and Reconnaissance (C4ISR) (the C4ISR framework, also called C4ISRAF, now known as Department of Defense Architecture Framework (DoDAF, CSISR 1998); the EIF (2004); IDEAS (2002); and the UK e-Government Interoperability Framework.

AIF (Athena, 2003) adopts a holistic approach of interoperability that allows a solid analysis of interoperability needs: It concerns concepts, formalisms, meta-models, and standards that help formalize the different levels of interoperability.

In 1998, the Architecture Working Group of the US Department of Defense (on C4ISR) defines the five levels of interoperability as follows:

- Level 0—isolated systems (manual extraction and integration of data)
- Level 1—connected interoperability in a peer-to-peer environment
- Level 2—functional interoperability in a distributed environment
- Level 3—domain-based interoperability in an integrated environment
- Level 4—enterprise-based interoperability in a universal environment

EIF is a generic framework jointly developed by the European Commission and the member states of the European Union (EU) to address business and government needs for information exchange (IDAbc, 2004). This framework defines three essential levels or dimensions of interoperability, respectively, namely, the technical, semantic, and organizational levels, as depicted in Figure 5.16. According to the EIF (2004), interoperability can occur at three levels:

1. Technical level, that is, data and message exchange
2. Semantic level, that is, information and service sharing
3. Organizational level, that is, business unit, process interactions across organization borders

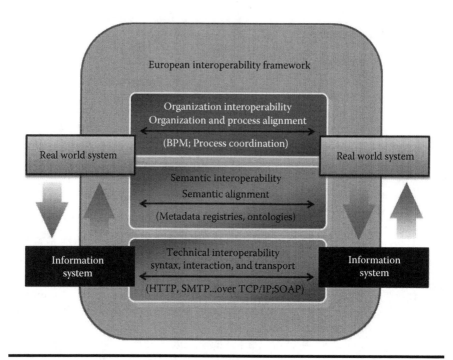

Figure 5.16 European Interoperability Framework. (Adapted from IDAbc, *EIF: European Interoperability Framework, Version 1.0.*, European Commission, Brussels, Belgium, 2004, available at http://ec.europa.eu/idabc/en/document/ 2319/5644.)

The IDEAS (2002) framework defines three levels: business (the business context and processes of organizations), knowledge (the definition of products, etc., in the organization), and ICT systems (applications and communication infrastructure), along with a transversal level of semantics to assure a mutual understanding of the three levels mentioned earlier.

According to these frameworks, Touzi et al. (2009) consider that the problem of interoperability mainly deals with the conceptual, technical, and organizational issues:

- At the conceptual level, the data, resources, and business processes of different information systems must be linked in spite of their heterogeneous structures and different interpretations. The problem is both syntactic and semantic.
- At the technical level, the aim is to reconcile the different applications, technologies, systems, and communication infrastructures used by the partners.

- At the organizational level, the business context of the collaboration must be explained: How do partners interact? Which data are exchanged? Which resources do they expose to others? Process and data models are examples of solutions for modeling interoperability at this level.

In general, enterprises can have multiple systems and applications. For example, many enterprises have numerous legacy systems and applications that are expected to continue their service. Integration is necessary when there is a need to improve interactions among units and systems in terms of material flow, information flow, or control flow. Integration is particularly important when mergers and acquisitions take place. As business environments, in recent years, have become increasingly distributed and heterogeneous across multiple organizational and geographical boundaries, there are strong demands to integrate various distributed enterprise systems and applications in order to enhance or increase enterprises' competitiveness. Many enterprises have invested heavily to integrate distributed enterprise systems and applications due to the continuous mergers and acquisitions, joint venture, outsourcing, corporate restructuring, infrastructure upgrades, adoption of mobile devices, embedded devices, and wireless sensors.

Since interoperable systems are central to enterprises, EAI has been proposed to help achieve quality integration. Originally, EAI was focused only on integrating enterprise systems with intraorganizational applications, but now it has been expanded to cover aspects of interorganizational integration (Concha et al., 2010). Nowadays, EAI facilitates the integration of both intra- and interorganizational systems. EAI encompasses technologies that enable distributed and heterogeneous enterprise applications to interact with one another across supply chain networks to help integrate many individual applications into a whole (Linthicum, 2000). It consists of methods and tools to coordinate various applications and to support the integration of both intra- and interorganizational systems. EAI solutions support the integration of business processes and data across a variety of enterprise systems and applications (Qureshi, 2005). EAI aims not only to connect the current system processes, but also to provide a flexible and convenient process integration mechanism. With EAI, intra- or interenterprise application systems can be integrated effectively and can ensure that different divisions, units, or even different enterprises can cooperate with each other (Xu, 2011). The objectives of EAI are to facilitate information exchange among business enterprises in a timely, accurate, and consistent manner and to support business operations in a manner that appears to be seamless. From a technology perspective, integrating distributed enterprise systems and applications can happen at different levels. By analogy with computer networks, researchers have found it useful to study the integration in terms of layers (Benatallah and Motahari-Nezhad, 2008) including the data layer, the business logic layer, the communication layer, and the presentation layer.

In recent years, enterprise interoperability has been recognized as a scientific research subject. As pointed out by Panetto and Cecil (2013), in spite of the significant research efforts that have been made, the proper theoretical foundations remain unavailable. Goncalves et al. (2013) have made a step further to establish the theoretical foundations of enterprise interoperability. IBoK has been proposed; it includes interoperability frameworks, interoperability theories, and interoperability models. The proposal on interoperability frameworks, interoperability theories, and interoperability models offers interesting insights into the current status of the research on interoperability and also provides insightful perspectives on the future research directions. Although the Enterprise Interoperability Science Foundation and IBoK (as first proposed by Goncalves et al. in 2013) are still in their embryonic stages, these are substantial efforts that have been made in recent years toward establishing the theory of interoperability and its application, which eventually will become one of solid cornerstones for a more comprehensive framework called IIIE, which will be introduced in Section 5.7.

5.6 Industrial Information Integration: Examples

5.6.1 Multilingual Semantic Interoperation in Interorganizational Enterprise Systems

Business process is an important research topic in IIIE, which concerns itself with the flow of business and manufacturing activities in intra- and interorganizational contexts, especially in the context of SCM. One of its major task is to facilitate business cooperation and collaboration between enterprises in the extended enterprise context. A technical challenge is the lack of business process interoperation across enterprises. Interorganizational business processes are often heterogeneous in the aspects of (1) maintaining syntactic consistency in structuring and (2) modeling business processes. For interorganizational systems, it is highly possible that users will adopt different standards for information creation and exchange according to their own contexts. Thus, a standardization approach is not always useful. An intelligent mediation approach attempts to solve the standardization problem based on agent technology. Nevertheless, since different industrial applications are mostly created and run in different contexts, rules of creating and using information by agents often have different semantic assumptions. This can lead to semantic conflicts between cross-context users for information exchange. To enable information exchange across heterogeneous information systems without misinterpretation, Guo (2009) provides a collaborative conceptualization approach, which resolves semantic conflicts of information exchange based on a collaborative vocabulary editing mechanism. The main task of this approach is to create semantically consistent concepts across heterogeneous information systems. Such concepts can be used to build

VSF Algorithm. Finding a validation set
Input: *L, C, a* 1-*to*-1 *relation* R_4: *D→C, a m-to-m relation* R_5:
C→L, and P ∈ *NSG as defined in Section III.*
Output: $P_3 \leftarrow \varnothing$
for (P.s_i){   /* P.s_i ∈ P */
        for($w_j.d_j.c_j \wedge w_k.d_k.c_k$){   /*$w_j{\cdot}d_j$, $w_k{\cdot}d_k$ ∈ s_i, j≠k* /
            $c_j \leftarrow R_4(w_j.d_j)$ /* $w_j{\cdot}d_j$ is a valued concept in w_j */
            $c_k \leftarrow R_4(w_k.d_k)$ /* $w_k{\cdot}d_k$ is a valued concept in w_k */
            if($\neg\varnothing \leftarrow R_5(c_j) \cap R_5(c_k)$) **then**
                $P_3 \leftarrow P_3 \cup w_j.d$
                $P_3 \leftarrow P_3 \cup w_k.d$ }
} /* building a set of validated sets */

Figure 5.17 Validation set finding (VSF) algorithm.

consistent business documents and processes for accurate information exchange and are both syntactically and semantically consistent for interorganizational enterprise systems. Furthermore, Guo et al. (2012) propose a concept-connected near synonym (NSG) framework for concept disambiguation. An NSG framework provides a vocabulary preprocessing process of collaborative vocabulary editing, which further ensures semantically consistent vocabulary for building semantically consistent business processes and documents between context-different information systems. Figure 5.17 shows the validation set finding (VSF) algorithm developed for evaluation purposes for the methods developed for interorganizational enterprise systems.

5.6.2 Agricultural Ecosystem Enterprise Information System

There is a need for integrated agricultural data processing and analysis for realizing the so-called digital agriculture, since existing agricultural and ecosystems information is stored in many disconnected locations using different formats and systems. To support agricultural operations and to help the integration of observation and prediction data for decision-making purposes in agriculture management, efforts are required to integrate relevant data into an integrated system such as an enterprise system to support agricultural management. Xu et al. (2008) propose an architectural framework that integrates agricultural and ecosystem information systems, and they have developed a deliverable enterprise system called the Agricultural Ecosystem Enterprise Information System (AEEIS) for agricultural and ecosystem use. The AEEIS system is equipped with enterprise systems and business intelligence capabilities to aid agricultural and ecological management. The AEEIS system provides a variety of functions including secure and automated exchange of data, data warehousing, data mining, OLAP, knowledge management, and information

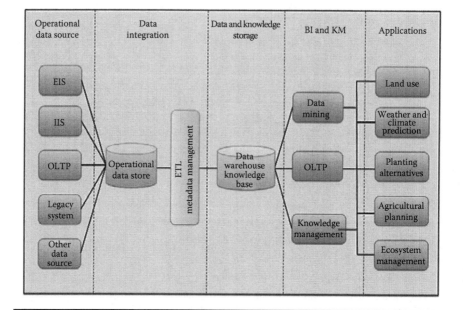

Figure 5.18 The structure of AEEIS.

dissemination. Figure 5.18 shows the structure of AEEIS, which consists of an operational database that integrates data from enterprise systems, an integrated information system, online transaction processing (OLTP), legacy systems, and other sources. In business intelligence and knowledge management, data mining, OLAP, and knowledge management components are available to process the data and information in order to produce business intelligence and knowledge. This allows the integration of disparate data sources into a single coherent framework. The operational database stores data regarding the characteristics of land use and plant species or crop varieties as well as their growth conditions. The operational database consists of two layers, with the land use database at the top and five vegetation databases at the bottom. AEEIS provides the ability to integrate existing, yet disconnected, data sources into a single system. In AEEIS, data are not only integrated but also analyzed, explored, and mined.

5.6.3 Water Resource Management Enterprise System

The management of water resources is important to many countries. Two topics of particular interest in water resource management are the allocation of water resources and flood protection. It is noted that flooding is one of the most frequent and the most damaging natural disasters in the world, causing 75% of all of the deaths and 40% of the total economic loss of all natural disasters combined. As a result, flood

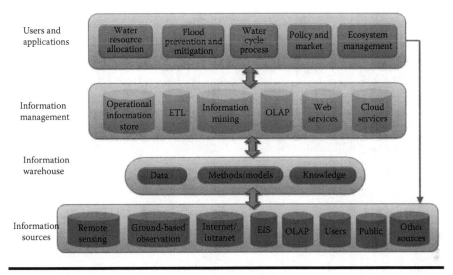

Figure 5.19 WRMEIS system architecture. (From Fang, S., Xu, L., Pei, H., Liu, Y., Liu, Z., Zhu, Y., and Zhang, H., An integrated approach to snowmelt flood forecasting in water resource management. *IEEE Trans. Ind. Inform.,* **10(1), 548–558, © 2014 IEEE.)**

forecasting is an active topic not only for researchers but also for the public and for governments. Fang et al. (2014) developed an integrated information system called the Water Resource Management Enterprise Information System (WRMEIS) based on the geoinformatics and enterprise systems to aid decision making for regional water resource management concerns. This system integrates a number of functions such as data acquisition, data management and sharing, modeling, and knowledge management. Figure 5.19 shows the structure of the WRMEIS, which consists of an information warehouse that integrates data, methods, models, and knowledge from remote sensing, ground-based observations, Internet, enterprise systems, OLTP, users, public, and other sources. At the top level of the system are users and applications, including most of the components of water resource management such as water resource allocation, flood prevention and mitigation, and simulation and forecasting of water cycle processes, as well as other topics such as policy analysis, land use and land cover change, and ecosystem management. The second level includes basic methods and technologies of information management such as operational information stores, extraction–transformation–loading (ETL), OLAP, and cloud services. At this level, operational information stores, data mining, cloud service, and online analytical processing allow the integration of disparate data sources into a single coherent framework. The third level is the information warehouse, which contains data, methods/models, and knowledge. The fourth level is information sources and includes remote sensing, ground-based observation, users, and so on.

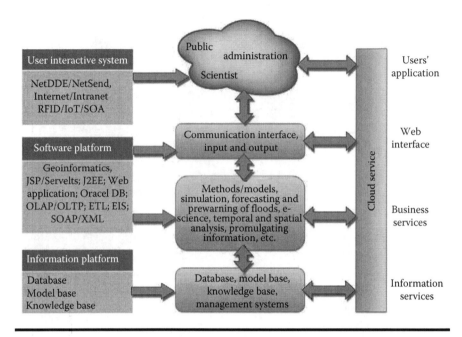

Figure 5.20 Main components of the SFFEIS. (From Fang, S., Xu, L., Pei, H., Liu, Y., Liu, Z., Zhu, Y., and Zhang, H., An integrated approach to snowmelt flood forecasting in water resource management. *IEEE Trans. Ind. Inform.*, 10(1), 548–558, © 2014 IEEE.)

Of particular note, a system called the Snowmelt Flood Forecasting Enterprise Information System (SFFEIS) within the WRMEIS structure has been implemented. It includes an operational database, ETL, a data warehouse, temporal and spatial analysis, simulation/prediction models, knowledge management, and other functions. The SFFIS has been developed using technologies including geoinformatics, enterprise systems, and cloud services. Figure 5.20 shows the main components of the SFFEIS. In the SFFEIS, there are three basic components: a user interactive system, a software platform, and an information platform (along with popular methods and technologies such as SOA, XML, RFID, IoT, ETL, and OLAP). They are integrated into this system within the framework of enterprise systems and cloud service. User applications, business services, and information services are basic functions of the SFFEIS.

Figure 5.21 shows the information acquisition facilities in the integrated system for early warning of snowmelt flood. Figure 5.22 shows the information storage facilities in the integrated system for early warning of snowmelt flood.

The system developed in this work has been tested and applied to a case in the Quergou River Basin, which is located in the middle part of the north-Tianshan Mountains in Xinjiang, China. The results show that water resource management is greatly benefited by using such an integrated system for detailed tasks.

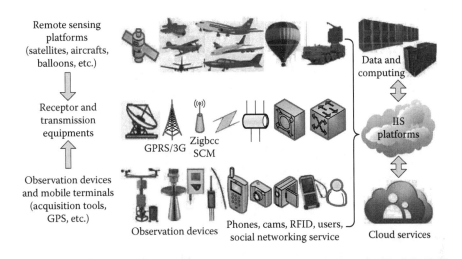

Figure 5.21 Information acquisition facilities in the integrated system for snowmelt flood early warning. (From Fang, S. et al., *Inf. Syst. Front.*, in press, 2014.)

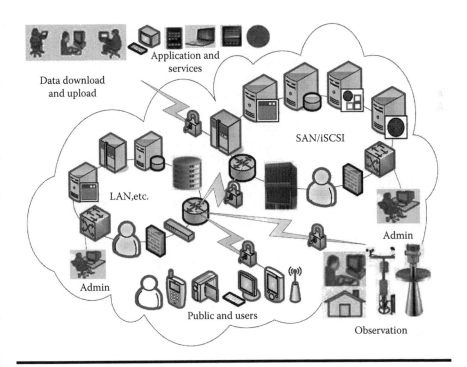

Figure 5.22 Information storage facilities in the integrated system for snowmelt flood early warning. (From Fang, S. et al., *Inf. Syst. Front.*, in press, 2014.)

The effectiveness of decision making can be improved by using integrated systems. The system is valuable for the acquisition, sharing, and management of multisource information in water resource management, and it provides a paradigm shift for future work (Fang et al., 2014).

5.6.4 Automated Assembly Planning System for Complex Products

Assembly planning has a significant impact on product delivery time, cost, quality, durability, as well as maintenance. Assembly planning is crucial to the success of a product, especially for complex products such as aircrafts, ships, automobiles, and aerospace products. The importance of digital assembly as a key component of digital manufacturing for assembly is increasing. In addition, the importance of the use of technology for product assemblies has been increasing, as industrial products are becoming more and more sophisticated. A complex product usually has a large number of parts. Xu et al. (2012) developed an integrated assembly planning system named AutoAssem for automate assembly planning. AutoAssem is dedicated to assembly planning for complex products. The novelties of AutoAssem include the implementation of some key technologies, including the automatic generation and extraction of multiple relational matrices, the assembly sequence planning with multiple algorithms, the automatic generation of an exploded view and assembly sequence, and the generation of interactive 3D assembly documentations.

AutoAssem has been developed as an integrated CAD system for assembly modeling; assembly planning for sequence, paths, and processes; and assembly evaluation and simulation. It consists of five main modules: assembly modeling, assembly sequence planning (ASP), path planning, visualization, and assembly simulation. Using hierarchical architecture, each of these five modules consists of several submodules at a lower level. AutoAssem has the ability to retrieve data regarding assembly sequence, paths, and processes from an integrated assembly plan. To perform a simulation, the system executes a dynamic gap analysis and generates the steps of assembling automatically.

AutoAssem provides a comprehensive solution to the assembly planning of complex products, in particular, for complex aircraft products. It is a completed enterprise system, and has been successfully employed in the assembly planning of many complex products including aircraft engines, car molds, naval power valves, CNC machine tools, and the general assembly of automobiles. Figure 5.23 presents the flowchart of assembly planning in AutoAssem.

5.6.5 Railway Signaling Enterprise System Based on IIIE

A railway signaling system is a safety-critical system that is used to ensure the fast and safe running of trains. Since the signaling system should be highly reliable, the development costs of such systems are usually high. The application data are usually

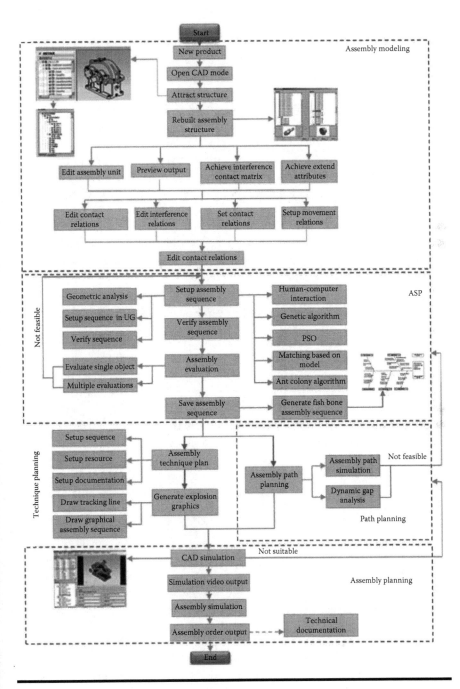

Figure 5.23 Flowchart of assembly planning in AutoAssem.

originated from various departments in a railway signaling system–integrated enterprise such as (1) the engineering department, which is in charge of site construction and installation and collects all the records on practical constructions; (2) the design department, which determines system requirements through communications with the end users and designs system architectures, interfaces, and station maps of the railway signaling system; (3) the system assurance department, which performs the analysis on reliability, availability, maintainability, and safety for the system; (4) the product technology department, which customizes specific systems based on the design of the project, conducts site debugging of the final configured system, and locates errors; and (5) the business procurement department, which purchases the equipment depending on the requests made by other departments.

So far, in most of the signaling system–integrated enterprises, the work of various departments is carried out independently, and the application data are exchanged through documents. Duplications that occur in this process result in an increased workload. The work done by different departments is sometimes not synchronized and causes parties to trip over each other and slow down the entire project. Inconsistent application data configurations can cause frequent rework and can make debugging more difficult. More importantly, the errors of these inconsistent configurations are not easy to find in the early testing stage, which can cause serious problems such as risks to safety. At present, application data configurations are usually conducted manually. Since the amount of the data is quite large and relationships are rather complicated, human errors are inevitable and can lead to further increase in the workload and can affect accuracy and efficiency. The configurations generated with IIIE will largely avoid human errors and will improve efficiency.

In addition, the tasks carried out by these departments are cooperative. In this context, each department should complete its responsible application data configurations based on the configurations of the upstream department. The configurations are further delivered to the downstream department. In this way, all of the integral application data configurations of the signaling system are obtained when the workflows between the departments are completed. It is necessary to provide an integrated platform for the application data configurations; this will enhance the effectiveness of development and implementation for the signaling system–integrated enterprises. In addition, with the introduction of automatic generation technology, application data can be configured more efficiently and accurately. In the study reported by Chen et al. (2013), the authors propose an enterprise system specially designed for signaling system–integrated enterprises, which integrates application data configurations and workflow. The automatic generation technology is introduced to make part of the application data configuration automatic. Figure 5.24 shows the data flows of the subsystems of the signaling system. Figure 5.25 shows the relationships between the workflows of departments and the application data configurations. Figure 5.26 shows the framework of the application data configuration tool. This framework is built on the platform of enterprise

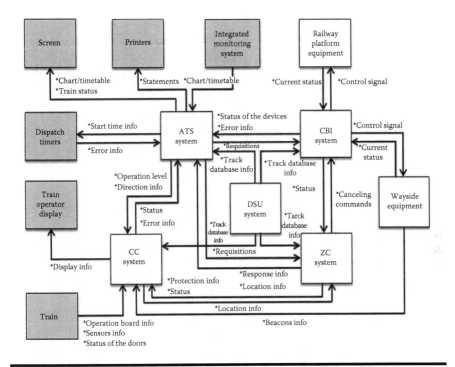

Figure 5.24 Data flows among the subsystems of the signaling system.

systems based on IIIE. It manages all the application data within the project life-time centrally and makes the automatic configurations of some application data a real entity.

5.7 IIIE: A New Discipline of Industrial Information Integration

As introduced in Chapter 1, it is well recognized that enterprise systems have had an important long-term strategic impact on global industrial development. Due to the importance of this subject, there has been a growing demand for research on enterprise systems to provide insights on issues, challenges, and solutions related to the design, implementation, and management of enterprise systems. In June 2005, at a meeting of the "IFIP" TC8 held at Guimarães, Portugal, the discipline frame-work of Enterprise Systems as a scientific subdiscipline called IIIE was proposed (Raffai, 2007; Roode, 2005). The IFIP TC 8 committee members (from many dif-ferent countries) intensively discussed the innovative and unique characteristics of IIIE as a scientific subdiscipline (Raffai, 2007). In this meeting, it was decided by the TC8 members that the IFIP TC8 First International Conference on Research

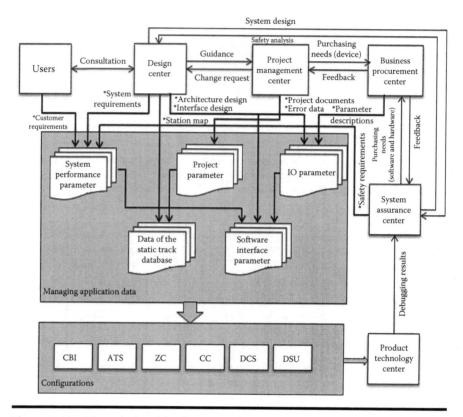

Figure 5.25 The relationships between the workflows of departments and the application data configurations.

and Practical Issues of Enterprise Information Systems (CONFENIS, 2006) would be held in April 2006 in Vienna, Austria. In August 2006, at the IFIP 2006 World Computer Congress held in Santiago, Chile, as the proposal was voted upon and endorsed by the Congress, the IFIP TC8 WG8.9 Enterprise Information Systems was formally established to promote worldwide academic interactions among both academics and practitioners in the area of enterprise information systems for advancing the concepts, methods, and techniques related to enterprise information systems. In October 2006, an Enterprise Information Systems Special Session was successfully held at the 2006 IEEE International Conference on SMC. In 2007, the IEEE SMC Technical Committee on EIS was established, focusing on the interface between engineering disciplines and industry information integration engineering. This is the first EIS TC established in IEEE.

Broadly speaking, IIIE is a set of foundation concepts and techniques that facilitate the industrial information integration process; specifically speaking, IIIE comprises methods for solving complex problems in developing information technology infrastructure for industrial sectors, especially in the aspect of information

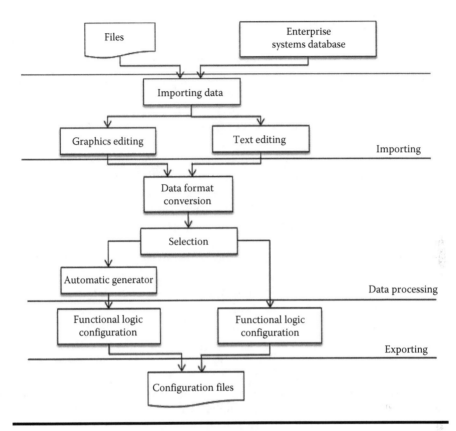

Figure 5.26   The framework of the application data configuration tool.

integration. IIIE has been proposed and studied through identifying its theoretical foundation, body of knowledge, frameworks, theories, and models at multiple levels. The key research questions addressed include (1) what is the scientific foundation that will provide IIIE with the disciplinary support at the levels of frameworks, theories, and models? and (2) at each level of IIIE (i.e., frameworks, theories, and models/techniques), how can real-world problem-solving support be provided? IIIE is an interdisciplinary discipline with the typical characteristics of a giant and complex system. According to the subsystems that make up a system, the number of subsystems involved, and the degree of complexity involved with the subsystems, the overall system can be categorized either as a simple system or as a giant system. If a system is made up of a huge number of subsystems, the system is referred to as a giant system. In addition, if a system has a numerous subsystems and layers and if the relationships among the subsystems and layers are complicated, the system is referred to as a complex giant system (Lin et al., 2013). As an interdisciplinary discipline, IIIE interacts with scientific disciplines such as mathematics, computer science, and almost every engineering discipline among the 12 engineering disciplines defined by the US National Academy of Engineering (http://www.nae.edu/MembersSection/Sections.aspx). The US National Academy of Engineering is organized into 12 sections, each representing a broad engineering category. IIIE interacts with almost every one of them in separate layers. In terms of scientific and engineering methods, at the methodological layer, IIIE interacts with computer science and engineering, industrial systems engineering, information systems engineering, and interdisciplinary engineering. In terms of developing and implementing enterprise systems in different industrial sectors, at the application layer, IIIE interacts with aerospace engineering, bioengineering, civil engineering, energy engineering, communication engineering, material engineering, and earth resources engineering. In addition to the scientific and engineering disciplines, IIIE also interacts with management and social sciences. For example, any effective business process relies on effective management. As a result, the perspectives that are included in workflow modeling and representation may include managerial perspective. Based on the definition of management defined by Xu and Xu (2011), in a broad sense, management is the most comprehensive science that covers all the disciplines. Judging from these, IIIE is defined as a complex giant system that can advance and integrate the concepts, theory, and methods in each relevant discipline and open up a new discipline for industry information integration purposes, which is characterized by its interdisciplinary nature. Figure 5.27 shows IIIE at the top level; relevant scientific, engineering, management, and social science disciplines at the second level; and application engineering fields at the third level. At the fourth level and the levels below, many relevant frameworks, theories, and models can be listed. Some existing frameworks as introduced in Chapter 3 are interdisciplinary themselves, just as enterprise interoperability is related to science, engineering, management, and social sciences (Goncalves et al., 2013). Thus, enterprise interoperability can be listed at a certain place in Figure 5.27. Figure 5.27 can be huge in size, in order to cover all of the details involved. For example, enterprise interoperability is involved with frameworks such as the AIF, Business Interoperability Parameters, the CEN/ISSS eBusiness Roadmap, C4IF, the IDEAS Interoperability Framework, the EIF, Levels of Conceptual Interoperability, LISI C4ISR, NC3TA, and the Organizational Interoperability Maturity Model (Goncalves et al., 2013).

In the following, we will introduce the main enabling technologies for IIIE, which include BPM, information integration and interoperability, EA and EAI, and SOA, although each of these is discussed in detail in different chapters and sections of this book: Business Process Management in Chapter 6; Information Integration in Chapter 7; Interoperability in Chapter 3 and Section 5.5; Enterprise Architecture in Chapter 4; Enterprise Application Integration in Chapter 8; and Service-oriented Architecture in Section 5.4 (Xu, 2011).

Rapid advances in industrial information integration methods have spurred tremendous growth in the use of enterprise systems. A variety of techniques have been used for probing IIIE so far. These techniques include BPM, workflow management, EAI, SOA, grid computing, and others. Many applications require a combination

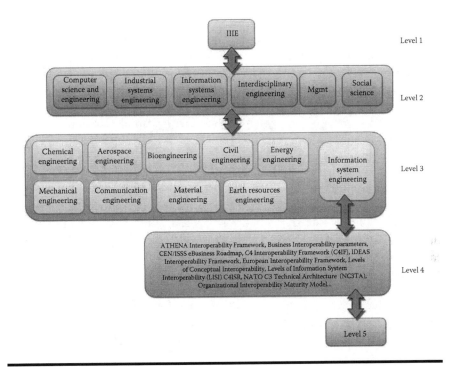

Figure 5.27   Discipline structure of IIIE.

of these techniques; this gives rise to the emergence of IIIE that requires techniques originated from different disciplines. At present, we are at a new breakpoint in the evolution of selected enabling technologies for IIIE. In this section, we introduce selected basic techniques that are significant in IIIE.

5.7.1 Business Process Management

IIIE enables the integration of business processes throughout an enterprise or extended enterprise with the help of BPM. BPM is an approach that is focused on aligning all of the aspects of an industrial organization in order to promote process effectiveness and efficiency with the help of information technology. Through business process modeling, BPM can help industries standardize and optimize business process, increasing their agility in responding to the changing environment for competitive advantage, accomplishing business process reengineering, and realizing cost reduction.

Process modeling is one of the most important topics in IIIE. The significance of process modeling to IIIE is obvious. The modeling, monitoring, and controlling of industrial processes is important, as it enables us to understand and optimize such processes. Manufacturing process modeling is a typical example. All process details in a manufacturing process, which relate to the desired outputs of the process, need to be understood. In general, a precise process model that relates processes is required. As such, modeling manufacturing processes is important as it enables manufacturers to understand the process and to optimize the process operation. Modeling industrial control is one example. Such modeling draws the domain expertise of multiple disciplines/subjects including information technology, process technology, factory automation, and industrial communication systems. Process modeling results can be applied to process automation and factory automation. The control and predictive capability of BPM also offer useful insights into quite a few engineering fields covered in Figure 5.27. As mentioned earlier, IIIE is interdisciplinary. Industrial process modeling can be listed somewhere in Figure 5.27 at a level below level 3; as such, the industrial process control itself is a complex interdisciplinary subject. To track process-related information and the status of each instance of the process as it moves through an organization, the concept of workflow becomes important. Workflow systems have been considered as efficient tools that enable the BPM, the business process reengineering, and eventually the automation of organizational business processes. Workflow management provides increased process efficiency through improved information availability, process standardization, task assignment on an automatic basis, and process monitoring using specific management tools (i.e., WfMS). Although workflow monitoring and management spans a broad continuum, the key idea of workflow management is to track process-related information.

When the first prototype of a workflow system was developed, the early idea of automation of business processes was initiated. Workflow management allows managing workflows for different types of processes, facilitating process automation, and providing predictive capabilities, and it enables organizations to maintain control over their processes. Business processes and their related workflow systems have gained greater interest since the early 1990s; research about enterprise business processes and workflows has become a prominent area that attracts attention both from academia and from industry.

A workflow consists of a number of tasks that need to be carried out and a set of conditions that determine the order of the tasks. The WMC defines a workflow as a computerized facilitation for the automation of a business process, in whole or in part. Three types of workflows are generally recognized in literature. A "production workflow" is associated with routine processes and is characterized by a fixed definition of tasks and an order of execution. An "ad hoc workflow" is associated with nonroutine processes, which could result in a novel situation. In an "administrative workflow," cases follow a well-defined procedure, but alternative routing of a case is possible. Compared with the other two types, production workflows correspond to critical business processes and possess high potential to add value to the organization. Hence, the administrative workflow is usually the focus of most studies on workflow modeling.

Workflow management has been considered an efficient way of monitoring, controlling, and optimizing business processes through information technology support and is playing an important role in improving an organization's performance through the automation of its business processes. The initial focus of business process modeling was the automation of business processes by using information technology. However, with the burgeoning global supply chain operation, today's enterprise systems are required to address more than merely the processes taking place within the walls of enterprises. Business process modeling is expected not only to automate business processes within the organization but also to enable organizations in the supply chain to automate interorganizational business processes. The main reason for this is that SCM has been rapidly developing worldwide, and the scope of enterprise integration and interoperation is extending from intra- to interorganizational. As such, more efforts have been focused on the integration of interorganizational systems to form interorganizational architecture. For this purpose, it is necessary to study both intra- and interorganizational business processes with a scientific approach. IIIE is required for addressing complex business processes taking place within and beyond the enterprise. Not only does the intraorganizational business process need to be addressed, but so does the interorganizational process. As e-business continues to prevail as an important trend, the embracing of interorganizational BPM and workflow management is considered to be one of the best prospects opened up for IIIE applications.

Today, workflow systems are increasingly applied to cooperative business domains such as SCM, and they are interorganizational. As such, workflow management needs to be completed on an interorganizational basis. Interorganizational BPM also provides enterprises the opportunity to reshape their business processes beyond their organizational boundaries. A changing business environment requires an organization to dynamically and frequently adjust and integrate both its intra- and interorganizational processes. Additional benefits of interconnecting business processes across systems and organizations include higher degrees of integration and the facilitation of the information and material flows.

Interorganizational workflows comprise intra- and interorganizational workflows. Wolfert et al. define intra- and interintegration and process and application integration in this way: "intraorganizational integration" overcomes fragmentation between organizational units; "interorganizational integration" integrates enterprises in the supply chain; "process integration" aligns tasks through coordination; and "application integration" aligns software systems to reach cross-system interoperability.

Process integration, as mentioned earlier, is one of main types of integrations and can be either intra- or interorganizational. Due to the closed connections and transformations between process management and workflow management, an intra- and interorganizational workflow management capability can enhance the performance of intra- and interorganizational integrations. Interenterprise workflow architecture supports the interoperations between independent enterprises. Meanwhile, an intra- and interorganizational workflow management capability can also enhance information sharing at both the intra- and interorganizational levels, eventually enabling all of the partners in the extended supply chain system to better collaborate, to optimize operations, and to gain competitive advantage.

WfMS defines, manages, and executes workflows through the execution of software. WfMS has become a standard solution for managing complicated processes in many business organizations since its appearance in the early 1990s. Despite a few failures associated with the introduction of WfMS, workflow technology has managed to become an indispensable part of enterprise systems. Workflow technology can be used to improve the business process and to increase performance, since the improvement can be quantified with respect to lead time, wait time, service time, utilization of resources, etc. WfMS can be employed as a repository of valuable process knowledge and can act as a vehicle for collecting and distributing knowledge across the supply chain. WfMS can also be used as a platform for knowledge sharing and learning in the supply chain and allows the knowledge workers in each organization to perform creative intellectual activities.

The complex nature of business processes, particularly processes spread across multiple organizations, presents technical challenges. Most traditional WfMSs assume one centralized enactment service, are able to support workflows only within one organization, and have problems in dealing with workflows crossing organizational boundaries. Practicing interorganizational workflow management requires coping with technical challenges.

Workflow research can be viewed in terms of three layers (Figure 5.8). The first layer pertains to issues about intraorganizational workflows, which link activities between different units within one organization. The second layer corresponds to interorganizational workflows, which cover distributed processes between different organizations, both of which comprise the interorganizational workflow. The third layer concerns the workflows in e-business settings. Effective management of business processes relies on sophisticated workflow modeling and analysis.

Among the modeling techniques, most of them have shown the capability in graphical representation and formal semantics in modeling workflows in an intraorganizational context. Currently, there is an urgent demand for translation between various models so that different WfMSs can interoperate with each other. This could lead to methods that will enable the integration of heterogeneous models within a unified framework.

The existing modeling techniques have advantages as well as disadvantages. Efforts regarding interorganizational workflow modeling are exploring better architectures in order to combine different organizational workflows while continuing to reconcile the differences. Some approaches have been specifically proposed for modeling interorganizational workflows, such as the routing approach and the interaction model. Some cognitive approaches have been proposed for the dynamic routing of information; meanwhile, new languages have been proposed to handle the routing of information among organizations.

In terms of evaluation, qualitative evaluation methods mainly focus on checking for structural soundness, which can usually be done through the validation and verification of workflows. Quantitative evaluation methods require the calculation of performance indices related to workflows. The existing techniques include computational simulation, the Markovian chain, and queuing theory.

At present, in the area of workflow management, there has been great interest in studying workflow modeling, workflow security management, and interorganizational workflow. Section 5.2 introduces SWSpec, a service workflow specification language that allows arbitrary services in a workflow to formally and uniformly impose requirements. System flexibility has been considered to be a major functionality of workflow systems. More research is needed for such functionality in order to provide sufficient flexibility for coping with complex business processes. Other topics for research include communication among multiworkflows in complicated business process, simplifying the workflow modeling process, and automating workflows. Existing techniques in process modeling still have limitations as they attempt to address some modeling issues. For example, business process models may contain numerous elements with complex intricate interrelationships. Efforts are needed to address how to properly capture such complexities.

5.7.2 Information Integration and Interoperability

SCM has experienced many changes in the past decade. The key change is that organizations in the supply chains are becoming more and more connected with each other. Meanwhile, supply chains are also becoming increasingly integrated with each other to form supply chain networks. Today's businesses of all sizes need to share data with suppliers, distributors, and customers. Information integration is significant not only for large-scale enterprise or for supply chain integration but also at the microscopic level. Compressed product development cycles and lifetimes and just-in-time stocking imply that SCM systems must be interconnected, and the applications composing the information systems of enterprises increasingly need to work together. As such, the demand for supply chain integration has been increasing.

As a consequence of such developments, enterprise systems are increasingly moving toward interorganizational integration as the benefits of interorganizational information sharing become obvious. An interorganizational system is aimed at providing a higher-level system related to activities that involve the coordination of business processes (both intra- and interorganizational) and is able to provide an integrated architecture to organizations within the supply chain. Now, more efforts have been focused on interorganizational systems, and more and more enterprises have moved toward interorganizational integration in order to support SCM. Interorganizational systems are able to allow communication between partners in the supply chain. Integrated enterprise systems can collect valuable SCM information for all of the related business processes across the supply chain. By using integrated SCM, businesses can better predict their markets, can better innovate in response to market conditions, and can better align their operations across supply chain networks. The DAMA project is an example. DAMA has developed an interenterprise architecture and a collaborative model for supply chains, which enables improved collaborative business across supply chains (Chapman and Petersen, 2000).

The integration of interorganizational systems is a complex task. Several frameworks have been proposed for information integration. Fox et al. indicate that at the core of the SCM system lays a generic enterprise model (Fox et al., 1993). Hasselbring proposes a three-layer architecture for integrating different types of architectures (2000). In Puschmann and Alt's framework, the data level is considered as a separate layer (2004). Giachetti's framework includes a typical characterization of different types of integration (2004). However, as indicated by Wolfert et al. (2010), the contents of these frameworks are not comprehensive, and an overall framework of information integration has yet to be developed.

Novel ES architectures with integration and interoperation capability have been emphasized. Grid computing is a new technology for distributed computing systems. The infrastructure of grid computing makes it possible to aggregate and share a large set of resources of different types, which are distributed geographically to form a single system. Grid computing may provide an effective tool for integrating enterprise systems. Research indicates that the characteristics of grid computing systems can meet the requirements of enterprise system integration for extended enterprises on the supply chain. A new grid-computing framework called Open Grid Service Architecture (OGSA) has been proposed to integrate grid resources into web services. Wang et al. (2008) indicate that this trend implies that the integration of web-based enterprise systems and grid computing infrastructure is possible. A novel global ES architecture based on OGSA, called GridERP, has been proposed to solve the problem of noneffective sharing of distributed resources and interoperability issues on the global deployment of enterprise systems, with the expectation of bringing enterprise system integration on the global supply chain to a higher level.

The current level of enterprise system integration may be limited by the sophistication of the relevant technologies or by the lack of new techniques, and the successful execution of SCM relies upon more sophisticated IIIE integration than what is currently available. It is expected that IIIE integration will attract more efficient and effective methods for automated SCM in which the seamless integration of interorganizational systems is highly expected. Among the new technologies, IoT have attracted much attention. The envisioned applications include information to be collected through IoT.

5.7.3 EA and EAI

To industrial organizations, an enterprise can be an organization, a part of a larger enterprise or an extended enterprise. An EA defines the scope of the enterprise, the internal structure of the enterprise, and its relationship with the environment. As it describes the structure of an enterprise, it comprises main enterprise components such as enterprise goals, organizational structures, and business process, as well as information infrastructure. An EA is generally considered an important aid for understanding and designing an enterprise. As information infrastructure is a component of EA and the term "enterprise" as used in EA generally involves information systems employed by an industrial organization, EA is highly relevant to IIIE, since IIIE concerns information flow within the entire industrial organization.

Enterprise architects use a variety of business models, conceptual tools, and analytical methods to describe the structure and dynamics of an enterprise. Artifacts are used to describe the logical organization of business processes and business functions, as well as information architecture and information flow. A collection of these artifacts is considered to be its EA. Software architecture, network architecture, and database architecture are partial components of an information architecture.

An EA's landscape is usually divided into various domains that allow enterprise architects to describe an enterprise from a number of important perspectives. One of the main domains in EA is the information domain. The important components in this domain include information architecture and data architecture. The other two domains with components that are also highly relevant to IIIE are the application domain and its component "interfaces between applications" and technology domain with its components as middleware, networking, and operating systems.

Representing the architecture of an enterprise correctly and logically will improve the performance of an organization. This includes innovations about the structure of an organization, business process reengineering, and the quality and timeliness of the information flow that represents material flows.

Enterprise integration has become a key issue for many enterprises looking to optimize business processes through integrating and streamlining processes both internally and with partners in the supply chain. It consists of plans, methods, and tools. Typically, an enterprise has existing legacy systems that are expected to continue in service while adding or migrating to a new set of applications. Integrating data and applications is expected to be accomplished without requiring significant changes to existing applications and/or data. To address this issue, a solution that can help achieve quality integration is referred to as EAI. Originally, EAI focused only on integrating enterprise systems with intraorganizational applications, but now it has been expanded to cover aspects of interorganizational integration. EAI facilitates the integration of both intra- and interorganizational systems. Major EAI-enabling technologies range from EDI to web services and XML-based process integration and provide a flexible, adaptable, and scalable EAI framework.

Solutions comprise the efficient integration of diverse business processes and data across the enterprises, the interoperation and integration of intra- and interorganizational enterprise applications, the conversion of varied data representations among involving systems, and the connection of proprietary/legacy data sources, enterprise systems, applications, processes, and workflows interorganizationally.

EAI entails integrating enterprise data sources and applications so that business data and processes can be easily shared. EAI must be able to integrate heterogeneous applications that are created with different methods and on different platforms. The integration of enterprise applications includes the integration of data, business processes, applications, and platforms, as well as integration standards. Through creating an integrative structure, EAI connects heterogeneous data sources, systems, and applications intra- or interorganizationally. EAI aims to not only connect the current and new system processes but also provide a flexible and convenient process integration mechanism. By using EAI, intra- or interorganizational systems can be integrated seamlessly to ensure that different divisions or even enterprises can cooperate with each other, even using different systems. A complete EAI offers functions such as business process integration and information integration. Through the coordination of the business processes of multiple enterprise applications and the combination of software, hardware, and standards together, enterprise systems can exchange and share data seamlessly in a supply chain environment.

In general, those enterprise applications that were not designed as interoperable can be integrated on an intra- and/or interorganizational basis. Legacy and newer systems can be integrated to provide greater competitive advantages. In EAI process, the constantly changing business requirements and the need for adapting to the rapid changes in the supply chain may require help from SOA, which was introduced in Sections 5.4 and 5.7.4.

The objective of EAI is to facilitate information exchange among business enterprises in a timely, accurate, and consistent fashion, in order to support business operations in a manner that appears to be seamless. EAI can facilitate the integration of both intra- and interorganizational systems and is capable of integrating both intra- and interorganizational applications.

5.7.4 Service-Oriented Architecture

SOA represents the latest trend in integrating heterogeneous systems. It has received much attention as an architecture for integrating platforms, protocols, and legacy systems, and it has been considered a suitable paradigm that helps integration, since it is characterized by simplicity, flexibility, and adaptability.

SOA represents an emerging paradigm for enterprises to use in order to coordinate seamlessly in the environment of heterogeneous information systems, enabling the timely sharing of information in the supply chain and developing flexible large-scale software systems. Some example applications include the InLife project (Ribeiro et al., 2009) and information integration based on SOA in agri-food industry (Wolfert et al., 2010). For an introduction to SOA, see Section 5.4.

In the last decade, enterprise systems have received increasing attention both from academic circles and from the industry worldwide, although the technical efforts on the early enterprise systems began in different countries almost at the same time during the last century. For instance, the development of enterprise systems in Western countries in 1980s and 1990s was actually progressing in parallel with the development of Automated Control Systems (ACS) in the former Soviet Union. ACS is equivalent to early enterprise systems in terms of its system characteristics. The only difference is in its technical nomenclature. Along with the establishment of the discipline of IIIE in 2005, the framework, theories, and methods of IIIE have been spreading throughout various countries. As Kataev et al. (2013) wrote, the major trend of enterprise systems development now is moving toward emphasizing fully integrated systems. They also wrote: "As the discipline framework of ES was for the first time systematically presented to both IFIP and IEEE in 2005, and IFIP TC 8 WG8.9 on ES and IEEE SMC Society Technical Committee on EIS (first ES TC in IEEE) were established in 2006 and 2007, respectively, the international research platform of ES was formally introduced into Russia." They concluded that enterprise systems development will continue to embrace cutting-edge technology and techniques, including the ACS and other technologies originally developed in Russia (such as CIM, based on polychromatic sets theory). This further shows that the research on IIIE as a new discipline of industrial information integration will continue to garner much attention.

References

ATHENA Consortium. 2003. Advanced Technologies for Interoperability of Heterogeneous Enterprise Networks and their Applications, FP6-2002-IST1, Integrated Project, 2003.

Benatallah, B. and Motahari-Nezhad, H. 2008. Service oriented architecture: Overview and directions. In Alfredo, F. and Egon, B. (Eds.), *Advances in Software Engineering, Lecture Notes in Computer Science*, Springer, Heidelberg, Vol. 5316, pp. 116–130.

Butner, K. 2009. *Blueprint for Supply Chain Visibility.* New York: IBM.

C4ISR. 1998. *C4ISR Architecture Framework*, Version 2.0. Washington, DC: Architecture Working Group (AWG), Department of Defense (DoD).

Chapman, L. and Petersen, M. 2000. *Demand Activated Manufacturing Architecture (DAMA) Model for Supply Chain Collaboration.* Albuquerque, NM: Sandia Laboratory.

Chaudhary, A., Saleem, M., and Bukhari, H. 2002. Web services in distributed applications: Advantages and problems. *Proceedings of IEEE Conference at Ghulam Ishaq Khan Institute of Engineering Sciences and Technology*, Islamabad, Pakistan.

Chen, D. and Doumeingts, G. 2003. European initiatives to develop interoperability of enterprise applications-basic concepts, framework and roadmap. *Journal of Annual Reviews in Control*, 27(3), 151–160.

Chen, X., Guan, A., Qiu, X., Huang, H., Liu, J., and Duan, H. 2013. Data configurations in railway signaling engineering-an application of enterprise systems techniques. *Enterprise Information Systems*, 7(3), 354–374.

Chen, Z. and Xu, L. 2001. An object-oriented intelligent CAD system for ceramic kiln. *Knowledge-Based Systems*, 14, 263–270.

Concha, D., Espadas, J., Romero, D., and Molina, A. 2010. The e-hub evolution: From a custom software architecture to a software-as-a-service implementation. *Computers in Industry*, 61(2), 145–151.

Debnath, N., Leonardi, M., Mauco, M., Montejano, G., and Riesco, D. 2008. Improving model driven architecture with requirements models. *Proceedings of Fifth International Conference on Information Technology: New Generations*, Las Vegas, NV, pp. 21–26.

EIF. 2004. *European Interoperability Framework*. White paper. Brussels, Belgium, February 2004. http://www.comptia.org.

Fang, S., Xu, L., Pei, H., Liu, Y., Liu, Z., Zhu, Y., and Zhang, H. 2014. An integrated approach to snowmelt flood forecasting in water resource management. *IEEE Transactions on Industrial Informatics*, 10(1), 548–558.

Fang, S., Xu, L., Zhu, Y., Liu, Y., Liu, Z., Pei, H., Yan, J., and Zhang, H. 2014. An integrated information system for snowmelt flood early warning based on Internet of Things. *Information Systems Frontiers*, in press. DOI: 10.1007/s10796-013-9466-1.

Fox, M., Chionglo, J., and Barbuceanu, M. 1993. *The Integrated Supply Chain Management System*. Toronto, Ontario, Canada: University of Toronto.

Giachetti, R. 2004. A framework to review the information integration of the enterprise. *International Journal of Production Research*, 42(6), 1147–1166.

Goncalves, R., Grilo, A., Agostinho, C., Lampathaki, F., and Charalabidis, Y. 2013. Systematisation of interoperability body of knowledge: The foundation for enterprise interoperability as a science. *Enterprise Information Systems*, 7(1), 7–32.

Gunzer, H. 2002. Introduction to web services, available at http://archive.devx.com/javasr/whitepapers/borland/12728jb6webservwp.pdf. October, 2013.

Guo, J. 2009. Collaborative conceptualization: Towards a conceptual foundation of interoperable electronic product catalogue system design. *Enterprise Information Systems*, 3(1), 59–94.

Guo, J., Xu, L., Xiao, G., and Gong, Z. 2012. Improving multilingual semantic interoperation in cross-organizational enterprise systems through concept disambiguation. *IEEE Transactions on Industrial Informatics*, 8(3), 647–658.

Hachani, S., Gzara, L., and Verjus, H. 2013. A service-oriented approach for flexible process support within enterprises: Application on PLM systems. *Enterprise Information Systems*, 7(1), 79–99.

Hasselbring, W. 2000. Information system integration. *Communications of ACM*, 43(4), 32–38.

He, W. and Xu, L. 2013. Integration of distributed enterprise applications: A survey. *IEEE Transactions on Industrial Informatics*, DOI: 10.1109/TII.2012.2189221.

Ho, L. and Lin, G. 2004. Critical success factor framework for the implementation of integrated-enterprise systems in the manufacturing environment. *International Journal of Production Research*, 42(17), 3731–3742.

Humphreys, P., Lai, M., and Sculli, D. 2001. An inter-organizational information system for supply chain management. *International Journal of Production Economics*, 70(3), 245–255.

Iacob, M. and Jonkers, H. 2009. A model-driven perspective on the rule based specification and analysis of service-based applications. *Enterprise Information Systems*, 3(3), 279–298.

IDAbc. 2004. *EIF: European Interoperability Framework, Version 1.0.* Brussels, Belgium: European Commission, available at http://ec.europa.eu/idabc/en/document/2319/5644. October, 2013.

IDEAS Consortium. 2002. IDEAS: Interoperability development for enterprise application and software roadmaps. IST-2001-37368.

Kataev, M., Bulysheva, L., Emelyanenko, A., and Emelyanenko, V. 2013. Enterprise systems in Russia: 1992–2012. *Enterprise Information Systems*, 7(2), 169–186.

Kleppe, A., Warmer, J., and Bast, W. 2003. *MDA Explained: The Model Driven Architecture: Practice and Promise.* Boston, MA: Addison-Wesley.

Langenwalter, G. 2000. *Enterprise Resource Planning and Beyond.* Boca Raton, FL: St. Lucie Press.

Li, B. and Li, M. 2009. Research and design on the refinery ERP and EERP based on SOA and the component oriented technology. *Proceedings of International Conference on Networking and Digital Society*, Guiyang, China, pp. 85–88.

Li, L. 2007. *Supply Chain Management: Concepts, Techniques and Practices.* Hackensack, NJ: World Scientific.

Lin, C., Chou, C., Lin, Y., Wu, I., and Chang, H. 2012. Apply model-driven architecture to re-conceptualization of BIM for extended usage. *Proceedings of International Workshop: Intelligent Computing in Engineering*, Herrsching, Germany, July 4–6, 2012.

Lin, Y., Duan, X., Zhao, C., and Xu, L. 2013. *Systems Science Methodological Approaches.* Boca Raton, FL: CRC Press.

Linthicum, D. 2000. *Enterprise Application Integration.* Reading, MA: Addison-Wesley.

Liu, J., Zhang, S., and Hu, J. 2005. A case study of an inter-enterprise workflow-supported supply chain management system. *Information and Management*, 42(3), 441–454.

Lorincz, P. 2007. Evolution of enterprise systems. *Proceedings of International Symposium on Logistics and Industrial Informatics*, Wildau, Germany, pp. 75–80.

Newcomer, D. 2002. *Understanding Web Services, XML, WSDL, SOAP, and UDDI.* Boston, MA: Addison-Wesley.

Panetto, H. and Cecil, J. 2013. Editorial information systems for enterprise integration, interoperability and networking: Theory and applications. *Enterprise Information Systems*, 7(1), 1–6.

Papazouglou, M. and Georgakooulos, D. 2003. Service-oriented computing: Introduction. *Communications of ACM*, 46(10), 24–38.

Puschmann, T. and Alt, R. 2004. Enterprise application integration systems and architecture-the case of the Robert Bosch Group. *Journal of Enterprise Information Management*, 17(2), 105–116.

Raffai, M. 2007. New working group in IFIP TC8 information systems committee: WG 8.9 working group on enterprise information systems. *SEFBIS Journal*, 2, 4–8.

Ribeiro, L., Barata, J., and Columbo, A. 2009. Supporting agile supply chains using a service-oriented shop floor. *Engineering Applications of Artificial Intelligence*, 22(6), 950–960.

Roode, D. 2005. IFIP General Assembly September 2005. Report from Technical Committee 8 (Information Systems), Gaborone, Botswana, August 27, 2005.

Quartel, D., Pokraev, S., and Dirgahayu, T. 2009. Model-driven development of mediation for business services using COSMO. *Enterprise Information Systems*, 3(3), 319–345.

Qureshi, K. 2005. Enterprise application integration. *Proceedings of IEEE 2005 International Conference on Emerging Technologies*, Islamabad, Pakistan, pp. 340–345.

Singh, M. and Huhn, M. 2005. *Service-Oriented Computing*. West Sussex, U.K.: Wiley.

Touzi, J., Benaben, F., Pingaud, H., and Lorré, J. P. 2009. A model-driven approach for collaborative service-oriented architecture design. *International Journal of Production Economics*, 121(1), 5–20.

Unger, T., Mietzner, R., and Leymann, F. 2009. Customer-defined service level agreements for composite applications. *Enterprise Information Systems*, 3(3), 369–391.

van Lessen, T., Nitzsche, J., and Leymann, F. 2009. Conversational web services: Leveraging BPEL light for expressing WSDL 2.0 message exchange patterns. *Enterprise Information Systems*, 3(3), 347–367.

Viriyasitavat, W., Xu, L., and Martin, A. 2012. SWSpec: The requirements specification language in service workflow environments. *IEEE Transactions on Industrial Informatics*, 8(3), 631–638.

Vyatkin, V. 2013. Software engineering in industrial automation: State-of-the-art review. *IEEE Transactions on Industrial Informatics*, 9(3), 1234–1249.

Wang, T., Su, C., Tsai, P., Liang, T., and Wu, W. 2008. Development of a GridERP architecture: Integration of grid computing and enterprise resource planning application. *Proceedings of 4th International Conference on Wireless Communications, Networking, Mobile Computing*, Dalian, China, pp. 1–4.

Wang, Z., Zhan, D., and Xu, X. 2006. Service-oriented infrastructure for collaborative product design in ETO enterprises. *Proceedings of the 10th International Conference on Computer Supported Cooperative Work in Design*, Nanjing, China, pp. 1–6.

Wolfert, J. et al. 2010. Organizing information integration in agri-food-a method based on a service-oriented architecture and living lab approach. *Computers and Electronics in Agriculture*, 70(2), 389–405.

Xu, L. 2011. Enterprise systems: State-of-the-art and future trends. *IEEE Transactions on Industrial Informatics*, 7(4), 630–640.

Xu, L., Li, Z., Li, S., and Tang, F. 2007. A decision support system for product design in concurrent engineering. *Decision Support Systems*, 42(4), 2029–2042.

Xu, L., Liang, N., and Gao, Q. 2008. An integrated approach for agricultural ecosystem management. *IEEE Transactions on Systems, Man, and Cybernetics, Part C: Applications and Reviews*, 38(4), 590–599.

Xu, L., Wang, C., Bi, Z., and Yu, J. 2012. AutoAssem: An automated assembly planning system for complex products. *IEEE Transactions on Industrial Informatics*, 8(3), 669–678.

Xu, R., Bai, J., and Wang, Y. 2010. The research and implementation of power application system integration based on enterprise service bus. *Proceedings of 2010 IEEE International Conference on Intelligent Computing and Intelligent Systems*, Beijing, China.

Xu, S. and Xu, L. 2011. Management: A scientific discipline for humanity. *Information Technology and Management*, 12(2), 51–54.

Chapter 6

Enterprise Process Modeling and Workflow Management

6.1 Introduction

From the perspective of IIIE, an "enterprise process" can be viewed as a mechanism for abstracting and modeling organizational processes and their various elements. The concept of "workflow" is closely related to the concept of enterprise process. A workflow represents a sequence of operations related to an element or a subelement in an organizational process. The relationship between enterprise process and workflow is similar to the relationship between a system and its subsystems.

Interest in enterprise processes and related workflow design engineering has risen since the early 1990s (van der Aalst and van Hee, 2014). Research in this area has resulted in the creation of a variety of new scientific and engineering methods and techniques. These methods and techniques provide opportunities for business and industry to gain a competitive edge by designing, redesigning, and managing their workflows.

In recent years, automation in the industry has been increasingly marked by the use of workflow technology. Meanwhile, emerging e-business and virtual enterprises have been leading the burgeoning trend of managing workflows across organizational boundaries. Connecting business processes within and across organizations can provide significant benefits in addition to a higher degree of integration.

The study of workflow has received much attention in the recent years, as workflow management plays an important role in improving an organization's competitiveness and competency through the abstraction and automation of business processes. The insights that an enterprise can glean from researching its workflows can offer it tremendous opportunities to reshape and redesign its business processes within as well as beyond the boundaries of its own organization.

The organizational workflow comprises intra- and interorganizational workflows. Managing interorganizational workflows is much more complex than managing intraorganizational workflows, since interorganizational workflows involve complex communications and interactions among and between the different systems of participating organizations; consequently, many technical issues can emerge during the integration process. Some of these issues include how to route information among participating organizations by using a standard language, how to realize the interoperability between distributed workflow systems, and how to prevent parties not involved in the partnership from learning its way of operation (core competency) and thus moving from collaborator to competitor (Eshuis and Grefen, 2008).

Up until now, the available workflow management methods have been mostly intraorganizational and have been based on centralized architectures that facilitate interoperability among existing workflow management systems. These existing technologies provide methodological foundations or starting points for the eventual realization of efficient interorganizational workflow collaborations.

Workflow management is considered as an efficient way of monitoring, optimizing, and controlling business processes and workflows in heterogeneous computing environments. "Workflow Management System" (WfMS) is a software system that supports the modeling, design, and execution of workflows. "WfMS" defines, executes, and manages workflows and, since its appearance in the early 1990s, has become a standard solution for managing complicated business processes in many organizations. Despite a few failures associated with the introduction of WfMS, workflow technology has managed to become an indispensable part of many businesses' enterprise systems.

WfMS has been successfully integrated into Enterprise Resource Planning (ERP) systems. In R/3 by SAP, the workflow module can be used to automate activities. WfMS plays an important role in developing cooperation among the participants in cooperative business domains, including CRM, supply chain management (SCM), and knowledge management (KM). WfMS helps businesses and industries adopt well-structured processes. It facilitates the conversion of a theoretical process that is merely defined at the information engineering level into a real-world executable process. It also facilitates business process reengineering and the associated workflow redesign. As such, WfMS has been accepted as one of the most highly successful types of systems in supporting cooperative enterprise operation, especially since it helps the businesses of today to operate across organizational boundaries, interacting with each other to face increasing competitive challenges. In simple words, WfMS has made the automated execution of the business process possible. Consequently, the worldwide commercial WfMS market has been growing at a steady pace.

This chapter is intended to introduce selected methods and techniques used for workflow abstraction and modeling. In Section 6.2, some basic definitions related to workflows are introduced. The techniques used for intraorganizational workflow modeling are introduced in Section 6.3. Workflows in interorganizational contexts are described in Section 6.4. In Section 6.5, we summarize the techniques for evaluating workflow technology, including the use of qualitative and quantitative analyses for evaluation purposes. Future research directions are discussed from both the technical and managerial points of view in Section 6.6.

6.2 Workflow Basics

The concept of workflow has been applied to model sequences of operations or tasks by many industrial and service organizations. According to the framework proposed by Basu and Kumar (Figure 6.1), workflow research can be viewed in terms of three layers (Basu and Kumar, 2002). The first layer pertains to the "intraorganizational workflows," which represent and link activities among different units within an organization. The second layer corresponds to the "interorganizational workflow," which represents and links distributed processes between different organizations. The intraorganizational workflows may involve interorganizational workflows. The third layer concerns the workflows in the e-business environment. Before considering the details, we will introduce some background about workflows in general.

Since its beginnings in the 1990s, workflow management technology has become increasingly more and more available. A large quantity of literature has been published regarding the topic of workflow since the 1990s. The early concept of the computer facilitation of the automation of business processes dates back to the 1970s, when the first prototype of a workflow system was developed in Zisman's PhD dissertation (Zisman, 1977). The Workflow Management Coalition (WMC) defines a workflow as a computerized facilitation for the automation of a business

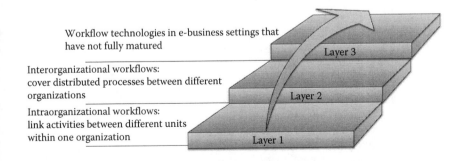

Figure 6.1 Workflow research in three layers.

process, in whole or in part. This definition implies several important characteristics of a workflow: (1) The workflow involves information; (2) it is processed by using resources; (3) the sequence of the processes is constrained by procedural rules; and (4) being a business process, it must achieve preset business goals (Yen, 2007). The concept of workflow is closely related to that of a business process; it consists of a number of tasks that need to be carried out and a set of conditions that determine the order of the tasks. Workflow management provides increased process efficiency through abstraction, design, standardization, automation, and monitoring, using tools such as WfMS. WfMS is very helpful in automating routine business processes. Once there is a dedicated automated system in place for scientific management of business processes, theoretically, such processes can be executed more efficiently (Lawrence, 1997). Meanwhile, the performance of the WfMS in the process of improving workflows and business processes can be evaluated with respect to aspects such as lead time, wait time, service time, flexibility, utilization of resources, and many other aspects.

Multiple definitions of workflow-related terms are in use in the literature, although increasing interest in workflow technology has raised a need to standardize concepts and practices. WMC has made efforts to standardize the process and has compiled a set of standards for WfMS. Three main types of workflows are generally introduced in literature (Leymann and Roller, 2000; van der Aalst, 1998). "Production workflow" represents routine processes characterized by a fixed definition of tasks and an order of execution. "Ad hoc workflow" represents nonroutine processes. In an "administrative workflow," a well-defined procedure is followed, but alternative routing is allowed. Compared with the ad hoc workflow and the administrative workflow, the production workflow corresponds to critical business processes in organizations and thus offers a high potential to add value. Hence, it is usually the focus of most studies on workflow modeling and design.

To ensure a better understanding of workflows, we will introduce some of the terminology related to workflows. A "process" is a topologically ordered set of procedures with states such as initiation, running, completion, and termination. A "workflow" represents a sequence of operations related to an element or a subelement in processes. In "systems perspective," a workflow can be decomposed into "subworkflows" systematically. Each workflow is characterized by one or more patterns as sequential, parallel, parallel merge, parallel-split, and parallel-merge-split. Workflows can also be classified into interactive workflows, event-driven workflows, etc.

A basic property of a workflow process is that it is work case based. A work case can be considered as an instance. A workflow instance includes specific workflow participants and instances of different processes within the workflow (Gudes and Tubman, 2002). It can be defined as the locus of control for a particular execution of a procedure (Kim and Ellis, 2001). The primary objective of a workflow system is to deal with "cases."

A "task" is an elemental process that represents a logical unit of work within the business process. It can be manual, semiautomatic, or automatic. The task executed in a process consists of input and output with the supporting resource, as defined by the function it performs. Workflow tools allow users to compose tasks into logical sequences. The number of tasks that a workflow can have can range from a small number into thousands. A "resource" represents generic resources, with examples such as a machine or even an organizational unit responsible for performing a task. Resources are often grouped into roles. A "role" represents a class of resources with similar characteristics; any one of them may perform a given task, as the organizational procedure requires specific roles to perform specific tasks. A role can also be defined as a designation of n participants that act. An actor is an entity that can fulfill roles and be associated with activities or procedures. There may be many resources in a role, and a resource may be a member of multiple roles. "Data information" refers to the data items involved in the process; this may be a number, a character string, or an image, for example, the input and output data of certain tasks. A task that needs to be executed for a specific case is named as a "work item." A "workflow procedure" is a set of work steps called activities, and the "order" defines which activities will be completed before another activity begins. An "activity" is a logic step in a process that requires resources to implement the operation. An activity can have states such as active, inactive, completion, and suspension. All of these can be specified in a workflow model, although workflows can vary in data, structure, size, constraints, computational requirements, and other resource requirements. Workflows can be computationally intensive; computation time required by workflows can vary.

The "routing" of a case along particular branches determines which tasks need to be performed and in which order. WMC has standardized four basic constructs in routing cases:

1. Sequential routing: Tasks are carried out one after the other, and usually there is a dependency relationship between them; the result of one task is entered into the next, as Figure 6.2a shows.
2. Parallel routing: Two or more tasks are performed simultaneously (or in any order) without affecting each other. The parallel tasks are initiated using an AND-split and resynchronized later using an AND-join building block, as Figure 6.2b shows.
3. Selective routing: For a choice between two or more tasks, the routing is often based on the rules and dataflow in the process; an OR-split is used to specify the choice between several alternatives, as Figure 6.2c shows.
4. Iterative routing: A task is performed several times before a satisfied output is generated; this, then, acts as the input to the next task. The iterative task will be checked to determine whether to "move on" or to "roll back," as Figure 6.2d shows.

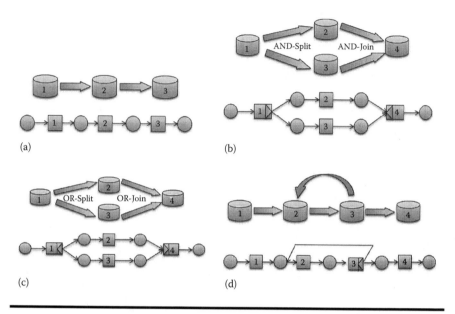

Figure 6.2 Basic constructs in routing. (a) Sequential routing, (b) parallel routing, (c) selective routing, and (d) interative routing.

Any process can be a combination of these four basic constructs. There are many ways to represent the workflow graphically; one of the most popular methods is to use Petri nets (PNs). In the 1990s, PNs were proposed as one of the main modeling techniques for workflows due to their strengths. They are graphical, expressive, and have a formal semantics that allow the distinction between the enabling of a task and the execution of a task in a WfMS (Prisecaru, 2008); as such, PNs are considered to be a formal method that has been successfully applied to workflow modeling. PNs are a mathematical tool, as well as a graphical tool, that permit explicit representation of the states and transitions of a system.

6.3 Intraorganizational Workflows

6.3.1 Modeling Perspectives

Workflows focus on the automation of complex processes in heterogeneous and distributed environments. Effective management of business processes relies largely on quality workflow modeling, analysis, and design. Over the last decade, workflow technology such as WfMS has been developed rapidly and has been increasingly used in many enterprises. So far, significant efforts have been made in modeling workflows in different perspectives. The following are a few perspectives from which workflows are commonly abstracted and modeled:

1. *Control flow/process perspective*: This perspective describes how tasks are executed and in which order (including sequencing, parallelism, and synchronization). This perspective also shows how a specific case is executed, according to the specified routing.
2. *Resource perspective*: This perspective defines the resources used in executing tasks; resources are specified, and the resources that are involved in a task are defined.
3. *Data/information perspective*: This perspective focuses on the flow of data between tasks in a workflow, the data/information entities involved, the structure of such entities, and their interrelationships.
4. *Task/function perspective*: This perspective describes the elementary operations performed while executing a task for a specific case.
5. *Operational perspective*: This perspective describes the elementary actions executed by tasks. In other words, this perspective describes the elementary actions/operations performed.

Some of the many questions related to a process can be answered from studying modeling efforts, based on these perspectives. For example, how does the organization work in general? How does a certain business process work? How is a process performing with certain resources? In a technique proposed by Sun et al., the data/information perspective is emphasized and formulated in two steps: first, the data flow matrix and the process data diagram are proposed to specify the data flow, and then a dependency analysis is utilized to conduct the data flow analysis (Sun et al., 2006). In another technique, the concept of resource workflow nets (RWFNs) based on nested PNs are proposed, in order to model both the resource perspective and the process perspective of a workflow independently; furthermore, they perform a synchronization function (Prisecaru, 2008). The metagraph approach integrates the informational, functional, and organizational perspectives within a single model by constructing views of metagraphs. Some methods focus only on a single perspective. Most workflow modeling efforts have been focused on the modeling of the process/control flow perspective of workflows, such as activity-based workflow modeling and object coordination nets. As a result, this direction is reflected in most workflow management systems. Practically, multiple perspectives are needed for modeling and design purposes.

Modeling efforts that integrate selected perspectives are not uncommon. Efforts have been made to summarize heuristic rules from multiple perspectives based on practical experiences, which can provide valuable help for designing business processes.

Workflow modeling methods based on PNs have been developed for representing resources in the process perspective. In resource-constrained workflow nets (RCWF-nets), the influence of resources on the processing of cases is considered, and the fundamental correctness requirements of RCWF-nets are emphasized. Colored PNs (CPNs) have been used to model an RCWF process. The model consists of both a task component to model tasks from the process perspective and a resource component to model the allocation of resources with different allocation methods from the resource perspective. Another approach proposes a special class of nested PNs to be used for the integrative modeling of processes and resource perspective of workflows and permits a distinction between them. Nested PNs are a special class of the PNs model. In this method, RWFNs are introduced as a special case of two-level nested PNs, in which the two perspectives are modeled as two separate object-nets: one object-net models the resource perspective and the other the process perspective. The dynamic behavior of the RWFN-net ensures collaboration between perspectives. In this method, the two object-nets synchronize whenever a task from the workflow net uses a role of the resource net, and they behave independently otherwise (Prisecaru, 2008). In another approach, a formal framework for process-oriented modeling and analysis was proposed, in which processes, tasks, resources, and other related concepts are specified in a formal language based on the sorted first-order predicate logic (Popova and Sharpanskykh, 2008).

Researchers believe that most of today's WfMS systems focus on the process dimension and oversimplify the organizational dimension. It is believed that there is a great need for modeling efforts focusing on organizational perspectives. In the framework proposed by Popova and Sharpanskykh, the organizational dimension is considered in addition to other perspectives. In particular, the performance-oriented view describes organizational structures, performance indicator structures, and the relations between them. From the aspect of the organizational perspective, organizational roles are included. The future trend of the research is to consider other perspectives in addition to the control flow perspective, especially the organizational perspective, which has received little attention given the potential impact of workflow automation on organizations as a whole.

6.3.2 Modeling Techniques

Many organizations have introduced workflow technology to support their business processes. This introduction is allowing this technology to grow quickly and to reach a wider market. There are many workflow products commercially available that implement various types of modeling techniques and tools.

The workflow models that most WfMS systems support are activity based and consist of elements such as objects, roles, and agents (Georgakopoulos et al., 1995). Among others, PNs are one of the most widely used techniques and are accepted mainly not only for providing a graphical representation but also as a mathematical method with a wide range of supporting tools. PNs are a class of modeling components that were originally proposed by Petri (1962). A classic PN is a four-tuple (P, T, F, M_0) where

1. $P = \{p_i: i = 1,\ldots,P\}$ is a finite set of places, representing possible states or conditions in the system.
2. $T = \{t_j: j = 1,\ldots,T\}$ is a finite set of transitions, describing events that may modify system states, $P \cap T = \emptyset$, $P \cup T \neq \emptyset$.
3. $F \subseteq (P \times T) \cup (T \times P)$ is a finite set of directed arcs. A binary relation represents the flow relation of the net, joining places, and transitions together.
4. $M_0: P \to \{1,2,\ldots\}$ is the initial marking in each place in the system.

Complete definitions and properties about PNs can be found in related literature and books.

In PN–based modeling, the condition is illustrated by place, the activity is illustrated by transition, and the state of a workflow is presented by the markings in each place. Despite the strength of PNs, it can be difficult to model some practical complex situations. As a result, PNs are still not being used frequently in practice. A major reason for this is that, although PNs have a good mathematical foundation and offer good graphical capability, many workflow patterns cannot be mapped into PNs (Cull and Eldabi, 2010). Therefore, many researchers have proposed that an extended version of the classical PN approach can be useful in many ways. For example, to address the complexity in modeling, the idea of a "worklet" in extended PNs was proposed. The worklet is a small workflow carried out by an organizational unit such as a department; it offers the modeler less difficulty both in modeling and in maintaining the models (Pudhota and Chang, 2005).

As a formal method, CPNs were proposed to define the workflow process. A CPN approach can deal with objects with various attributes. It can help define comprehensive families of workflow process specifications in order to control the number of business processes for different types of cases (Liu et al., 2002). Temporal workflow system components have been used to specify temporal constructs and to standardize the representation of temporal constraints (Chinn and Madey, 2000). A timed PN approach can analyze temporal behavior through temporal representation and reasoning for workflow management. To specify the timing constraints, modeling efforts have also been made to convert a directed network graph workflow model into a timing constraint workflow net, which is an extension of a WF-net (Li et al., 2003). Another extension of classical PN approach is the information control net (ICN) (Ellis, 1999), which is a simple but mathematical rigorous formalism. These nets use AND and OR nodes as the building blocks to model procedures. Apart from these extensions, some researchers utilize the strong representation and formal basis of PNs, combining them with other approaches to form new models. For example, Yet Another Workflow Language (YAWL) takes PNs as a starting point and then adds mechanisms to allow for a more direct and intuitive support. By doing so, some complex patterns, for example, synchronizations or nonlocal withdrawals, can be mapped onto PNs (van der Aalst, 2000). YAWL has applications in several industrial sectors including the automotive industry, defense, healthcare, and utilities (van der Aalst and ter Hofstede, 2005). Dataflow language (DFL) is a formal workflow modeling approach based on PNs and nested relational calculus, in which the former is responsible for the organization of the processing tasks and the latter is responsible for handing the collection of data items (Hidders et al., 2008). Interorganizational logical workflow (ILWN) was proposed for modeling and analyzing real-time cooperative systems based on timed PNs and temporal logic. Although more functions are available, this method can efficiently mitigate the problem of state explosion (Du et al., 2007). A significant amount of researches on PNs have so far provided support for the PN-based WfMS tools. Some examples of products that exploit PNs should be mentioned, such as ExSpect.

Currently, service workflows appear in several forms within an organization where services are used as a building block to streamline business processes and its concept has been realized in enterprise system solutions and in decentralized collaborative environments such as grids, virtual organizations, and cloud computing, in which services become a fundamental element for collaboration. As an example of extended version of the classical PN approach in service sectors, service workflow net (SWN) (Viriyasitavat et al., 2012; Xu et al., 2012) is seven-tuple $\mathcal{M} = (P, T, R, f, i, o, l)$, where P, T, R, f, and l definitions from basic PN are given as follows:

1. P is a set of services connected to a task in T
2. T is a set of tasks
3. $R \subseteq (P \times T) \cup (T \times P)$ represents connected flows
4. $f : \left(P \rightarrow \left(\psi \times \{\text{split}, \text{join}\} \right) \right) \cup \left(T \rightarrow \left(\psi \times \{\text{split}, \text{join}\} \right) \right)$ is a function with the formula (ψ) expressing connection either a split or a join type
5. $l : P \rightarrow A \cup \{\tau\}$ is a function where τ is null and A is a set of service properties

Instead of precisely representing a specific type of workflow instances, SWN is a PN variant to represent workflow structure of services and tasks, capturing all possible executions. This is because the activation of tasks and services at runtime cannot be known in advance due to nondeterministic choices from external inputs. In SWN, the coordination of tasks and services is controlled by a basic set of workflow connectives, AND, OR, and XOR. Despite being restricted in the basic ones, they are general to address all possible ways of activations.

A graph theory–based technique called "metagraph" applies a specialized graphical structure to capture the relationships between workflow tasks (Basu and Blanning, 1994, 2000). A metagraph is an ordered pair $S = (X, E)$, where $X = \{x_i, i = 1, \ldots, I\}$ is a finite generating set and $E = \{e_k, k = 1, \ldots, K\}$ is a set of "edges," with each edge as an ordered pair $e_k = (V_k, W_k)$ in which $V_k \in X$ is the "invertex" of the edge e_k and $W_k \in X$ is the "outvertex." A metagraph $M(B, C)$ from a source $B \subseteq X$ to a target $C \subseteq X$ is a set of edges, such that

1. There is a set of simple paths $\{h_m(x'_m, x''_m), m=1...M\}$ with $x'_m \subseteq B, x''_m \subseteq C$, then $\forall m$, $M(B,C) = \bigcup_{m=1}^{M} Set(h_m(x'_m, x''_m))$

2. $\left(\dfrac{\bigcup_{l=1}^{L} V_l}{\bigcup_{l=1}^{L} W} \right) \subseteq B$

3. $C \subseteq \bigcup_{l=1}^{L} W_l$

All of the important components in a workflow can be presented in a metagraph. Each information element in a workflow can be represented as an element of the generating set X (e.g., x_0, x_1, x_2, x_3, x_4, x_5 in Figure 6.3), so that a collection of information elements can be represented as a vertex ($<x_2, x_3>$), with each task presented as an edge (e.g., e_1, e_2, e_3, and e_4). Then a metagraph can be used to represent the tasks that comprise a collection of related workflows in a business process. This approach makes it possible to examine different but related aspects of workflow systems in a metagraph. Except for the graphical visualization of the processes, the advantage of this approach is that it provides a powerful supporting mechanism for a formal analysis about interactions within and between related workflows.

The idea of an event-driven model can be applied to workflows. The event-driven process chain (EPC) method in the workflow model uses event–condition–action rule-based language. The formalism describes the control flow of a process as a chain of events and tasks (Keller et al., 1992). A method was proposed to convert an EPC-based workflow model to a Workflow Intuitive Formal Approach (WIFA) model, which can formally verify the construct of workflows; meanwhile, it facilitates the interaction between heterogeneous interorganizational workflows (Tsai et al., 2006). It is known for its application in business process reengineering with SAP R/3 ERP systems. Adding roles and data objects to the ECA model, an event–role–object–condition–action approach was proposed. This approach describes a general framework for implementing dynamic routing and operational control mechanisms (Kumar and Zhao, 1999).

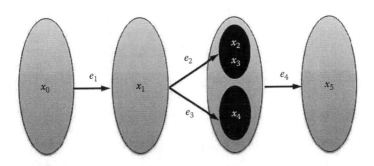

Figure 6.3   Representing information elements and task in metagraph.

In modeling workflows, the object-oriented approach has also been adopted. Some WfMS products are object oriented for their workflow specifications. One of the main advantages is that objects provide a closer semblance with reality and are less prone to change than are function-oriented methods (Barros and ter Hofstede, 1998). The typical modeling tool adopted for these types of products is the Unified Modeling Language (UML). UML is a graphical and visual modeling language. It is a tool used to analyze and design object-oriented systems. It is diagrammatic in nature and is composed of different types of diagrams: use case diagram, class diagram, sequence diagram, collaboration diagram, statechart diagram, activity diagram, etc. Some of the diagrams are related to enterprise modeling and process analysis, including use case, sequence, collaboration, and activity diagrams. For example, an activity diagram is used to model the flows circulating between activities within a process (Glassey, 2008). In some cases, PNs are incorporated into UML, where they function as the activity diagram. The composition of many different techniques enables the strong expressiveness of UML; meanwhile, an application may use one or more formalisms, which can result in many different models. However, a consistent standard for object-oriented workflow modeling has not yet been formed.

Some other approaches for workflow management are available. A cognitive approach for implementing real-time routing and control of dynamic and complex process management was proposed, in which business logic (which involves process routing, operational constraints, exceptional handling, etc.) is used to determine which tasks to execute in a certain situation. This method is intended to provide flexibility and adaptability in a complex cross-organizational environment (Wang and Wang, 2006). A knowledge-based workflow management system (K-WfMS) is proposed to increase the adaptability of WfMS against organizational changes (Lee et al., 1999). In this approach, a workflow is defined by a set of business rules and the knowledge of the organizational structure. A workflow is represented as task sequences in K-WfMS, and tasks are assigned according to the responsibilities of organizational roles. To an enterprise, accumulated workflow models are valuable knowledge assets, and it is important to manage them systematically for reuse purposes (Kim et al., 2002). A proposal was made to create a document-based workflow modeling support system to help search the appropriate workflow models for organizations and to help them improve the productivity of workflow management by utilizing domain-dependent case search and case-based reasoning techniques.

The comparisons of the select techniques used are given in Table 6.1. They are evaluated under the criteria of theoretical foundations, the degree of visualization, and the representation. It is believed that all modeling tools should offer the necessary constructs needed for analysis and design purposes. Table 6.1 is not intended to promote the best technique, but it tries to provide some analysis and some evaluations to help determine which techniques might be more suitable to certain situations. In addition, there are other techniques that have been developed but have no

Table 6.1 Contrast of Workflow Modeling Techniques

	Formal Basis	Visualization	Representation	Perspectives					Examples
				Process	Resource	Task	Data	Operation	
Activity-based modeling	✓	✓	✓	✓	—	—	—	—	Petri net
Graph-based modeling	✓	✓	—	—	✓	✓	✓	—	Metagraph
Event-driven modeling	—	—	✓	✓	—	✓	—	—	EPC
Object-oriented modeling	—	✓	✓	✓	✓	✓	—	—	UML
Rule-based modeling	—	—	✓	—	✓	✓	—	—	KWM

Note: "✓" represents "strongly involved," while "—" indicates "weakly or not involved" in the table.

formally defined semantics or theoretical basis (Dalal et al., 2004), such as IDEF. Technically speaking, they are considered to be intuitive, as the interpretation may shift depending on the application domain, the characteristics of the business processes, and the perspectives of the modeler.

Among the techniques introduced earlier, most have shown their basic capability of graphical representation and formal semantics in modeling workflows in an intraorganizational context. Currently, there is an urgent demand for the translation between these models so that different workflow management systems can interoperate with each other for industrial applications. This is an interesting research topic that has not received much attention until recently. Research could lead to the development of methods that are capable of integrating heterogeneous models into a common framework.

6.4 Interorganizational Workflows

WfMS models and applications were originally developed for intraorganizational purposes. In this context, WfMS applications are used to specify the cooperation within the organization. A typical WfMS application is composed of n tasks, which need to be completed by n departments within the organizations.

Interorganizational workflows are different from intraorganizational workflows largely due to their feature of crossing organizational boundaries. In the information-based economy of Internet and IoT, with the rapid increase in the number of workflow processes in which multiple organizations are involved, it is important to study, model, and verify interorganizational workflows, as workflow interoperability in such a complex and dynamic environment is critical (van der Aalst and ter Hofstede, 2000). Interorganizational workflow systems are an important component of enterprise systems, which support interorganizational collaboration. Modeling, analysis, and design methods of interorganizational workflows are the key issues in developing such workflow systems. The importance of the methods and techniques for modeling interorganizational workflows has been emphasized by many authors since the late 1990s.

Chronologically, in 1998, documentary PNs, a variant of high-level PNs, were proposed for use in modeling (Bons et al., 1998). In 1999, MSCs were proposed for use in specifying the interaction between organizations (van der Aalst, 1999). In 2000, interorganizational workflow nets were proposed to model loosely coupled interorganizational workflows (van der Aalst, 2000). In 2001, the X-transaction model was developed to support interorganizational workflow management. This model realizes a flexible intra- and interorganizational rollback effect (Wang et al., 2008). In 2002, XRL, which is based on XML language, was defined and proposed for use in specifying interorganizational workflows (Verbeek et al., 2004). An approach for designing interorganizational workflows was proposed, which can support the cooperation of partners while preserving the autonomy of the partner organizations. In 2003, the Public-to-Private approach for constructing interorganizational workflows based on the concept of inheritance was proposed (van der Aalst, 2003). In 2007, logical workflow nets (LWNs) and interorganizational logical workflow nets (ILWNs) were proposed (Du et al., 2007); this method can be used for modeling and analyzing real-time cooperative systems based on time PNs, workflow techniques, and temporal logic. In 2008, an approach for modeling interorganizational workflows based on nested PNs was proposed (Prisecaru and Jucan, 2008). Meanwhile, an object-oriented technique for analyzing the soundness of interorganizational workflows that focuses on interactive messages in an interorganizational context was proposed (Sun and Du, 2008). Jiang et al. (2008) summarize a few characteristics of the workflows in the interorganizational context: (1) "Autonomic versus collaborative" relates to each enterprise as an independent economic entity, although they should cooperate with each other to realize the common goal; (2) "Distributed versus interrelated" relates to the geographically distributed enterprises that are integrated into an overall workflow. One process may be part of another, or some processes can be intercrossed with each other; (3) "Stable versus dynamic" relates to the workflows of each partner being relatively stable, while the whole interorganizational workflow should remain flexible and dynamic to be able to adapt to rapidly changing market demands. In 2009, the modeling power of LPNs and PNs with inhibitor arcs (IPN) was verified.

Various forms of interoperations between different workflows have been studied (van der Aalst, 2000):

1. *Capacity sharing*: Tasks are executed by resources in distributed organizations under the control of one workflow manager. These coordinated organizations are basically in the same industry.
2. *Chained execution*: The workflow is split into a few subprocesses, and each is executed by a different business partner, one after the other. This is similar to the sequential routing in the intraorganizational workflow, except that the resources are located in various organizations.
3. *Subcontracting*: This is common in the industry, as one partner subcontracts some phases in the workflow to other partners. The workflow is hierarchical, while the contracting organization is at the top level.
4. *Case transfer*: Interorganizational workflows are partitioned in the case dimension (vertical partitioning), in which each organization has the same workflow description. When one organization has too many cases in the process, some can be transferred to other organizations to balance the workload.
5. *Loosely coupled*: Interorganizational workflows are partitioned in the process dimension (horizontal partitioning), in which each organization deals with one or more parts of the entire process. A protocol is used to communicate between these business partners according to the routing of the process.

Of the five forms of interoperation, capacity sharing is the simplest, and the only one utilizing centralized control, whereas the other forms use a decentralized control. Loosely coupled architectures are considered complex and dynamic. WMC evaluates these architectures with respect to criteria such as coordination efforts at design-time/runtime and system performance and concludes that the loosely coupled architecture is the most applicable one for interorganizational applications.

Based on the pattern of business partners interacting with each other within the context of the overall workflow, interorganizational workflows are classified into "distributed workflow" and "outsourced workflow" (Schulz and Orlowska, 2004). Distributed workflow indicates that all tasks of the workflow of each partner are implemented inside the scope of the workflow (see Figure 6.4a) with each task having its own input and output, thus making it an indispensable part of the entire workflow. It is important for the different systems to coordinate their interactions with each other by means of routing tasks and dependencies. In an outsourced workflow, one or more activities of an existing workflow are implemented outside of the scope of the workflow by an external task or workflow. The existing activities in the workflow are placeholders for external activities (see Figure 6.4b). To some extent, the structure of "chained execution" and "loosely coupled" workflows reflects the examples of a distributed workflow; "subcontracting" can be viewed as outsourced workflow.

WMC has worked out some standards regarding interoperability, such as the Wf-XML standard (WMC, 2000). This is based on traditional centralized architectures and uses the Workflow Process Definition Language that has no formal

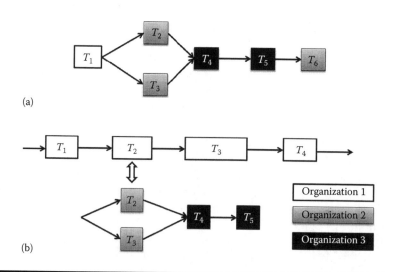

Figure 6.4   (a) Distributed workflow and (b) outsourced workflow.

semantics. This has played an important part in dealing with the technical issues regarding connecting systems, but it is limited in its ability to address some of the fundamental problems related to interorganizational workflows or highly dynamic workflows. Actually, the semantics and architectures should be defined before attempting to solve the technical issues (van der Aalst, 2000).

In modeling interorganizational workflows, many problems may occur. Most of them are related to exploring distributed architectures for combining workflows of business partners as well as to learning how to reconcile and reconstruct different workflow models. The reason that this is problematic is because the process information of each partner is hidden from other partners. While the partners are able to exchange transactional data, it is difficult for them to exchange the detailed process-level information, which is crucial for close interaction and cooperation in e-business. In spite of this, a stream of researches has begun to address these problems, which will be introduced later in the chapter.

6.4.1 Workflow Modeling between Organizations

Enterprise systems is becoming the new frontier of business information systems, leading to the trend of focusing on the cross-functional business processes that link various divisions and organizations. Most traditional workflow management systems assume one centralized enactment service and are able to support workflows inside an organization. For workflows crossing organizational boundaries, it is critical to ensure that problems such as inconsistency do not arise due to the complex communications across different organizations. The inherently hybrid nature of business processes, particularly as such processes spread across multiple organizational entities, presents real challenges. The concept of how workflows can be merged was proposed (Sun et al., 2006), in which the process of combining one workflow schema into another and removing redundant steps while keeping all of the necessary ones is critical. The current research is offering systematic approaches for building complex workflows from simple ones appropriately.

WMC has summed up various forms of interoperability and has suggested some standards on supporting workflow interoperability; however, more work is needed before it enables smooth interactions between organizations, as it aims to address only the technical issues and not the content of the coordination structure. Many researchers have been trying to explore architectures for combining the workflows of different partners as well as ways to coordinate the coexistence between them. In this section, these researches are classified into (1) interaction models, that is, to suggest patterns for cooperating partners to interact with each other; and (2) routing approaches, that is, to provide support for the routing of workflows across organizations in a standardized and flexible way. It should also be noted that this mainly refers to the information perspective only.

6.4.2 Interaction Models

In cross-enterprise collaboration, the extent to which each participating enterprise should expose its internal processes is an interesting research topic. On one side, each partner would not want to expose its processes, in order to preserve its privacy, autonomy, and competence; on the other side, successful collaboration among multiple enterprises obviously requires knowledge about each other. In general, the interorganizational workflows interconnect individual workflows, and the interfaces of such workflows are required to be known by the virtual enterprise. The Vega Project develops a platform to support the interaction of users and systems from different organizations that form a virtual enterprise (Schulz, 1998). It distributes a workflow service following a black-box approach in which the individual workflow models are not required to be made public to other partners in the virtual enterprise. The WISE Project proposes a framework for a virtual business process by using the process interfaces as provided by cooperating enterprises (Lazcano et al., 2000). It introduces a centrally coordinated approach in which interorganizational workflows are executed by a central workflow engine and collaborating partners do not need to communicate directly with each other. On the contrary, in the monitored–nested model, the sequence and states of activities within a provider's process are all observable by service consumers (Kuechler et al., 2001). Therefore, cooperating enterprises can obtain more process information about one another to improve coordination and efficiency. However, trading partners can see the whole structure of internal processes, since this model does not provide process abstraction.

In terms of the relationship between private and public workflows, selectable and scalable exposition of private workflows in a public context was proposed (Schulz, 2002). The intent is to leave a private workflow unchanged and to relate it to a process-based interface that can be adapted to satisfy specific business requirements. The workflow-view method provides a way to support the contradicting requirements of public visibility versus privacy mentioned earlier (Chiu et al., 2004). It is an approach to selectively hide the details of private workflows even as it allows interactions between trading partners in a gray box mode (i.e., they can access each other's internal information to some extent). A workflow can be described at different levels of granularity for its hierarchical composition; the use of granularity depends on the level of trust between involved organizations, which provides the opportunity for the creation of a workflow view, which, in turn, becomes a subset of the primary workflow. The access control at the view level is exercised according to the rights assigned to the roles played by the external partners. Based on the workflow-view approach, Schulz and Orlowska (2004) discuss the entities of a distributed workflow architecture that provides execution support for mediated and unmediated interorganizational workflows and argue that this architecture supports the view-based interorganizational workflow model. This architecture allows for business interactions between organizations without requiring them to have full

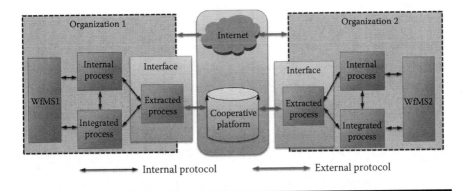

Figure 6.5 Architecture of view-based interorganizational workflow.

knowledge of the structure of each other's processes and without exposing confidential data to their partners.

With a resemblance, the process-view-based coordination model was proposed to address the issues of managing B2B workflows; a process to conceal sensitive information is developed while maintaining the parts essential for cooperation (Liu and Shen, 2003). This approach is executed through three phases: the base process phase, the process-view phase, and the integration phase. The process view, which is an aggregate abstraction of a base process, can be viewed as an external interface of an internal process. The process-view and workflow-view approaches are similar to a large extent (see Figure 6.5 for a general framework about view-based workflow modeling), in which both are motivated by views of object databases. The small difference found between them is that the process view is generated through order-preserving rules and aggregate abstraction, while the workflow view is derived as a subset of the workflow and depends on the access rights of the external partners. In order to construct a tailored process view on private processes for different partners, a customized process view was proposed in which activities from the noncustomized view that are not requested were hidden or omitted (Eshuis and Grefen, 2008) and in which noise from the process views provided can be removed.

The Crossflow project develops architecture for workflow interactions across organizations (Grefen et al., 2000). The Crossflow architecture, which is WfMS independent, builds upon proxy gateways that protect core services of each organization from exposure and introduces a business contract to manage the WfMS cooperation between service providers and consumers. There is no central instance that has a comprehensive view of multiorganizational business processes, even though there is a requirement for direct communication between partners to execute a coordinated workflow. Similarly, business contracts that encapsulate a formal commitment as a set of obligations and that coordinate and control the interaction between business workflows are presented (Weigand and Heuvel 2002). These contracts are statements of shared purpose that comprise mutual obligations and authorizations that reflect the legally binding agreements between trading partners. They are realized through the contract specification language (XLBC) to formally link the Component Definition Language specification of workflow systems. In another approach, workflow view is used as a fundamental support system for e-service workflow interoperability, then a contract model is proposed based on workflow views to simplify the process of developing interorganizational workflows regarding contracts (Chiu et al., 2002).

An approach based on MSCs and PNs is proposed to describe how PN-based workflows are used in a distributed environment. MSCs, with their functions including recording messages that are exchanged and ordering events associated with sending and receiving messages, are used to specify the interaction between organizations. As mentioned earlier, PNs are used to model intraorganizational workflows. In one method, the loosely coupled workflows are mapped into PNs, and the soundness and consistency with the MSC of the interorganizational workflows are checked through PN-based tools such as Woflan (van der Aalst, 2000).

A Common Open Service Market infrastructure is proposed for the enactment of interorganizational workflow (Merz et al., 1997). Its infrastructure comprises a number of mobile agents. The control flow within each agent is managed by a CPN-based workflow engine. The workflow automation through the agent-based reflective process approach is proposed for the service-based crossorganizational workflow framework, which applies agent technologies to problems related to dynamism and distribution of workflows (Blake and Gomaa, 2005). It considers not only the complex workflow-oriented interactions among services, but also the complex interactions of agents internal to the architecture, and it integrates current industry-standard development processes and modeling techniques. A mobile agent-based workflow management system was also proposed to support inter- and intraorganizational workflow management systems in a grid environment (Nichols et al., 2006) while allowing mobile agents to improve the capacity of WfMS by alleviating a recognized bottleneck in the architecture.

Table 6.2 shows the development of the interaction models. The future trend that can be seen is that each partner in the crossorganizational workflows would expose part of its process to its partners in order to facilitate the cooperation. Regarding how to extract a subset process from the internal process, many thoughts and mechanisms have been proposed, including the contract view, the process view, and the workflow view. Whatever mechanism is used, different views of the private process are presented to relevant partners at the interface side. In most of the current studies, these views are generally generated and disclosed by the provider without the involvement of external partners, with no choices but accept them passively. What if the provided view is not appropriate or is not enough? What if they require additional information about their partners? The work of Eshuis and Grefen (2008) made progress in this direction, although what

Table 6.2   Interaction Models

Interaction Model	Specification	Exposure	Reference
Vega model	It follows a black-box approach where the individual workflow models are not required to be made public to other partners	No exposure	Schulz (1998)
WISE model	The individual workflows do not connect with each other directly, but are linked with interorganizational workflows, which are executed by central engine	No exposure	Lazcano et al. (2000)
Monitored–nested model	Trading partners can see the entire structure of each other's internal processes to improve coordination efficiency	Total exposure	Kuechler et al. (2001)
Crossflow model	It introduces business contract to manage cooperation between participating partners, but still requires direct communication to execute crossorganizational workflow	Contracted exposure	Grefen et al. (2000); Weigand and van den Heuvel (2002)
Workflow-view model	It selectively hides details of private workflows and allows the interactions between trading partners under access control	Controlled access	Chiu et al. (2002, 2004)
Process-view model	An aggregate abstract of internal process is viewed as the interface to conceal sensitive information, but to reveal the essential parts for cooperation	Aggregate abstract	Liu and Shen (2003, 2004)

they propose to do in the phase of customization is only to hide and combine some redundant activities in the process view.

In the view-based interaction models, the views are generated from the standpoint of the providers, without the knowledge of the requirements of the external partners (Liu and Shen, 2003, 2004). The requesting partners should also play active, rather than passive, roles in the interaction, since what is provided in the process view is not enough for an effective interaction. Then an emphasis needs to be placed on determining how to mirror the opinions and requests in the interface.

WMC and related literature have presented a few forms of workflow interoperation to be used between organizations; however, the existing interaction models have not paid enough attention to the forms of interorganizational workflow. It is desirable to link the implementation strategy with the interoperation form, as well as to develop specified interaction models for certain forms.

6.4.3 Routing Approaches

An interenterprise system architecture based on the Internet was proposed (Liu et al., 2006). It is mainly composed of a few workflow-supported SCM systems in different organizations that are connected to each other through some form of integrated interfaces. The integrated interface uses HTML to transfer messages between enterprises; the messages are coded in XML. HTML acts as the standard for transferring web pages between user interfaces; XML serves as the standard for exchanging data and semantic information between organizations. Possible future improvements include a standardized or homogeneous language to route process information across organizations, as support for routing is a major gap existing in most of the architecture for interorganizational workflows.

To address this problem, the idea of a routing slip was proposed (Kumar and Zhao, 2002). The routing slip is a simple sequence of addresses to which a document must be sent. It is described or specified using XRL and is stored in the data type definition (DTD) file. Complementing the stream of research, a framework for implementing an interorganizational workflow was proposed, which includes an XRL on the basis of eXtensible Markup Language (XML) syntax (van der Aalst and Kumar, 2003), in order to provide support for the routing of workflow information among partners for Internet-based e-business services. XRL utilizes XML syntax to define DTD, which consists of a set of elements or tags for describing workflow applications. Furthermore, a prototype workflow management system named XRL/flower, based on both XML and PNs, was developed. XML tools can be deployed to parse, check, and handle XRL documents, which are automatically translated into PN constructs. Standard Generalized Markup Language (SGML) is used to capture the process of generating and manipulating structured documents based on the SGML standard (Weitz, 1998). SGML net also has a PN structure as a base layer and uses document templates to manipulate document routing in the workflow.

A framework for managing dynamic routing in workflow applications was proposed, in order to overcome the rigid and exacting style of routing (Kumar and Zhao, 1999). The framework is composed of three techniques: (1) the workflow control tables specify absolute static authorizations that govern the rights of access to documents; (2) the sequence constraints specify the temporal order in which documents are accessed, as well as the ability to override the workflow control tables; (3) the event-based rule supersedes both the workflow control table and the sequential constraint and represents a sophisticated way of managing the workflow and handling of special situations. The proposed model is complex and appropriate for an interorganizational environment, which is characterized by different types of business processes and involves flexible routing patterns.

6.5 Workflow Analysis

After defining the process specifications that capture interactions and routing details between components in the workflows, it is vital to ensure the correctness and efficiency of the workflows. An erroneous workflow may lead to many problems. The later that an error is detected, the more it will cost to correct it. The problems resulting from an incorrect interorganizational workflow are even more difficult to recover because of the distributed control issue. Hence, it is useful to analyze each workflow thoroughly, prior to its implementation. Workflow analysis is differentiated between qualitative analysis and quantitative analysis.

6.5.1 Qualitative Analysis

Qualitative analysis is mainly concerned with the logical correctness of the defined process; that is, the absence of anomalies such as deadlocks (i.e., a case is blocked and no longer proceeds through the process) and livelocks (i.e., a case becomes stuck in a never-ending loop). The main objective of qualitative analysis is to prove that the model is valid and to find answers for questions such as the following: Does each process end improperly? Is the sequence of the states correct? Are there any nonsolvable conflicts for allocating resources to different tasks (Salimifard and Wright, 2001)?

The qualitative analysis is also seen as a structural analysis. It is generally found in two aspects: validation and verification. Validation means testing for semantic completeness to ensure that the workflow behaves as it was intended to do. An approach based on integer programming was proposed to check whether the modeled behavior through PNs and the observed behavior in the frequency profile could match each other (van der Aalst, 2006). By providing necessary conditions for the match, research can determine whether the workflow modeled with PNs is validated. Verification requires checking the syntactic soundness of a workflow and eliminating redundancies. By doing this, some typical problems in workflow specifications can be identified, and their complexity can be addressed. Among these are initiation problems, termination problems, and equivalence problems. A method of using PN-based analysis techniques to verify the typical process control specification was proposed. To illustrate the applicability of the approach, a verification tool called Woflan was developed. It has also been applied to the interorganizational contexts (Verbeek et al., 2004). It combines the techniques of XRL and Woflan together to create a powerful tool set for designing, verifying, and implementing interorganizational workflows. The XRL/Woflan uses XSLT to transform XRL specifications to specific PNs. By doing this, it allows users not only to design routing constructs but also to determine whether the workflow is correct based on the PN representation.

Another stream of qualitative analysis focuses on the task-allocation mechanism in a workflow system. Generally, there are two basic mechanisms available: the push mechanism wherein a work item is pushed to a single resource, and the pull mechanism wherein a resource pulls items from the pool of work items. Most state-of-the-art workflow systems such as Staffware consider the pull mechanism as the basic paradigm to prevent unknown or unanticipated situations. However, even as it provides more flexibility, the pull mechanism may lead to inefficiency. To address this problem, an approach was proposed that allows balancing of quality and performance; that proposed mechanism also allows to fine-tune the trade-off between push and pull strategies according to the needs of the application (Kumar et al., 2002). An adaptive transaction protocol for the pervasive workflow management system that provides the dynamicity and flexibility required in the interorganizational environment was also proposed (Montagut et al., 2008). This concept relies on business partners that are assigned to tasks using an algorithm wherein the selection process is based on functional and transactional requirements that are specified in the protocol and defined at the workflow design phase.

6.5.2 Quantitative Analysis

Quantitative analysis is mainly concerned with the performance of the defined process. It focuses on calculating the performance indices such as average completion, level of service, and resource utilization. The technique mainly used is simulation through experimenting with the specified workflow under the assumption of a specific behavior of the environment; as a result, some of the key performance indicators can be estimated. It also proves to be a good way of validating the initial measurements. A place/transition model was proposed for studying the performance of the process (Desel and Erwin, 2000). Through simulation, the extended model can obtain some performance indices associated with time or cost. An approach integrating simulation modeling and analysis capabilities within the workflow management system was developed to study the general performance and reliability of WfMS (Miller et al., 1995). However, some researchers point out that some challenges have to be faced when it comes to mapping an administrative business process onto a simulation model, for example, resources are not available all the time (Reijers and Mansar, 2005).

Another technique for performance analysis is the Markov Chain. A Markov decision process was proposed to model web-based workflow composition (Doshi et al., 2005). The model contains the possible states of a case and the probabilities of transitions between them. Thus, various properties are established using this approach (e.g., what are the chances of a case taking a particular route through a process?). By expanding the Markov chains with cost and time aspects, a range of performance indicators can be generated. The queuing theory, in which the emphasis is placed upon performance indicators as waiting times, completion times, etc., can also be used for the analysis of workflow systems. Workflow performance can be analyzed through queuing models such as the layered queuing model, which represents software and hardware components and their interactions at multiple layers in a predictive model.

A method called multidimensional workflow-net was proposed. This method incorporates multiple timed workflow-nets (TWF-net) with organizational and resource information to model workflows (Li et al., 2004). To facilitate performance analysis, a decomposition algorithm was developed to decompose the TWF-net into a set of subsets, each of which describes the routing path of a type of transaction instances. By doing this, some performance indices (such as lower bound of average turnaround time) can be easily calculated. Soundness is an essential property of interorganizational logical time workflow nets (ILTWNs) and logical time workflow nets (LTWNs) to indicate the correctness of ILTWNs and LTWNs and can be used (1) to verify whether the structure and dynamic behavior of ILTWNs and LTWNs are consistent with the requirement specifications of the modeled system and (2) to test whether the system terminates at an acceptable state. As determining soundness is EXPSPACE hard (van der Aalst, 1999), soundness has been explored for its subclass as T-fair ILTWNs (Liu et al., 2012). The results reduce the complexity of the analysis based on the static structure of ILTWNs.

Comparatively, quantitative analysis is relatively more difficult to enact before the workflow is actually implemented, since the workflow systems are relatively complex and can be far too dynamic to be predicted and described precisely. However, based on the results of the analysis, some measures for improving the workflow can be suggested.

6.5.3 Empirical Study

As a totally different methodology, empirical study can also be used to analyze the performance of workflow systems. With a quality method of data collection, Reijers and Mansar (2005) evaluate the performance of workflow systems, which are implemented in 16 different business processes of 6 organizations, with respect to lead time, wait time, service time, and utilization of resources. The results show that WfMS positively affects the identified performance indicators. Survey questionnaires of users' opinions concerning workflow have also been used to evaluate workflow systems. The findings reveal the degree of user satisfaction concerning the effects of the WfMS system on workflow.

6.6 Future Directions

In terms of modeling and analyzing workflows across organizational boundaries, additional problems can be encountered, which lead to interesting research questions. Although many studies have been conducted in the area, much work still remains to be done. In this section, some research opportunities in two main categories are summarized and discussed: technical and managerial aspects.

6.6.1 Technical Aspects

1. *Improvement in modeling techniques*: Results of the evaluations of some widely used types of modeling techniques show that each has strengths as well as weaknesses. For example, PNs provide easy-to-use graphical capability and strong mathematical foundations with supporting tools; however, they are mainly used in modeling the control flow perspective. Without formal semantics, KWM introduces a set of cognitive and flexible rules that can be used to determine which activities are to be executed by which resource (Lee et al., 1999). Although many modeling techniques work smoothly on a theoretical basis, it can be observed that there is a gap in practice: Some conceptual models are hardly ever used in practical implementation. There can be several explanations for such gap between conceptual modeling and practical implementation: either the divergent terminology uses a different degree of technical details or the heterogeneous modeling methods that were proposed are not in use. In the long run, these issues can be resolved, at least partially through standardizing terminology and developing more systematic models. Traditional methods for improving business process modeling such as workflow mining and process retrieval still require much manual work. To address this, based on the structure of a business process, a method called workflow recommendation technique was proposed to provide process designers with support for automatically constructing the new business process that is under consideration (Li et al., 2014). With the help of the minimum depth-first search (DFS) codes of business process graphs, an efficient method for calculating the distance between process fragments and selecting candidate node sets for recommendation purpose was developed. In addition, a recommendation system for improving the modeling efficiency and accuracy was implemented. Experiment results have proved its effectiveness in practical applications. An important and interesting research issue would be to develop next-generation modeling techniques that can maximize existing methods' strengths while avoid weaknesses. Of course, this is challenging.

2. *Multiple perspectives/dimensions in modeling efforts*: Most workflow models are developed based on different perspectives such as the control flow perspective, the data/information perspective, the resource perspective, and the

operational perspective. The existing workflow models are typically control flow oriented, while real-world workflows tend to be multiple perspective oriented. Survey research also indicates that businesses choose to combine one or more perspectives for their modeling, analysis, and design efforts. Although it has been realized that multiple perspectives/dimensions are important, however, few existing techniques have been developed to integrate perspectives/dimensions. However, combining some of the aforementioned perspectives/dimensions and/or proposing a hybrid modeling approach have been given attention in recent years (Wieczorek et al., 2009).

3. *WfMS as a vehicle for enterprise integration*: As one of the components of complex enterprise systems, interorganizational WfMS systems offer the tools for enterprise integration, such as the facilitation of the activities of multiple enterprises to be streamlined, controlled, and monitored in order to prepare for integration. What are expected are the advanced techniques for integrating distributed workflows semantically, logically, and physically. More technical and business parameters must be included for building comprehensive workflows.

4. *Resilient WfMS for exceptional handling*: Resilience is a concept that is rooted in socioecological systems and systems science. It has been applied in different areas and called engineering resilience, ecological resilience, etc. (Holling, 1973; Zhang, 2010). Due to the distribution of individual WfMSs, more unforeseen exceptions could possibly occur (e.g., nonrecoverable damage to one partner due to a natural disaster). An interorganizational workflow management system should incorporate a resilient mechanism to establish a robust cooperative environment. For the purpose of interorganizational workflow modeling and the verification of incident command systems, a system called WIFA was proposed (Rosca and Wang, 2007). However, more research would need to be pursued. The goal is to apply the concept of resilience to designing workflow management systems.

5. *Batch processing function and passing value indeterminacy*: Many existing methods are unable to describe the batch processing function of cooperative systems or model passing value indeterminacy (Liu et al., 2012). Batch processing function means that a batch of data of the same type from different organizations needs to be processed before a specified deadline. Passing value indeterminacy means that all arrival data be processed at the same time, and late arrival be processed in the next workflow cycle; meanwhile, the efficiency of the workflow must be optimized. Research indicates that formal models of real-time cooperative systems need to be developed in order to model and analyze both the batch processing function and the passing value indeterminacy.

6. *Modeling and analysis of workflow linked to SCM*: In order to promote the competitive advantage of SCM, features such as leanness and agility are considered important. Currently, the majority of workflow systems are developed

from the perspective of information modeling without considering the actual requirements of SCM. The gap between the modeling of workflows and supply chain characteristics is substantial. Recently, a method for modeling and analysis of interorganizational workflow systems in the context of LSCs using PNs was proposed (Ma et al., 2011). This method describes the assumed conditions of interorganizational workflow net according to the idea of LSC and then discusses the standardization of collaborating business process between organizations in the context of LSC. The concept of LTPNs was defined through combining labeled PNs with time PNs. The concept of LTWNs was also defined based on LTPNs. CLTWNs were then defined based on LTWNs. The notion of OR-silent CLTWNS and a verifying approach were proposed for the soundness of LTWNs and CLTWNs. The purpose of this research was to establish a formal method for the modeling and analysis of workflow systems for LSC. This study initiates a new perspective of research on interorganizational workflow management and promotes operation management of LSC in real-world settings.

7. *System complexity*: As executing complex processes in heterogeneous computing environments is a major characteristic of SCM, we recognize that the size and the complexity of workflow systems will be escalating as more supply chain–wide interorganizational workflows are implemented. As a result, more dynamic changes need to be dealt with. Research is needed to develop new workflow methods that will prove to be workable in practice. This may be achieved through applying systems theory. Workflow technology will continue to improve as system complexity is being explored.

8. *New technology*: Although workflow technology has been in use since its development in the 1990s, recently, there has been an increasing momentum for the research and development of WfMS and their applications, due to the increasing demand fueled by the underlying advances in computing technologies, notably cloud computing, grid computing, and service computing. Grid computing involves coordinating and sharing computing resources across the web globally. As such, it makes the computing power that was previously available only to selective organizations available to almost everyone. New workflow tools have been developed in a grid environment. Cloud and grid computing enable users to use workflow tools to share data and resources. Workflow scheduling on the grid becomes challenging. Scheduling of workflows on the grid is a complex optimization problem.

9. *Discrepancies between the actual workflow and the design workflow*: As a component of enterprise systems, WfMS are configured on the basis of workflow models. There are discrepancies between the workflows that were designed and included in the systems and the workflows perceived in practice. Although WfMSs offer generic models describing workflow cases that can be configured to support specific processes, more efforts are needed to reduce the discrepancies between the actual workflow and the design workflow.

10. *Scientific workflow management*: The term "scientific workflows" was first coined by Vouk and Singh (1997) in 1996. In recent years, scientific workflows have emerged as a new tool for scientific researchers to use to study complex scientific processes for enabling and accelerating many scientific discoveries. Similar to the workflow is the computerized facilitation of business or industrial processes, a scientific workflow is the computerized facilitation or automation of a scientific process. A scientific workflow management system defines, executes, monitors, and manages scientific workflows in which the execution order is driven by a logical representation of a scientific workflow in a particular domain. Although scientific workflows leverage existing workflow techniques, they deviate from them, as well; since they offer a different set of requirements raised from a wide range of science and engineering domains, due to the intertwining involvement of science and engineering in business and industry. However, the integration of techniques in scientific workflows and industrial workflows will become a new research frontier in *IIIE*.

6.6.2 Managerial Aspects

1. *WfMS as a platform for knowledge sharing and learning*: The knowledge workers in each organization are encouraged to perform creative intellectual activities, which allow for dynamic changes to be captured and stored in a WfMS. As such, the WfMS can be considered as a repository of valuable process knowledge and can act as a vehicle for the collection and distribution of that knowledge. Also, an additional workflow needs to be constructed to realize the benefits of knowledge sharing in the crossorganizational context.
2. *Performance analysis*: In contrast to qualitative analysis, quantitative analysis has received limited attention even though performance analysis results have proven to be vital in determining the efficiency of the constructed workflow. The systems methods related to the process based on a systems approach can be developed. Thus, some decisions can be made about whether the process can be implemented or whether the process needs improvement.
3. *Social and organizational environment*: We recognize that the ultimate success of a workflow system is highly dependent upon the social and organizational environment in which the WfMS is being implemented.
4. *Factors leading to the adoption of WfMS*: Installing a WfMS inevitably changes the business process and work activities; therefore, it can have a major impact on the organization's people and organizational culture. Although there have been a few studies declaring the positive impact of WfMS, it may also create negative impacts, including overly rigid procedures and a reduction in the motivation of workers. Therefore, it is of great significance to explore the successful factors leading to the adoption of WfMS. By doing so, some measures can be taken to increase the acceptance of WfMS in organizations.

5. *Incentive mechanism for participating partners*: In an interorganizational work-flow, each partner is expected to complete its workflow requirements well. In addition, to elevate the efficiency of the overall workflow, participants are required to cooperate more actively. Appropriately designed incentive mechanisms need to be in place to provide stimulation.

6.7 Summary

Workflow systems have been considered efficient tools that enable the automation of organizational business processes. In an attempt to maximize efficiency and effectiveness in various industry sectors, increasing importance and attention is being placed on workflow technology. This change takes on yet greater importance as more organizations are moving toward using workflow technology. This chapter's topic is bringing workflow research to a wider community related to enterprise systems. With the emergence of e-business and e-commerce, one important trend continues to appear: workflows should be operated across multiple organizations so that companies have the opportunity to reshape their business processes beyond their organizational boundaries. This chapter introduces the characteristics of workflows as well as selected methods and techniques used in modeling organizational workflows. The techniques used for workflow modeling are classified and discussed. Each of these techniques has its own strengths as well as weaknesses. For example, PNs are a well-established process modeling tool, based on their solid mathematical foundation and their graphical representation capability. With the development of PNs, several of their extensions have been developed such as time (Wang, 1998), colored, and stochastic PNs (Kusiak and Yang, 1992). However, it is difficult to apply PNs to describe batch processing functions and the indeterminacy situation. As batch data with the same type of information from different organizations need to be processed before a specified deadline, and the arrival times of these data are indeterminate, PNs with weights of flow arcs may describe the batch processing function, but cannot easily represent the data arrival indeterminacy as the weights are not less than 1 (Du and Guo, 2009). PNs with inhibitor arcs (IPNs) have the modeling power of a Turing machine (Suzuki and Lu, 1989). As such, although IPNs are able to model the batch processing function and the data arrival indeterminacy, the models of systems with these properties are very complex (Du and Guo, 2009).

Although there have been many formalisms in the approach, there has not been a standardization in the area of workflow modeling. Interorganizational workflow modeling delves deeper into exploring the architectures for combining different organizational workflows, while it continues to reconcile the relationships between each of the systems. Some approaches are specifically proposed for modeling inter-organizational workflows, which are introduced in this chapter from two separate aspects: the first is the interaction model and the second is the routing approach.

These methods emphasize the need for providing support for workflow within and among organizations. Some cognitive approaches are suggested to support the dynamic routing of information. New languages (e.g., XRL and SGML) are proposed to handle the routing of information among organizations. It also appears that a selective, yet partial, exposure of private workflow data to external partners would be an interesting research topic. As for how to extract these data from internal processes, more research would need to be pursued.

Qualitative, quantitative, and empirical methods for analyzing workflows have been introduced in this chapter. The qualitative analysis mainly focuses on checking the structural soundness, which can usually be done through the validation and verification of workflows. The quantitative analysis requires the calculation of the performance indices related to the workflows. The existing techniques include computational simulation, Markovian chain, and queuing theory. Finally, some future research topics are explored in both the technical and managerial aspects.

Workflows have been proving their success in automating business and industrial processes. In the future, we can see that large-scale workflow systems will become more complex. This means that research on ES and its component WfMS is becoming more and more important to lead industries in technological directions that are more efficient, effective, and competitive.

References

Barros, A. P. and ter Hofstede, A. H. M. 1998. Toward the construction of workflow-suitable conceptual modeling techniques. *Information Systems Journal*, 8(4), 313–337.

Basu, A. and Blanning, R. W. 1994. Model integration using metagraphs. *Information Systems Research*, 5(3), 195–218.

Basu, A. and Blanning, R. W. 2000. A formal approach to workflow analysis. *Information Systems Research*, 11(1), 17–36.

Basu, A. and Kumar, A. 2002. Research commentary: Workflow management issues in e-business. *Information Systems Research*, 13(1), 1–14.

Blake, M. B. and Gomaa, H. 2005. Agent-oriented compositional approaches to services-based cross-organizational workflow. *Decision Support Systems*, 40(1), 31–50.

Bons, R., Lee, R., and Wagenaar, R. 1998. Designing trustworthy interorganizational trade procedures for open electronic commerce. *International Journal of Electronic Commerce*, 2(3), 61–83.

Chinn, S. J. and Madey, G. R. 2000. Temporal representation and reasoning for workflow in engineering design change review. *IEEE Transactions on Engineering Management*, 47(4), 485–492.

Chiu, D. K. W., Cheung, S. C., Tills, S., Karlapalem, K., Li, Q., and Kafeza, E. 2004. Workflow view driven cross-organizational interoperability in a web service environment. *Information Technology and Management*, 5(3–4), 221–250.

Chiu, D. K. W., Karlapalem, K., Li, Q., and Kafeza, E. 2002. Workflow view based e-contracts in a cross-organizational e-services environment. *Distributed and Parallel Databases*, 12(2), 193–216.

Cull, R. and Eldabi, T. 2010. A hybrid approach to workflow modelling. *Journal of Enterprise Information Management*, 23(3), 268–281.

Dalal, N. P., Kamath, M., Kolarik, W. J., and Sivaraman, E. 2004. Toward an integrated framework for modeling enterprise processes. *Communications of the ACM*, 47(3), 83–87.

Desel, J. and Erwin, T. 2000. Modeling, simulation and analysis of business processes. In van der Aalst, W. M. P. et al. (Eds.), *Business Process Management. Lecture Notes Computer Science 1806*. Heidelberg, Germany: Springer-Verlag, pp. 129–141.

Doshi, P. et al. 2005. Dynamic workflow composition using Markov decision processes. *International Journal of Web Services Research*, 2(1), 1–17.

Du, Y. Y., Guo, B. Q. 2009. Logic petri nets and equivalency. *Information Technology Journal*, 8(1), 95–100.

Du, Y. Y., Jiang, C. J., and Zhou, M. C. 2007. Modeling and analysis of real-time cooperative systems using Petri nets. *IEEE Transactions on System Man Cybernetics Part A: Systems and Humans*, 37(5), 643–654.

Ellis, C. 1999. Workflow technology. In Beaudouin-Lafon, M. (Ed.), *Computer Supported Co-operative Work* (Trends in Software). Wiley & Sons, New York, NY, pp. 29–54.

Eshuis, R. and Grefen, P. 2008. Constructing customized process views. *Data & Knowledge Engineering*, 64(2), 419–438.

Georgakopoulos, D., Hornick, M., and Sheth, A. 1995. An overview of workflow management: From process modeling to workflow. *Distributed and Parallel Databases*, 3(2), 119–153.

Glassey, O. 2008. A case study on process modeling-three questions and three techniques. *Decision Support Systems*, 44(4), 842–853.

Grefen, P., Aberer, K., Ludwig, H., and Hoffner, Y. 2000. Cross-flow: Cross-organizational workflow management in dynamic virtual enterprise. *International Journal of Computer Systems Science & Engineering*, 15(5): 277–290.

Gudes, E. and Tubman, A. 2002. AutoWF: A secure web workflow system using autonomous objects. *Data & Knowledge Engineering*, 43(1), 1–27.

Hidders, J. et al. 2008. DFL: A dataflow language based on Petri nets and nested relational calculus. *Information Systems*, 33(3), 261–284.

Holling, C. S. 1973. Resilience and stability of ecological systems. *Annual Review of Ecology and Systematics*, 4, 1–23.

Jiang, P., Shao, X., Qiu, H., and Li, P. et al. 2008. Interoperability of cross-organizational workflows based on process view for collaborative product development. *Concurrent Engineering*, 16(1), 73–86.

Keller, G., Nuttgens, M., and Scheer, A. W. 1992. Semantic process modeling on the basis of event-driven process. Technical Report, University of Saarland, Saarbrücken, Germany.

Kim, J., Suh, W., and Lee, H. 2002. Document-based workflow modeling: A case-based reasoning approach. *Expert Systems with Applications*, 23(2), 77–93.

Kim, K. and Ellis, C. 2001. Performance analytic models and analyses for workflow architectures. *Information Systems Frontiers*, 3(3), 339–355.

Kuechler, W., Vaishnavi, V. K., and Kuechler, D. 2001. Supporting optimization of business-to-business e-commerce relationships. *Decision Support Systems*, 31(3), 363–377.

Kumar, A., van der Aalst, W. M. P., and Verbeek, E. M. W. 2002. Dynamic work distribution in workflow management systems: How to balance quality and performance. *Journal of Management Information Systems*, 18(3), 157–193.

Kumar, A. and Zhao, J. 1999. Dynamic routing and operational controls in workflow management systems. *Management Science*, 45(2), 253–272.

Kumar, A. and Zhao, J. 2002. Workflow support for electronic commerce applications. *Decision Support Systems*, 32(3), 265–278.

Kusiak, A. and Yang, H. 1992. Modeling design cycles with stochastic Petri nets. *Proceedings of Winter Annual Meeting of the American Society of Mechanical Engineers*, Anaheim, CA.

Lawrence, P. 1997. *Workflow Handbook 1997*. Chichester, U.K.; New York: Wiley & Sons.

Lazcano, A., Alonso, G., Schuldt, H., and Schuler, C. 2000. The WISE approach to electronic commerce. *International Journal of Computer Systems Science and Engineering*, 15(5), 343–355.

Lee, H. B., Kim, J. W., and Park, S. J. 1999. KWM: Knowledge-based workflow model for agile organization. *Journal of Intelligent Information System*, 13, 261–278.

Leymann, F. and Roller, D. 2000. *Production Workflow: Concepts and Techniques*. Upper Saddle River, NJ: Prentice Hall.

Li, J., Fan, Y. S., and Zhou, M. 2003. Timing constraint workflow nets for workflow analysis. *IEEE Transactions on Systems, Man, and Cybernetics-Part A: Systems and Humans*, 33(2), 179–193.

Li, J., Fan, Y. S., and Zhou, M. C. 2004. Performance modeling and analysis of workflow. *IEEE Transactions on Systems, Man, and Cybernetics-Part A: Systems and Humans*, 34(2), 229–242.

Li, Y., Cao, B., Xu, L., Yin, J., Deng, S., Yin, Y., and Wu, Z. 2014. An efficient recommendation method for improving business process modeling. *IEEE Transactions on Industrial Informatics*, 10(1), 502–513, 2014.

Liu, D. R. and Shen, M. X. 2003. Workflow modeling for virtual processes: An order-preserving process-view approach. *Information Systems*, 28(6), 505–532.

Liu, D. R. and Shen, M. X. 2004. Business-to-business workflow interoperation based on process-views. *Decision Support Systems*, 38(3), 399–419.

Liu, D. S. et al. 2002. Modeling workflow processes with colored Petri nets. *Computers in Industry*, 49(3), 267–281.

Liu, J. X., Zhang, S. S., and Hu, J. M. 2006. A case study of an inter-enterprise workflow-supported supply chain management system. *Information and Management*, 42(3), 441–454.

Liu, W., Du, Y., and Yan, C. 2012. Soundness preservation in composed logical time workflow nets. *Enterprise Information Systems*, 6(1), 95–113.

Ma, J., Wang, K., and Xu, L. 2011. Modelling and analysis of workflow for lean supply chains. *Enterprise Information Systems*, 5(4), 423–447.

Merz, M., Liberman, B., and Lamersdorf, W. 1997. Using mobile agents to support inter-organizational workflow management. *International Journal of Applied Artificial Intelligence*, 11(6), 551–572.

Miller, J. et al. 1995. Simulation modeling within workflow technology. *Proceedings of 1995 Winter Simulation Conference*, Arlington, VA, pp. 612–619.

Montagut, F., Molva, R., and Tecumseh Golega, S. 2008. The pervasive workflow: A decentralized workflow system supporting long-running transactions. *IEEE Transactions on Systems, Man and Cybernetics Part C-Applications and Reviews*, 38(3), 319–333.

Nichols, J., Demirkan, H., and Goul, M. 2006. Autonomic workflow execution in the grid. *IEEE Transactions on Systems, Man and Cybernetics Part C-Applications and Review*, 36(3), 353–364.

Petri, C. A. 1962. Kommunication mit Automaten, PhD thesis, Bonn University, Bonn, Germany.

Popova, V. and Sharpanskykh, A. 2008. Process-oriented organization modeling and analysis. *Enterprise Information Systems*, 2(2), 161–193.

Prisecaru, O. 2008. Resource workflow nets: An approach to workflow modeling and analysis. *Enterprise Information Systems*, 2(2), 101–124.

Prisecaru, O. and Jucan, T. 2008. Interorganizational workflow nets: A Petri net based approach for modelling and analyzing interorganizational workflows. *Proceedings of the EOMAS Workshop*, June 16–17, Montpellier, France, pp. 64–78.

Pudhota, L. and Chang, E. 2005. Coloured Petri net for flexible business workflow modeling. *Proceedings of Third IEEE International Conference on Industrial Informatics*, New York, NY, pp. 633–638.

Reijers, H. A. and Mansar, S. L. 2005. Best practices in business process redesign: An overview and qualitative evaluation of successful redesign heuristics. *Omega*, 33(4), 283–306.

Rosca, D. and Wang, J. 2007. Inter-organizational workflow modeling and analysis of incident command systems. *Proceedings of IEEE International Conference on Network, Sensing and Control*, London, U.K., pp. 218–223.

Salimifard, K. and Wright, M. 2001. Petri net-based modeling of workflow systems: An overview. *European Journal of Operational Research*, 134(3), 664–676.

Schulz, K. 2002. Modeling and architecting of cross organizational workflows, PhD thesis, University of Queensland, Brisbane, Queensland, Australia.

Schulz, K. 1998. Implementation of the distributed workflow service. Technical report, ESPRIT 20408, Vega D303, 1998.

Schulz, K. and Orlowska, M. E. 2004. Facilitating cross-organizational workflows with a workflow view approach. *Data & Knowledge Engineering*, 51(1), 109–147.

Sun, H. and Du, Y. 2008. Soundness analysis of inter-organizational workflows. *Information Technology Journal*, 7, 1194–1199.

Sun, S. X., Zhao, J., and Nunamaker, J. F. 2006. Formulating the data-flow perspective for business process management. *Information Systems Research*, 17(4), 374–391.

Suzuki, I. and Lu, H. 1989. Temporal Petri nets and their application to modeling and analysis of a Handshake Daisy Chain Arbiter. *IEEE Transactions on Computers*, 38(5), 696–704.

Tsai, A., Wang, J., Tepfenhart, W., and Rosca, D. 2006. EPC workflow model to WIFA model conversion. *Proceedings of IEEE International Conference on Systems, Man, and Cybernetics*, 2006, Taipei, Taiwan, pp. 2758–2763.

van der Aalst, W. M. P. 1998. The application of Petri nets to workflow management. *The Journal of Circuits, Systems and Computers*, 8(1), 21–66.

van der Aalst, W. M. P. 1999. Interorganizational workflows: An approach based on message sequence charts and Petri nets. *Systems Analysis, Modelling, Simulation*, 34(3), 335–367.

van der Aalst, W. M. P. 2000. Loosely coupled interorganizational workflows: Modeling and analyzing workflows crossing organizational boundaries. *Information and Management*, 37(2), 67–75.

van der Aalst, W. M. P. 2003. Inheritance of interorganizational workflows: How to agree to disagree without losing control? *Information Technology and Management*, 4(4), 345–389.

van der Aalst, W. M. P. 2006. Matching observed behavior and modeled behavior: An approach based on Petri nets and integer programming. *Decision Support System*, 42(3), 1843–1859.

van der Aalst, W. M. P. and Kumar, A. 2003. XML-based schema definition for support of interorganizational workflow. *Information Systems Research*, 14(1), 23–46.

van der Aalst, W. M. P. and ter Hofstede, A. H. M. 2000. Verification of workflow task structures: A Petri-net-based approach. *Information System*, 25(1), 43–69.

van der Aalst, W. M. P. and ter Hofstede, A. H. M. 2005. YAWL: Yet another workflow language. *Information System*, 30(4), 245–275.

van der Aalst, W. M. P. and van Hee, K. 2004. *Workflow Management: Models, Methods, and Systems*. Cambridge, MA: MIT Press.

Verbeek, H. M. W., van der Aalst, W. M. P., and Kumar, A. 2004. XRL/Woflan: Verification and extensibility of an XML/Petri-net-based language for inter-organizational workflows. *Information Technology and Management*, 5(1–2), 65–110.

Viriyasitavat, W., Xu, L., and Martin, A. 2012. SWSpec: The requirement specification language in service workflow environments. *IEEE Transactions on Industrial Informatics*, 8(3), 631–638.

Vouk, M. and Singh, M. 1997. Quality of service and scientific workflows. In *Quality of Numerical Software: Assessment and Enhancements*, Boisvert, R. (Ed.), Springer, New York, pp. 77–89.

Wang, J. 1998. *Time Petri Nets: Theory and Application*. Boston, MA: Kluwer Academic Publishers.

Wang, T., Vonk, J., and Grefen, P. 2008. Towards a contractual approach for transaction management. *Enterprise Information Systems*, 2(4), 443–458.

Wang, W. and Wang, H. 2006. From process logic to business logic-a cognitive approach to business process management. *Information and Management*, 43(2), 179–193.

Weigand, H. and van den Heuvel, W. J. 2002. Cross-organizational workflow using contracts. *Decision Support System*, 33(3), 247–265.

Weitz, W. 1998. Workflow modeling for Internet-based commerce: An approach based on high level Petri nets. *Lecture Notes in Computer Science*, Springer, New York, Vol. 1402, pp. 166–178.

Wieczorek, M., Hoheisel, A., and Prodan, R. 2009. Towards a general model of the multicriteria workflow scheduling on the grid. *Future Generation Computer Systems*, 25(3), 237–256.

Workflow Management Coalition. 2000. Workflow standard interoperability Wf-XML binding, WFMC-TC-1023, Version 1.0.

Xu, L., Viriyasitavat, W., Ruchikachorn, P., and Martin, A. 2012. Using propositional logic for requirements verification of service workflow. *IEEE Transactions on Industrial Informatics*, 8(3), 639–646.

Yen, V. 2007. A node-centric analysis of metagraphs and its applications to workflow models. *Enterprise Information Systems*, 1(2), 139–159.

Zisman, M. D. 1977. Representation, specification and automation of office procedures, PhD thesis, University of Pennsylvania, Wharton School of Business, Philadelphia, PA.

Zhang, W. 2010. Guest editor's foreword. *Enterprise Information Systems*, 4(2), 95–97.

Chapter 7

Enterprise Information Integration Modeling and Integrating Information Flows

7.1 Data and Information Integration

The growing complexity of business environments poses challenges to any enterprise seeking to maintain competitiveness in today's rapidly changing and evolving environment. It used to be that automated SCM was the purview only of very large organizations. However, today's businesses of all sizes need to share data with customers, suppliers, and distributors. As such, more and more enterprises are embarking upon both intra- and interorganizational business initiatives, such as PDM, PLM, CIM, and ESs. Enterprises are increasingly working together; thus, the demand for supply chain integration is rising. Such integration has been made possible by the advancement of information technologies, especially information integration techniques. At present, information integration is significant not only for large-scale enterprises or for supply chain integration but also at a microscopic level. For example, information integration is one of the most important aspects of maintaining high efficiency in product development.

An intraorganizational ES is aimed at providing a higher-level system, which is related to activities that involve the coordination of business processes within the organization and which is able to provide an integrated architecture to the organization for enhancing its organizational performance. It is a system to streamline the data flow among the different functions in an enterprise. Modern business needs drive enterprises to more frequently link and share internal business processes and applications with external business partners. In such an environment, an enterprise must integrate with the business environments and must efficiently utilize information integration technology in order to be able to share data and information with all related parties. Interorganizational systems enable communication between the partners in the supply chain. Such systems can collect valuable SCM information for all related business processes across the entire supply chain. Using integrated SCM, businesses can align their operations across global networks. In recent years, efforts have been focused on the integration of interorganizational systems, and consequently, more and more enterprises have moved toward an interorganizational integration that supports SCM through ESs, in order to respond to the rapidly changing environment. The DAMA project is just such an endeavor. DAMA has developed an interenterprise architecture and collaborative model, which enables improved collaboration in business across supply chains.

Enterprises are facing the challenge of information integration. Information integration is related to business integration, enterprise integration, application integration, and ES integration. The integration of ESs and the development of a complex enterprise integration solution are complex and difficult tasks for most enterprises. As Figure 7.1 shows, enterprise integration is built upon business integration. It is a journey that an enterprise undertakes to interconnect solo functions and processes in order to streamline its organizational processes. The objective of ES integration is to provide the right information at the right place and at the right time and thereby to enable communication among people, machines, and

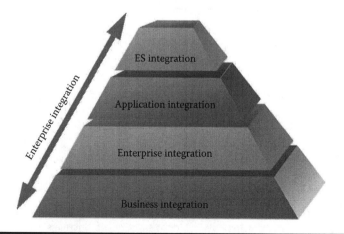

Figure 7.1   From business integration to enterprise systems integration.

systems to ensure their efficient cooperation and coordination. Several frameworks have been proposed for information integration. Fox et al. propose a generic enterprise model, which is the core of the SCM system (Fox et al., 1993). Hasselbring proposes a three-layer architecture that integrates different types of architectures (Hasselbring, 2000). Giachetti's framework includes a characterization of the different types of integration (Giachetti, 2004). In Puschmann and Alt's framework, the data level is considered as a separate layer (2004). In recent years, grid computing has emerged as a new technology for use by distributed computing systems. The infrastructure of grid computing allows for the aggregation and sharing of a large set of resources of different types that are distributed geographically to form a single system. Grid computing is expected to provide an effective tool for integrating ESs. Research indicates that the characteristics of grid computing systems meet the requirements of ES integration for extended enterprises on the global supply chain. A new grid-computing framework called OGSA has been proposed to integrate grid resources into web services. This trend implies that integration between a web-based ES and a grid computing infrastructure is possible. A novel ES architecture based on OGSA, called GridERP, has been proposed to solve the problem of the noneffective sharing of distributed resources and the interoperability issue on the deployment of ESs, with the expectation of bringing ES integration on the global supply chain to a higher level.

Despite the many efforts that have been made, information integration at present is still limited by the sophistication of the relevant technologies or by the lack of techniques. A successful execution of SCM will reply more upon sophisticated information integration than that is currently available. It is expected that information integration will attract more attention in automated SCM, in which seamless integration with external systems has always been emphasized. Researchers pointed out that there was limited research on integration frameworks. Wolfert et al. indicate that the contents of existing integration frameworks are not comprehensive and that an overall framework of information integration has not yet been developed (Wolfert et al., 2010). This is especially true as the explosion of the Internet and IoT has caused a secondary explosion in the type and the amount of information available to enterprises. The use of IoT-based business models significantly promotes the need for developing newer frameworks for information integration. Researchers have discussed ways to efficiently use wireless sensors for data acquisitions in information integration. Among the new technologies, RFID and the IoT have attracted much attention. RFID is a contactless and low-power wireless communication technology, which has applications across many areas of the supply chain. One of the envisioned applications includes information to be collected from a network of RFID sensors and IoT combined. Both RFID and IoT are considered as IT infrastructures; as such, it is reasonable to predict that both RFID and IoT will play a major role as newer frameworks for information integration are developed in the near future.

7.2 RFID: An Emerging Information Architecture

RFID refers to the identification and tracking using radio waves. RFID is an automatic identification solution that streamlines both identification and data acquisition. In recent years, RFID has become very popular in material flow systems, logistics, and SCM, as a technology used for automatic identification and for data capture. Integrating technologies such as RFID can help improve the effectiveness of the information flow in a supply chain, as SCM attempts to effectively obtain real-time information and to enhance dynamic management and control via information sharing from the involved partners in the supply chain. Using RFID, the partners in the supply chain will be able to access information, based on the data shared through RFID.

RFID readings can increase data accuracy and transfer speed, which, in turn, can help SCM. As such, RFID technology has been adopted in SCM for the purpose of improving tracing capability, since traceability is inherently linked with supply chain integrity (Kumar and Budin, 2006). The application of RFID technology, the most cutting-edge technology for supply chain traceability and integrity, has enabled enterprises to facilitate real-time traceability. In the construction industry, RFID can be assisted with personal digital assistants (PDAs) to enable on-site engineers to integrate work processes seamlessly at job sites. Wang et al. presented a web-based portal system that incorporates RFID to improve the efficiency and effectiveness of on-site data acquisition and information sharing in the supply chain (2007). This system not only improves the efficient acquisition of data on-site using RFID but also provides a monitor to control the construction progress. This study demonstrates the effectiveness of an RFID-based SCM application in the construction industry as it responds efficiently and enhances the information flow in the supply chain environment. Real-time monitoring and control is always desirable. Kumar and Budin gave two examples of massive recalls that could have been avoided if tracing capabilities had been used, since tracing capabilities would enable recalls of partial products instead of massive recalls (Kumar and Budin, 2006). In the food industry, RFID can greatly reduce the number of recalls as well as their negative impacts (Magina and Vlachos, 2005; Pang, 2013). In the Meatrac project, the design idea was to incorporate new communication technology, RFID, and control technology in order to provide full traceability. The achievement of the project not only is limited to tractability and traceability in real time but also creates possible applications for other purposes (Mousavi et al., 2005). Zang and Fan have implemented an event-processing mechanism in ESs based on RFID, which includes architecture, data structures, optimization strategies, and algorithm. They implemented an event-processing mechanism based on RFID in ESs, which can also be applied to the manufacturing industry (Zang and Fan, 2007). With the adoption of RFID, event processing fits well into ESs in terms of its facilitation of event aggregation into high-level actionable information and the event response to improve the responsiveness. RFID systems are one of the most promising technologies to improve management across the supply chains. RFID shows great promise in SCM, since it allows the nearly autonomous tracking of and passing of information. RFID technology can also be assisted by other technologies to make process integration as seamless as possible. As early efforts in integrating RFID into comprehensive enterprise information integration frameworks, industrial information integration models based on RFID and IoT have been proposed (Hsu and Wallace, 2007; Pang, 2013).

Rapid advances in industrial information integration methods have also spurred growth in the use of WSNs. With the applications of WSNs, event processing can fit well in ESs in terms of its facilitation of event aggregations into high-level actionable information and event response to improve the responsiveness. Significant progress has been made in the research on WSN in recent years, and WSN applications are increasing. However, Bonivento et al. (2007) explain that the lack of system-level design methodologies is one of the main obstacles in adopting RFID and WSN systems. Other scholars draw a similar conclusion, noting that limited progress has been achieved on the design of higher-layer protocols for industrial applications of WSNs, which are of critical importance. In spite of this, both RFID and WSNs are becoming the enabling technologies of the IoT.

7.3 IoT: An Emerging Internet-Based Information Architecture

7.3.1 Introduction

The Internet has changed business and personal lives in past years and continues to do so. The IoT becomes a foundation for connecting things, sensors, actuators, and other smart technologies. IoT is a term that has been introduced in recent years to describe objects that are able to communicate via the Internet. Haller et al. have provided the following definition: "A world where physical objects are seamlessly integrated into the information network, and where they, the physical objects, can become active participants in business processes. Services are available to interact with these 'smart objects' over the Internet, query their state and any information associated with them, taking into account security and privacy issues" (Haller et al., 2009).

IoT is an emerging Internet-based information architecture that can be employed to facilitate information flow in global supply chain networks. With the assumption that objects have digital functionality and can be identified and tracked automatically, IoT can dramatically streamline how the supply chain will be managed. The significance of IoT to SCM, especially the new supply chain information transmission approach based on RFID and IoT, has been emphasized and proposed. As IoT will have an impact on global supply chain networks, many new opportunities in applying RFID with IoT (together) to SCM are emerging.

IoT is an application of the Internet. IoT gives immediate access to information about the physical world and the objects and leads to innovative services with high efficiency and productivity. The IoT is gradually becoming a buzzword. The history of IoT has been discussed by Van Kranengurg and Bassi (2012). Two arguable pioneers who coined the term of Internet of Things are Bill Gates in his book titled *The Road Ahead* in 1995 and Kevin Ashton and Neil Gershenseld at MIT in 1999 for their research at Auto-OD Center (http://postscapes.com/internet-of-things-history; K. Ashton, "That 'Internet of Things' thing," http://www.rfidjournal.com/article/view/4986). This concept was further elaborated and noted that the era of the pervasive and ubiquitous computing is coming. The first IoT conference was held in Europe in 2006–2008. Over 50 member companies, including Bosch, Cisco, Fujits, Google, Intel, SAP, and Sun, formed an alliance to launch the Internet Protocol for Smart Object and to enable the IoT. Coordination and Support Action for Global RFID-related Activities and Standardisation is a European Framework 7 project. It was considered as the international effort concerning regulations, standardization, and other requirements for realizing IoT. During that time, IoT was not yet a tangible reality but was rather the prospective vision of a number of technologies that, combined together, could drastically change the way in which our society functions in the coming decade.

As reported by Pretz in 2013, the next generation of Internet should be a things-connected network, wherein things are wirelessly connected via smart sensors and are able to interact without human intervention (2013). Some basic technologies have been already developed and applied in the automotive, healthcare, food supply chains, and transportation. Information sharing is one of the features of IoT that is able to help build global collaboration in the industry. Efforts have been made on the development, standardization, security, and application aspects of IoT. In the development of IoT, the integration of devices equipped with intelligent sensors with cloud-based Internet is a challenge, as it involves basic technology, interfaces, and new standards.

As mentioned earlier, the IoT was first coproposed by Kevin Ashton to refer to uniquely identifiable interoperable connected objects with RFID technology, which then was being rapidly developed. It has been agreed that the IoT was born in 2008–2009. Intelligent sensing and wireless communication technologies have become part of the IoT, and new challenges and research horizons have emerged. Currently, the definition of IoT is still in the stage of evolution, that is, it is subject to different perspectives and new technology. In 2005, ITU described IoT as a "dynamic global network infrastructure with self-configuring capabilities based on standard and interoperable communication protocols where physical and virtual 'Things' have identities, physical attributes, and virtual personalities and use intelligent interfaces, and are seamlessly integrated into the information network" (ITU 2005). The IoT can be considered as a superset of connecting devices that are uniquely identifiable by existing near-field communication (NFC) technologies. The words "Internet" and "Things" describe a worldwide interconnected network

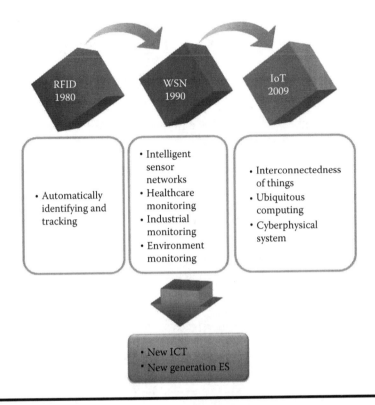

Figure 7.2 IoT, related technology, and their impact on new ICT and enterprise systems.

based on sensory, communication, networking, and information processing technologies, which might be the new version of ICT. Figure 7.2 shows the evolution of IoT and the closely related technologies.

The IoT was initiated by the use of RFID technology, which is increasingly being used in logistics, pharmaceutical production, retail, SCM, and diverse industrial sectors. The emerging wirelessly intelligent sensory technologies have significantly extended the sensory capabilities of devices, and therefore the original concept of IoT has been extended to ambient intelligence and autonomous control. To date, a number of technologies are involved in the IoT, such as RFID, intelligent sensing, WSN, NFC, wireless communications, cloud computing, and others. All of these make the IoT becoming an evolving computing concept.

There are several different definitions for IoT, depending on the technology and implementation of the ideas that are moving forward. However, the basic idea of the IoT is that objects have a unique way of identification in virtual representation, which makes it possible that all of the things around us can be identified by the things around them. In IoT, for all things, data can be exchanged (if needed), and data can be processed according to predefined schemes. IoT can be of significance in a variety of industrial sectors, including the manufacturing sector.

7.3.1.1 IoT-Oriented Infrastructure for Manufacturing Systems

A manufacturing system consists of system components and their related subsystems. Each component has an information unit for controlling the component. For manufacturing enterprises, ESs are the keys to success, since the advancement of information technology is revolutionizing the manufacturing systems. The impact of IoT on the manufacturing systems in the context of ESs will be significant:

- *Ubiquitous computing for manufacturing enterprises*
  Everything can be connected, so that the data can be acquired promptly and can be readily shared by all of the units. This makes it possible to integrate manufacturing resources with a much broader scope, including the resources within the organization and the interorganizational resources from partners in the supply chain.
- *Facilitating systems-level optimization*
  The manufacturing system is targeted for a system-level balance between flexibility and efficiency. On one hand, the manufacturing system is modularized, and each module is optimized at the module level for its specified function; on the other hand, the selection of modules and the assembling topologies offer the flexibility of a system configuration to meet various functions at the system level. The topology of system configuration can be optimized with the widely available information over the system.
- *Real-time full control*
  Traditional ESs are good for planning and scheduling, which are more at the macro level. Such systems will be integrated with the real-time control systems at the micro level. Online data acquisition systems are used to provide not only real-time control at the machine level but also feedback at the macro level about any changes (and/or uncertainties), so that the plans and schedules can be adjusted to accommodate the changes and uncertainties promptly. A good example is the change trend of the SAP software tool. In the MES within the ERP manufactured by SAP, with the emergence of IoT, it is expected to integrate with the online process control eventually (Logeais, 2008).
- *Customer-empowered by real-time data availability*
  Besides the privilege of comparing products from different vendors, IoT allows customers to personalize the product requirements and to place and change orders in real time based on their needs (Pang, 2013). The satisfaction level of customers can be greatly enhanced.

  IoT can be aligned with the architecture of a manufacturing system. Figure 7.3 shows the relationship between the IoT and future manufacturing systems.

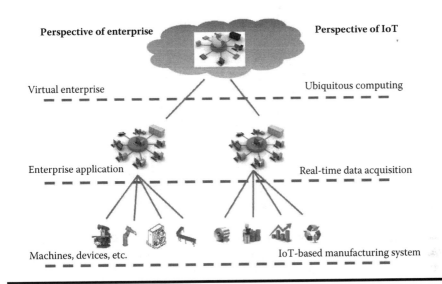

Figure 7.3 IoT for future manufacturing.

The future of the Internet will consist of heterogeneous connected devices that will further extend the borders of the physical entities and virtual components worldwide. IoT will empower the connected things with new capabilities and will create many new real-world applications.

7.3.2 Enabling Technologies

The characteristics of IoT include (1) the pervasive sensing of objects, (2) the integration of heterogeneous hardware and software, and (3) the extremely large scale of its nodes. The existing literature has introduced the state of the art of the enabling technologies for IoT including architecture frameworks and solutions, commercial products of communications, standardizations, modeling techniques, communication protocols, identification and resolution frameworks, objects platforms, security, and privacy. To clarify the progress made on protocols, algorithms, and proposed solutions and to determine what are the unsolved issues, Atzori et al. (2010) surveyed key enabling technologies for IoT in communication, identification, tracking, wired and wireless sensors, and distributed intelligence for smart objects.

7.3.2.1 Ubiquitous Computing

The Internet can be described as a ubiquitous infrastructure. IoT will change the ways of managing and operating production, distribution, transportation, and services. IoT can be applied to capture the status of an entire enterprise and its business processes more accurately. IoT is also known for its ubiquitous computing and ambient intelligence. IoT is a concept in which the virtual world of information technology integrates seamlessly with the real world of things (Uckelmann et al., 2011).

Guinard et al. (2010) indicated two trends with respect to the devices. They are as follows: (1) the hardware is getting smaller, less expensive, and more capable; and (2) the software industry is moving toward service-oriented approaches, and especially for the business software, new complex applications are based on the collaboration of other services. In IoT, smart things/objects are expected to be active participants in communication and business processes; they can be enabled to interact among themselves, reacting autonomously to the sensed environment and trigger actions and services with or without direct human intervention (Vermesan, 2009).

The trends for smart devices are as follows: (1) applying data analytics, modeling and simulation, and fusion and computation, for the scientific analysis of multidisciplinary data; (2) integrating devices, systems, and organizations for information sharing, real-time monitoring, and business process management; (3) using anything/anytime/anywhere communication to sense, capture, measure, and transfer data for dynamic planning and scheduling for short-term and long-term activities; and (4) vertically, improving upon the performance of an individual device. They are becoming versatile, powerful, and intelligent to deal with the changes and complexity of today's business world. Horizontally, a simple device without the functions of computation can be integrated; abundant information can be acquired for real-time decision making.

7.3.2.2 RFID

RFID and WSN are the cornerstones of IoT. RFID was initially introduced in retail and logistics applications. A simple RFID system is composed of an RFID and an RFID reader tag. Due to its capacity to identify, track, and trace, the RFID system is increasingly being used. RFID systems can provide precise real-time information about the involved devices, so they are successfully used in various industrial sectors. They can simplify operations processes, improve efficiency, and reduce costs. RFID technology has been used in the transportation sector to improve fuel efficiency. RFID technologies have also been used for data acquisition and information sharing in SCM. RFID's industrial applications are increasing significantly, especially in manufacturing, logistics, and SCM. Based on incomplete recent data, of all RFID-based applications, approximately, 56% were used for access control, 29% for supply chain, 25% for motorway tolls, 24% for security control, 21% for product control, and 15% for asset management. The next generation of RFID technology will focus on the item-level RFID usage and RFID-aware management issues. Although RFID technology is successfully used in many areas, however, there still are some problems that need to be solved:

- *The RFID tag reading collision*, which includes the collisions between RFID readers or RFID tags and multiple reads of the same RFID tag
- *The signal interference problem*, which includes the interference in RFID systems and the interference from other radio devices
- *Information privacy*, which involves customer privacy and the confidentiality of RFID tags that can be scanned by authorized RFID scanners
- Lack of common RFID standards
- Integration of RFID and intelligent wireless sensors

The emerging intelligent sensory technologies, including those related to infrared, γ-ray, pressure, vibration, electromagnetic, biosensor, and x-ray, can deliver all of the acquired information connected to IoT for analysis purposes. Several hardware solutions for RFID-based sensors have been proposed. The integration of data acquired by intelligent sensors with data delivered by RFIDs can make IoT capable of facilitating industrial and services information processing and the further deployment of services in extended applications.

7.3.2.3 Wireless Sensor Networks

The IoT is a network of all physical objects with identifiable IDs, which is based on the Internet. WSN is attractive for many applications due to the fact that no wire connections for communication are required. WSN has been applied in the supply chains so that real-time data can be obtained for a variety of purposes. Most of the ongoing WSN research focuses on energy-efficient routing, aggregation, and data management algorithms. The deployment, application development, and standardization aspects have received less attention, to date. One of the major challenges is to optimize the deployment of the large-scale systems and the integration of data from a large quantity of sources. The wireless network is an important infrastructure: (1) IPv6 makes it possible to connect an unlimited number of devices; (2) Wi-fi and WiMax provide high-speed and low-cost communication; (3) ZigBee/bluetooth/RFID provides the communication in low-speed and local communication; and (4) the mobile platform offers anything/anytime/anywhere communications.

7.3.2.3.1 RFID and WSN

RFID technology has advanced tremendously in recent years, as is evidenced by RFID's applications in various industrial sectors. Due to the needs of the continuous monitoring and controlling of everything, everywhere, computing systems are required to be ubiquitous and distributed. Some researchers have predicted that the computing technology's next revolution will be the widespread use of wireless computing and communication devices in the framework of IoT. ES-related applications (especially extended enterprise integration applications) involve a large number of enterprise databases that belong to different organizations, and these databases increasingly rely on real-time data sources such as RFID and WSN, since RFID and WSN are essentially distributed systems.

RFID and WSN represent two complementary technologies. RFID is capable of detecting and identifying objects that may form challenges to conventional sensor technologies. RFID tags are less expensive when compared to sensor nodes. It is preferable to use RFID tags instead of the sensor nodes of a WSN in many places; however, an RFID tag does not include information about the states of the targeted objects or the environment. Integrating with a WSN enables RFID readers and tags to have intelligence and allows an RFID network to be operated in multihop fashion. Many applications have been proposed to use RFID and WSN for various purposes including using RFID and WSN for power facility management and integrating RFID into service infrastructures to improve traceability. Researchers have investigated the use of RFID and WSN with ZigBee; electronic labels were attached to the sensors to integrate the RFID with the WSN. Systems have been developed to monitor food supply chains (Pang, 2013). Researchers have argued that an integration of RFID and WSNs is of great value in ensuring the accurate and timely localization and tracking of objects.

7.3.2.4 Cloud Computing

Advances in automatic identification, wireless communications, intelligent sensing, and distributed data processing have narrowed the gap between the notion of ubiquitous computing and the world of networked sensing and intelligent "things." Despite the consensus about the great potential of the concept and the significant progress in a number of enabling technologies, there is a general lack of an integrated vision on how to realize it.

Cloud computing is a large-scale and low-cost processing method, which is based on an IP connection for computation and storage. In 2001, Google CEO Eric Schmidt proposed the Cloud Computing concept. In 2003, Google used cloud computing based on the Google appEngine, which was a large-scale application of PaaS. In 2006, Amazon announced the elastic computing server (EC2) as a successful application of IaaS mode. In 2007, China Mobile launched a plan to promote cloud computing.

The characteristics of cloud computing include the following:

- Ubiquitous network access: use is available through standard Internet-enabled devices
- Location-independent resource pooling: processing and storage demands are balanced across a common infrastructure with no particular resource assigned to any individual user
- On-demand self-service: individuals can set themselves up without needing anyone's help

- Rapid elasticity: consumers can increase or decrease capacity at will
- Pay per use: consumers are charged fees based on their usage of a combination of computing power, bandwidth use, and/or storage

All of these characteristics are necessary to produce an enterprise private cloud capable of achieving compelling business value. Cloud computing includes savings on capital equipment and operating costs, reducing supporting costs, and significantly increasing business agility. All of these enable enterprises to improve their profit margins and competitiveness in the markets they serve.

Cloud computing is needed to address the dynamic and exponentially growing demands for real-time reliable data processing of IoT including interoperable service-oriented middleware and architectures to share real-world data among heterogeneous devices; networking technologies for wired and wireless networking to interconnect "things"; and decision-making application services that store, integrate, and process dynamic data streams from devices with limited computational capacity on a real-time basis. Cloud infrastructures also provide the storage and computing capabilities to address the IoT application services' needs to process big data. However, cloud computing is still far from mature, even without the considerations of standardization and interoperation. Figure 7.4 shows the characteristics, delivery, and deployment models of cloud computing.

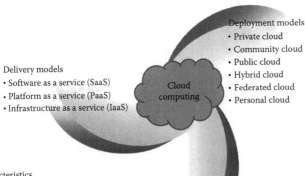

Delivery models
- Software as a service (SaaS)
- Platform as a service (PaaS)
- Infrastructure as a service (IaaS)

Cloud computing

Deployment models
- Private cloud
- Community cloud
- Public cloud
- Hybrid cloud
- Federated cloud
- Personal cloud

Characteristics
- *On-demand self-service*: individuals can set themselves up without needing anyone's help
- *Ubiquitous network access*: available through standard Internet-enabled devices
- *Location independent resource pooling*: processing and storage demands are balanced across a common infrastructure with no particular resource assigned to any individual user
- *Rapid elasticity*: consumers can increase or decrease capacity at will
- *Pay per use*: consumers are charged fees based on their usage of a combination of computing power, bandwidth use and/or storage

Figure 7.4 Characteristics, delivery, and deployment models of cloud computing.

7.3.2.5 More on the Enabling Technologies of IoT

7.3.2.5.1 Communication Technologies in IoT

Physical objects involve very diverse hardware specifications in aspects of communication, computation, memory, data storage capacity, and transmission power. All of these things can be well organized through networks that involve all kinds of communication technologies. For example, devices can be well organized by using a gateway for the communication over the Internet. The Internet as a fundamental component of the IoT can be seen as a black box that can be studied through formal models such as stochastic queuing delay models, stochastic packet loss models, and others.

IoT involves a number of heterogeneous networks, such as WSNs, wireless mesh networks, mobile networks, and WLAN, with the service layer collecting information about all the things through the networks. These networks help things in IoT perform complex functions such as information exchange, computation, and others. The reliable communication between the gateway and the things can be helpful for centralized decision making through IoT. The gateway is able to run complicated optimization algorithms locally by exploiting its network knowledge. Therefore, the computational complexity is shifted from the things to the gateway, and a global optimal route and computational parameter values for the gateways can be obtained. This is feasible as the resulting complexity is affordable for standard gateway hardware capabilities.

Hardware capabilities and the communication requirements among different types of devices can be very different. From a hardware perspective, things can have very different memory, communication, and computation capabilities. Things can have very different quality of service (QoS) requirements in terms of delay, energy consumption, and reliability. For example, minimizing the energy usage for communication/computation purposes is a major constraint for those battery-powered devices without efficient energy-related techniques. On the contrary, this energy constraint is not critical for devices with a power supply connection.

IoT would also greatly benefit from the existing protocols in Internet such as IPv6, since this would make it possible to directly address any number of things needed through the Internet. The commonly used communication protocol and standards include the following:

- RFID (e.g., ISO 18000 6c EPC class 1 Gen2)
- NFC, IEEE 802.11 (WLAN), IEEE 802.15.4 (ZigBee), IEEE 802.15.1 (Bluetooth)
- Multihop Wireless Sensor/Mesh Networks
- IETF Low-power Wireless Personal Area Networks (6LoWPAN)
- M2M
- Traditional IP technologies, such as IP, IPv6

Details of the communication technologies can be found in Table 7.1.

Table 7.1 Communication Technologies in IoT

Communication Protocols	Transmission Rate	Spectrum	Transmission Range
RFID	424 kbps	135 KHz 13.56 MHz 866–960 MHz 2.4 GHz	>50 cm >50 cm >3 m >1.5 m
NFC	100 kbps–10 Mbps	2.45 GHz	
ZigBee	256 kbps/20 kbps	2.4 GHz/900 MHz	10 m
Bluetooth	1 Mbps	2.4 GHz	10 m
BLE	10 kbps	2.4 GHz	10 m
UWB	50 Mbps	Wide range	30 m
Wi-Fi	50–320 Mbps	2.4/5.8 GHz	100
WiMax	70 Mbps	2–11 GHz	50 km
UMTS/CDMA/ EDGE/MBWA	2 Mbps	896	~

7.3.2.5.2 Networks Involved in IoT

There exist many cross-layer protocols for WSNs, wireless mesh networks (WMNs), and ad hoc networks (AHNs). However, they cannot be applied to the IoT for several reasons. First, the heterogeneity of the IoT means that things have largely diverse hardware capabilities, different QoS requirements, and individual characters. On the contrary, in WSN, nodes usually have very similar hardware specifications, common communication requirements, and a shared goal. Second, the Internet is involved in the IoT network architecture, from which it inherits a centralized and hierarchical architecture. In comparison, in WSN, WMN, and AHN, highly flat network architectures are considered, in which nodes communicate in a multihop fashion, and the Internet is not involved.

7.3.2.5.3 Service Management in IoT

Service management in IoT refers to the implementation and management of quality IoT services that meet the needs of users or applications. SOA can promote encapsulation of services; by doing this, the details of services, such as the implementation of services and the protocols used, can be hidden from the concept of services. This makes it possible to decouple between components in the system and therefore to hide the heterogeneity from the service consumers. SOA-IoT allows applications to use the heterogeneous objects as compatible services. On the other hand, the dynamic nature of IoT applications requires IoT to provide reliable and consistent service.

There are a variety of service management architectures contributing to the IoT. The OSGi platform is able to provide applications with a dynamic SOA architecture that can effectively enable the deployment of smart service. The advances in software industry show that the OSGi is an effective modular platform for service deployment in many applications. In IoT, the service composition based on the OSGi platform can be implemented by Apache Felix iPoJo.

7.3.2.5.4 Dynamic Service Composition in IoT

IoT is a subset of the future Internet; every virtual and physical object can communicate with every other object giving seamless service to other entities. Millions of devices in IoT need to be interoperable. The service-oriented IoT can make each component offer its functionalities as standard services, which might significantly increase the efficiency of both devices and networks that are involved in IoT. In order to best organize the services that the real objects provide, each service can find a virtual responding element in IoT.

In a service-oriented IoT, services can be created and deployed according to the steps as (1) developing services' composition platforms, (2) abstracting the devices' functionalities and communication capabilities, and (3) provisioning of a common set of services. The IoT makes it possible to build each real object a mirror in the IoT.

A service is a collection of data and associated behaviors that accomplish a particular function. A service may reference other primary or secondary services and/ or a set of characteristics that make up the service. Services can be categorized into two types: "primary service" and "secondary service." The former denotes services that expose the primary functionalities at an IoT node, which can be seen as the basic service component and can be included by another service. A secondary service can provide auxiliary functionality to the primary service or to other secondary services. In IoT, each service may consist of one or more characteristics that define the service data structure, permission, descriptors, and other attributes.

In the newly released Bluetooth SIG specification, a service can be easily described with XML for easy exchange with other middleware. Following is an example of "Health Thermometer Service." The service provides the measurements of a health thermometer by a universally unique identifier (UUID— 0x1809) as shown in Figure 7.5; however, the user does not need to know how the measurement is acquired by the sensor.

In IoT, a characteristic consists of the following three segments:

1. Characteristic declaration, which describes the properties of characteristic value (read, write, indicate, etc.), characteristic value handle, and characteristic value type (UUID)

```xml
<?xml version="1.0" encoding="utf-8"?>
<service uuid ="1809">
  <uri>org.bluetooth.service.health_thermometer</uri>
  <description>Health Thermometer Service </description>
  <characteristic uuid ="2a1c", id = "xgatt_temperature_celsius">
    <description> Celsius temperature </description>
    <properties indicate = "true" />
    <value type = "hex"> 0000000000</value>
  </characteristic>
</service>
```

Figure 7.5   An illustrated example of service in IoT.

Table 7.2   Characteristic of a Service

Handle	Type	Permissions	Value
39	0X2800 (Service UUID)	Read	E0:FF
40	0X2803 (Characteristic UUID)	Read	10:29:00:E1:FF

2. Characteristic value, which contains the value of a characteristic
3. Characteristic descriptor, which provides additional information about the characteristic

Table 7.2 shows an example of a characteristic.

7.3.2.5.4.1 Integration of Different Service Technologies — The service-oriented IoT aims at developing an effective architecture for service operations in an IoT, which extends the preexisting architectures of IoT and takes the unique characteristics of service-oriented IoT into consideration. The knowledge about services in the service-oriented IoT should be well represented to support discovery, detection, classification, composition, and testing.

The architecture of IoT can contain three basic layers: the "application layer," the "network layer," and the "sensing layer." (1) The "application layer" provides the functionalities that are built on top of an implementation of the IoT. The application layer is connected to the process modeling components for IoT-aware business processes that can be executed in the process execution components; (2) the middle layer contains three basic components: "service entity arrangements," "virtual entity and information," and "resources." The arrangement and access of IoT services to external entities and services are organized by the "service entity arrangements" component. The "virtual entity" (VE) component contains the functionality that associates VEs to relevant services as well as a means to search for such services. The "resources" module provides the functionalities required by services for processing information and for notifying application software and services about events related to resources and corresponding virtual entities; (3) the "sensing layer" involves sensing devices, such as RFID tags, sensor nodes, which can collect, record, and process observations and measurements. The "network layer" is able to access the sensing layer with device-level API, which facilitates the information exchange between the applications and the environment.

7.3.2.5.5 Security and Privacy Challenges in IoT

In IoT, security and privacy are challenges. In order to integrate the sensing layer devices into the IoT, it is necessary to develop effective security technology, which provides security and privacy protection for IoT activities. The applications in IoT may face challenges such as physical attacks to RFID tags and data stores; integrity of codes protection; and the availability of things. In RFID systems, a number of security schemes and authentication protocols have been proposed to cope with security threats. Juels (2006) proposed a "block tag" to prevent unauthorized tracing. On the other hand, low-cost symmetric-key cryptography algorithms, such as Tiny Encryption Algorithm and Advanced Encryption Standard, have been proposed to protect the information exchange in IoT. It is reported that low-cost RFID tags have implemented some asymmetric key cryptography algorithms, such as Elliptic Curve Cryptography. On the other hand, the security protocols developed for WSN can be integrated into IoT as an intrinsic part of IoT.

In IoT, numerous applications can be involved. Therefore, it is necessary to develop a high-degree security mechanism. The challenges in security and privacy protection include resilience to attacks, data authentication, access control, and client privacy, among others.

7.3.3 Standards

Figure 7.6 introduces the technologies related to IoT. Numerous technical standards that are vital to the success of IoT have been proposed by institutions in different countries. Among these, middleware, interfaces, and open standards in IoT are very important. The main research objectives in this area include (1) designing policies and distributed architectures; (2) improving privacy and individual protection; (3) improving the trustworthiness, acceptability, and security of IoT; (4) standardization; (5) improving the development of fundamental technologies such as MEMS and ubiquitous localization; (6) applications based on IoT, such as healthcare systems, and environment monitoring. IoT-related standards have received much research attention around the world.

Developing standards for IoT requires consideration of the efficiency and availability of specifications. Currently, a number of institutions are working on the primary standards in IoT. Globally, the International Telecommunication

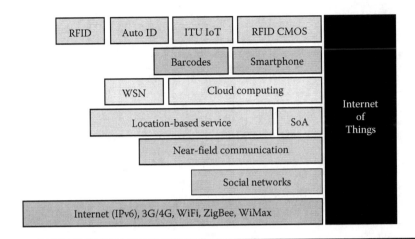

Figure 7.6   IoT-related technologies.

Union (ITU), EPCglobal, International Electro-technical Commission (IEC), International Organization for Standardization (ISO), and IEEE have developed a set of standards. In Europe, the European Telecommunications Standards Institute (ETSI) and the European Committee for Electro-technical Standardization (CEN/CENELEC) have released a set of standards regarding the fundamental techniques in IoT. In China, the China Communications Standards Association and the China Electronics Standardization Institute are working on the standards of semipassive RFID and ultra-high frequency (UHF) band RFID. In the United States, the American National Standards Institute is working on the standards of IoT. The lack of standards may decrease the competitiveness of IoT products. A global collaboration regarding standards is needed to deal with the lack of homogeneity among standards bodies. Formal contracts such as the World Standards Cooperation (WSC) are expected to be able to govern the relationships between the international standards bodies and regional standards bodies.

The coordination among standards is very important at the early phase of IoT development, since it can help the developers and users to determine the best technical protocols for dynamic applications and services in IoT. On the other hand, standardization of the technologies used in IoT is urgent as it can accelerate the spread of IoT technology. Table 7.3 lists the standards involved in IoT.

7.3.4 Current Research

In the last decade, RFID-based technology has been widely used in logistics, retailing, and pharmaceutics. Since 2010, advances have been made in intelligent sensors, wireless communication, and sensor network technologies. As a new-generation networking technology, IoT will connect objects and systems through

Table 7.3 List of Standards Involved in IoT

Technologies	Standards
Communication	IEEE 802.15.4(ZigBee) IEEE 802.11(WLAN) IEEE 802.15.1(Bluetooth, low-energy Bluetooth) IEEE 802.15.6 (Wireless body area networks) IEEE 1888 IPv6 3G/4G UWB
RFID	RFID tag ISO 11784 RFID air interface protocol: ISO 11785 RFID payment system and contactless smartcard: ISO 14443/15693 Mobile RFID: ISO/IEC 18092 ISO/IEC 29143 ISO 18000-1—Generic parameters for the air interface for globally accepted frequencies ISO 18000-2—for frequencies below 135 kHz ISO 18000-3—for 13.56 MHz ISO 18000-4—for 2.45 GHz ISO 18000-6—for 860–960 MHz ISO 18000-7—for 433 MHz
Data content and encoding	EPC global electronic product code, or EPC™ EPC global physical mark up language EPC Global Object Naming Service (ONS)
Electronic product code	Auto-ID: Global Trade Identification Number (GTIN), Serial Shipping Container Code (SSCC), and the Global Location Number (GLN)
Sensor	ISO/IEC JTC1 SC31 and ISO/IEC JTC1 WG7 Sensor Interfaces: IEEE 1451.x, IEC SC 17B, EPC global, ISO TC 211, ISO TC 205
Network Management	ZigBee Alliance, IETF SNMP WG, ITU-T SG 2, ITU-T SG 16, IEEE 1588
Middle	ISO TC 205, ITU-T SG 16
QoS	ITU-T, IETF

automatically communicating and exchanging data. The IoT spans a huge number of applications in which billions of things can be connected. However, the success of the IoT depends on standardization, which will provide interoperability, compatibility, reliability, and effective operations on a global scale. In order to best provide services to end-users or applications, IoT technical standards should be developed, which will define the specifications of information exchange, processing, and communications between things.

In order to take a leadership position, the UK government has launched a £5m project about IoT's foundation in technology and innovation; meanwhile, investment and policies at a national and local level have been developed. In the EU, the IoT European Research Cluster FP7 (http://www.rfid-in-action.eu/cerp/) has hosted a number of integrated projects on the fundamental technologies of IoT, such as "Internet of Things Architecture (IoT-A)," which has mainly designed the reference model and architectures. In these projects, the applications and end-users provide the precise requirements to drive the theoretical work. Meanwhile, the ETSI is working on the policy making. In China, the "Sensing China" project, which assumes that if everything around had an identification tag that could collect data and could be accessed through the Internet, it would be possible to keep track of the status of the things in IoT and monitor any number of parameters, was officially launched in June 2010 by the government. In the United States, IBM and the Information Technology & Innovation Foundation released a report in 2009 that claimed that the new ICT technology development could be an effective way to improve the existing information technology infrastructure and would have a positive impact on productivity and innovation. The US government has focused ICT strategies on energy efficiency, broadband technology opportunities, rural utilities service, and healthcare information technology. Meanwhile, South Korea has launched RFID/USN and a "New IT Strategy" program to advance its IoT infrastructure development. The government of Japan launched "u-Japan x ICT" and "i-Japan strategies" in 2008 and 2009, which aim at deploying the IoT into all aspects of daily living.

7.3.5 Applications

7.3.5.1 Introduction

The IoT paradigm aims to bring an intelligent interconnection of objects in the physical world through information-sensing devices using network protocols and information systems, so that objects can communicate and interact with each other. IoT is not only an IT infrastructure but also, more importantly, an information system that is applied to sense the environment and to identify, position, track, and monitor objects. The front end is "sense, understand, and control," and the back end is for feedback and control. It is very important to make efforts toward new applications. IoT enables information gathering, storing, and transmitting to be available for things equipped with IoT tags. IoT tags have been widely used in the management of the supply chain, which is involved in the retailing industry, smart shelf operations, healthcare, the food industry, the logistic industry, the travel and tourism industry, library applications, and many others (Pang, 2013). Currently, IoT has been already deployed in these ways:

- For the user: A large number of hardware and software components (RFID tags, mobile phones, social networks, mobile apps, etc.) have been developed for consumers that allow users to access additional information regarding products.
- For the manufacturer: Increasingly, products are produced using unique identification technologies such as barcodes, RFID tags, intelligent sensors on personal electronic devices, home appliances, and so on, which can monitor the products before they even reach the users.
- For traditional industries: IoT has been used to increase the effectiveness of traditional industries by introducing new means of information exchange and processing techniques.

Researchers have discussed how to adapt the IoT to customers' requirements: IoT has been proposed to be applied to many applications. The possibilities of IoT to make people's life easier and to automate many of our current tasks are huge. IoT originated from a vision strongly coupled with RFID-tagged objects. Many new businesses in retail, logistics, food, healthcare, energy, smart home, and transportation have been developed due to the IoT (Pang, 2013). New applications arise in areas such as logistics, healthcare, rescue, and environmental preservation. Software engineering will provide innovative infrastructures, fulfilling the requirements of IoT including heterogeneity (different objects, sensors, protocols and applications, dynamicity (arrival and departure of objects) and evolution (support for new protocols and sensors). Gama et al. (2012) applied service-oriented middleware that tries to leverage the existing IoT architectural concepts to bring more flexibility and dynamicity. To improve tailing dam safety, a tailing dam monitoring and prealarm system based on IoT and cloud computing was accomplished with the capacities of real-time monitoring of the saturated line, impounded water level, and dam deformation (Sun and Zhang, 2012). Pang (2013) suggested that IoT is a promising solution for a personalized healthcare system. IoT accommodates updating in order to reach a more accurate treatment of chronic diseases such as diabetes. IoT allows developing solutions to the convenience the patients, physician, and nurses. Heer et al. (2011) used building automated control as an example to illustrate the challenges in a secure IP-based IoT with the focus on security protocols. The requirements of IoT architecture have been discussed from the perspective of security. Alam et al. (2011) discussed the secure access provision to IoT-enabled services and interoperability in different administrative domains. Welbourne et al. (2009) developed a community-oriented infrastructure at the University of Washington.

It created a microcosm for IoT for use in studying applications, systems, and social issues that are likely to emerge in a realistic setting. The system is used to empower users by facilitating their understanding, management, and control of RFID data and privacy settings. Large-scale applications are emerging in many industries from smart grid to real-time transportation management and optimization.

From the perspective of application, IoT can be viewed as an integrated information system based on the Internet to achieve high efficiency in industrial sectors such as transportation, healthcare, and many others. The nature of such an information system shows (1) massive amounts of data in real time; (2) ambient intelligence, since the integration and fusion of information possess a synergized effect; (3) multidisciplinarity; and (4) dynamics. The IoT is of high importance on economy and society. For its use to become more ubiquitous, the development of information technology infrastructures plays a key role. It can be foreseen that the IoT will greatly contribute to addressing many needs including social needs.

7.3.5.2 Industrial Deployment

Table 7.4 shows the possible applications of IoT in industrial sectors.

IoT can reduce the gap between the components in the current digital economy, in which a services-centric economy is realized through network transactions. Meanwhile, business processes can benefit from the IoT at three levels: (1) the level of individuals, such as consumer service and individual service; (2) the level of enterprises, such as business process and SCM; and (3) the level of industry, which involves industrial and economic development.

Enterprises that adopt IoT technology can benefit from more competitive products, more profitable and greener business models, resource optimization, and real-time information processing in business activities. The globally connected IoT can provide enterprises with integrated logistic service networks. Manufacturers can benefit from global availability, cost-efficient purchasing, and inventory management. The IoT-based infrastructure enables the business partners to seamlessly integrate enterprise resources.

7.3.5.3 Social Internet of Things

Recently, the idea to integrate the IoT with social networks has been proposed in Fielding and Taylor (2012), and a new paradigm named "Social Internet of Things (SIoT)" has been proposed to describe a world in which things around us can be intelligently sensed. It is predicted that SIoT can effectively perform things and service discovery and can improve the scalability of IoT in human social networks. Meanwhile, the privacy and protection in social networks can be implanted into IoT in order to improve the security of IoT. The ideal of SIoT as reported in Miorandi et al. (2012) is motivated by the popular social networks: Facebook, Twitter, and micro-blogging, which permeate everyday life. The SIoT has attracted

Table 7.4   Possible Applications of IoT in Industrial Sectors

Industrial Deployment	Applications
Logistics and SCM (supply chain management)	Goods position monitoring Theft prevention Container monitoring in SC SC events monitoring
Access control	NCF Access control system E-home Security infrastructure
Control of industrial processes	Intelligent quality control system Process parameters monitoring Sensor-network-based control Inventory tracking Machine condition monitoring
Learning, education, and training	Collaborative applications in education Augmented cognition application Multimedia-based collaboration
Civil protection	Conditions monitoring for bridges, tunnels, and gymnasiums Early warning systems for risk detection
E-home	Intelligent building energy conservation system
Intelligent agriculture	Temperature, humidity, and soil conditions monitoring system Crop disease monitoring system Microclimate control
Intelligent transportation and traffic	Traffic monitory system Vehicle management system Traveler information system
Environment protection	Pollution detection Chemical gas leakage detection system Weather observation Air-conditioning monitoring system

attention from many areas including e-business and e-learning. The homophily (Fielding and Taylor, 2012) method to establish higher levels of trust can be helpful in optimizing relationships between things. In SRI (2008) and Welbourne et al. (2009), the authors discuss the combination of social relationships into the future of the Internet. Hernandez-Castro et al. (2008) discussed the integration of IoT into existing social networks such as Facebook and Twitter. Fielding and Taylor (2012)

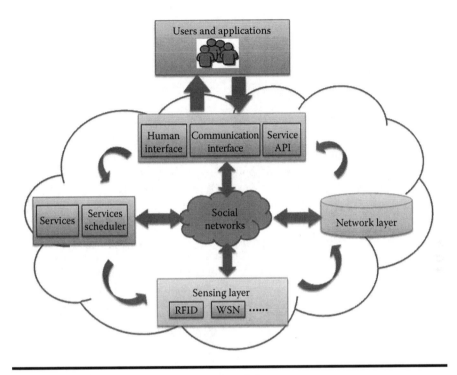

Figure 7.7 A proposed architecture of Social IoT.

investigated the potential of SIoT to support novel applications and networking services for the IoT in more effective and efficient ways. An integration scheme of social networking into IoT has been described, and the system architecture for the implementation of an SIoT has been developed, as Figure 7.7 shows.

7.3.5.4 Healthcare Applications

Healthcare is an important application area for IoT, which can enhance service quality and reduce costs. In healthcare, a number of medical sensors or devices can be used to monitor medical parameters such as body temperature, blood glucose level, blood pressure, and heart rate. Advances in sensors, wireless communication, and data processing technologies are the driving forces of implementing IoT into healthcare systems. The emerging wearable body sensor networks can be used to continuously monitor patient activities or medical parameters. IoT can provide healthcare systems with the interconnection of such heterogeneous devices and can help in obtaining a comprehensive picture of health parameters (Pang, 2013).

IoT that interconnects with wearable biosensors can help healthcare services with patient parameter monitoring, daily activity tracking (steps walked, calories burned, exercises performed, etc.), and the care of the aging. It can be foreseen that IoT-equipped intelligent medical sensors can significantly enhance the quality of life and can even prevent the onset of health problems. The low-cost medical sensor technology is able to wirelessly connect with other things in IoT, which can make it possible to develop implantable wireless identifiable sensors to monitor the health parameters of a patient's life.

Recent BLE-based technologies can help with interconnection in the healthcare, security, and home entertainment industries. BLE-based technologies enable a device to communicate with objects that are integrated with respective chips, such as mobile phones, watches, mobile body sensors, home appliances, and personal computers. Through this, all things can be connected with IoT.

On the other hand, the development of mobile devices and mobile health applications creates a huge market for IoT in the healthcare sector. The individual mobile health applications that assist cardiology practices to measure blood pressure, and diabetes treatment facilities to record blood glucose, have been developed. A new concept named the Health Internet of Things has been proposed (Fielding and Taylor, 2012). It is based on the use of sensor technologies and wireless networks for the monitoring of medical parameters. IoT technologies can be used to improve assisted living solutions. Medical devices that connect to IoT (including medical sensors and wearable sensors) can be used to gather the healthcare information that can be transmitted to remote medical centers.

7.3.5.5 IoT in Infrastructures

IoT solutions have also been developed in many infrastructure areas: smart cities, environmental monitoring, smart homes/building, and so on. In smart buildings, IoT can help to improve the quality of building management and can reduce the consumption of resources. In recent years, the term "Smart Cities" has been proposed to denote the cyber-physical ecosystem emerging through the deployment of intelligent sensors and novel services over citywide scenarios. In China, the "Sensing China" project was launched in June 2010; it assumes that if everything around us could have an identification tag that could broadcast information to the Internet, people could keep track of the status of the things through IoT and could monitor any number of parameters. With the successful deployment of IoT in a community or in a city, the huge impacts of IoT on all aspects of life can be foreseen.

7.3.5.6 Security and Surveillance

As described earlier, every physical object in IoT can find a responding counterpart that can provide services for applications or users. Each object should be well addressed and labeled in IoT. The interconnection between things might bring unprecedented convenience as well as security issues. Therefore, strong security protection is necessary in IoT to avoid attacks and malfunctions. In traditional networks such as the Internet, security protocols and privacy assurance are widely used to protect privacy and communication security. However, traditional lightweight cryptography and security protocols are not secure enough for IoT. Current security protocols and mechanisms must be improved and then integrated into IoT in order to provide better protection.

On the other hand, a strong legal and technical framework is also essential. Due to the complex nature of IoT, the protection of billions of intelligent things is very difficult. Things might face a lot of threats such as data leakage, identity theft, and threats coming from external networks. Therefore, IoT must provide strong security protection for all components at all stages, from the sensing layer to the application layers, from identification to services provision, and from RFID tags to IT infrastructure governance. Information should be secured from the beginning of its existence to the end of its life cycle. In IoT, heterogeneity greatly affects the security protection of networks; it is easy to suffer threats from communication channels, cryptography, and route selection.

Information privacy is one of the most sensitive subjects in IoT. The availability of data in IoT makes it difficult to protect information generated by personalized services. To design a proper data and privacy protection mechanism, many factors need to be taken into consideration. User authentication/authorization can involve numerous technologies, such as access control and trust management.

7.3.5.7 Data Cleaning in IoT Applications

Data cleaning is used to eliminate redundant data while maintaining the integrity of the original data. Some progress has been made on data filtering and cleaning technologies. Researchers have proposed data cleaning algorithms in which different steps of cleaning are applied based on the raw data. Researchers have also proposed a data cleaning strategy based on the time correlation, in which probability models are used, and they were mainly developed for solving data leakage problems. In general, such algorithms are developed for handling the unreliability of RFID data caused by data leakage and repeated readings. While, in such algorithms, the data redundancy issue was not the focus, some other studies have discussed ways to handle redundant data. Data cleaning, transformation, and loading techniques based on the mathematical theories have also been proposed.

7.3.5.7.1 RFID and WSN

7.3.5.7.1.1 RFID — The purpose of RFID is to obtain identity data from the targeted object through the radio frequency signal. As such, RFID has been used for the automatic recognition of objects. In general, an RFID system consists of two parts: an electronic "tag" and a "reader." An electronic tag is composed of a

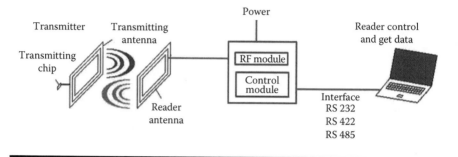

Figure 7.8   Working principle of RFID.

chip and an antenna. The tag communicates with the reader based on the principle of inductance coupling or electromagnetic reflection. The reader is the device that reads the identity label of a targeted object. An antenna can be attached to a tag, to a reader, or to an interface to the reader via a coaxial cable. As shown in Figure 7.8, an RFID operates as follows:

1. The reader encodes the message that is prepared to be sent. The message is loaded onto a high-frequency carrier signal and then is sent out by antenna.
2. When a tag enters into the working range of a reader, it receives the message. The corresponding embedded circuit chip on the tag will process the message with rectification, modulation, decoding, and decryption and then will verify it based on the command request, the password, and the authority.
3. If it is a reading command, the control logic circuit will read relevant information from its local memory. The data are encrypted, coded, modulated, and then sent back to the reader via antenna. When the reader receives the message, it will send it for processing via demodulation, decoding, and decryption.
4. If it is a writing command, the control logic circuit will apply working voltage in the electronic tag. During the verification, if the corresponding password and authority cannot be passed, the message will be returned.

The incoming sensing and RFID message is processed through accessing the event database. For a separate query request, data aggregation, filtering, and mapping are involved.

7.3.5.7.1.2 Wireless Sensor Network — A WSN consists of a large number of sensor nodes. Sensors are inexpensive, low in power consumption, and versatile. They are small in size and can communicate with each other over short distances. WSNs have been applied in many different areas including the military, the environment, and the healthcare. There are two leading international standards for

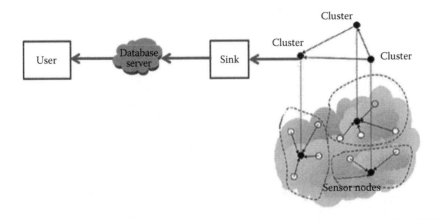

Figure 7.9   Three-layer logic architecture of a WSN.

low-power wireless communications: Bluetooth (802.15.3) and ZigBee (802.15.4). As shown in Figure 7.9, under the system architecture of Bluetooth or ZigBee, a WSN consists of three layers: the sensor node layer, the cluster layer, and the sink node (reader) layer. Under the standard of IEEE 802.15.4, a sensor node and an RFID can be treated equally in a sense that both of them can build peer-to-peer networks randomly to aid in network communication. In integrating a WSN with an RFID system, both of the readers and sensor nodes will use wireless transceiver chips that are compatible with IEEE 802.15.4.

Bluetooth allows for the creation and maintenance of short-range personal area network (PAN). Bluetooth transfers data at the rate of 1 Mbps; the range of Bluetooth device is about 10 m. However, the main disadvantage of the Bluetooth technique is its high energy consumption; usually, Bluetooth cannot be used by sensors that are powered by a battery. On the other hand, a ZigBee-based WSN is a security network for short-distance communication based on the IEEE 802.15.4 standard. It has been gradually becoming more used by low-rate wireless personal area networks (LR-WPANs), and it is becoming a media access control (MAC) standard. ZigBee offers powerful networking capabilities. It is capable of supporting three types of self-organizing wireless networks: star, network, and cluster structures. ZigBee technology has the following characteristics: (1) it has low power consumption, (2) it is inexpensive due to opening standard, (3) it supports communication in a limited range, (4) it provides a low rate and short latency, (5) it has high capacity, and (6) it possesses high security. ZigBee also offers anti-interference ability.

7.3.5.7.2 Integration of RFID and WSN

The integration of RFID and WSN is shown as Figure 7.10. The integration of ZigBee sensors with RFID technology makes it possible to identify the targeted

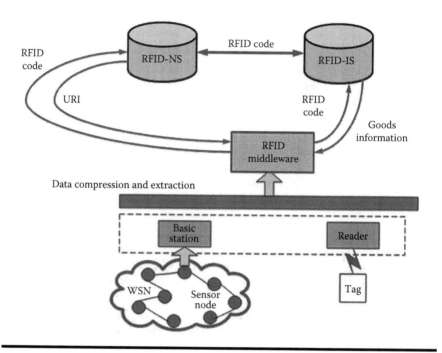

Figure 7.10   Integration of RFID and WSN for IoT.

object globally and perceive the statuses of the object in a real-time basis. The integration of RFID and WSN can expand the scope of applications of the two technologies; this has been identified as the future trend for IoTs.

7.3.5.7.3 System Architecture

A five-layer system architecture has been proposed to illustrate the integrated information framework in Figure 7.11. To fully take advantage of RFID and WSN technologies, system architecture contains five logical layers: "physical layer," "data link layer," "network layer," "transport layer," and "application layer." The functions at each layer are explained in the following text.

1. *Physical layer*: The functions in the physical layer include channel selection, signal monitoring, and sending and receiving messages. The design goal in this layer is to minimize energy consumption and increase link capacity.
2. *Data link layer*: This layer ensures that the physical data can be transmitted correctly; it also relates to the system spectrum efficiency and secures communication among equipment. Based on the IEEE802 standard, the data link layer consists of two sublayers: logical link control (LLC) and MAC. LLC ensures the security and reliability of transmission. MAC provides the

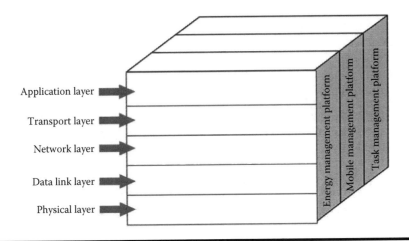

Figure 7.11 Architecture for integration of RFID and WSN.

service interface, including point-to-point communication that allows the upper layer to access physical channels. To reduce power consumption and to adopt the caching mechanism, the MAC sublayer in IEEE 802.15.4 is based on the 802.11 wireless LAN standards for CSMA/CA access.

3. *Network layer*: The functions at this layer include packet routing and congestion control. Packet routing answers the question of which adjacent node should the current node send its packet to, in order to get it as quickly as possible to its eventual destination. The selection about packet routing may be based on the shortest time, the minimized number of hops, or the shortest distance. Congestion occurs when a node is carrying so much data that the quality of its service deteriorates.

4. *Transport layer*: The functions at this layer are flow control, error control, and quality control, to allow for reliable transmission service. The routing protocols are based on ad hoc technology; it has similar characteristics (i.e., low energy consumption) to the bottom layer of the IEEE802.15.4 standard.

5. *Application layer*: The functions in this layer ensure secure data transmission and provide required services for the given application.

7.3.5.7.4 Data Filtering Algorithm

In this section, a data cleaning algorithm is introduced.

7.3.5.7.4.1 Assumptions and Definitions

It is assumed that there are m readers, $m \times n$ fixed reference tags with attributes that are known, and there are l tags for arbitration. Each reader contains n reference tags that can be detected and located in the working area of the reader. To apply the algorithm, the following concepts and definitions are defined.

$G_{j,k}$= the signal intensity of reference tag k, which belongs to reader j, $j \in (1,...m)$, $k \in (1,...n)$

$W_{i,j}$= the signal intensity of tag i waiting for arbitration on reader j, $i \in (1,...l)$, $j \in (1,...m)$

Definition 7.1 $D_{i,j}$ is the Euclidean distance between the signal intensity of tag i waiting for arbitration on reader j and the signal intensity of the m reference tags of reader j. If the value of $D_{i,j}$ is smaller than that of reader j, it more likely belongs to the reader of the tag i.

$$D_{i,j} = \sqrt{\sum_{k=1}^{n}(W_{i,j} - G_{j,k})^2}, \; i \in (1 \cdots l), j \in (1 \cdots m) \tag{7.1}$$

Definition 7.2 A *triple d (Rid, Tid, τ)* is a set of original data. *Rid* is denoted as unique identification of the reader. *Tid* is denoted as the unique identification of the tag. τ is a timestamp that the tag is detected. Through a triple d, the mapping relationship of the RFID system between physical device and logic identity can be built.

Definition 7.3 "Cross redundant data" are produced by some readers, which are located at the same working area at certain moments. Cross redundant data satisfy the following two conditions:

1. For the two arbitrary triples d_i *(Rid_i, Tid_i, τ_i)* and d_j *(Rid_j, Tid_j, τ_j)* from the same cross space, the condition of $(Rid_i \neq Rid_j) \cap (Tid_i = Tid_j)$ is satisfied.
2. At the same time, the relation of $(\tau_i = \tau_j) \cup (|\tau_i - \tau_j| \leq \sigma)$ is also satisfied, where σ is the time threshold.

Definition 7.4 "Affiliation" is the possibility that tag i responds to reader j within the induction scope of reader j at the moment t; further, it is denoted as follows:

$$F(i,j,t) = \frac{1}{3}\sum_{k=0}^{2} p(i,j,t-k) \; i \in (1 \cdots l), j \in (1 \cdots m) \tag{7.2}$$

$p(i, j, t)$ is the frequency that reader j links to tag i at the moment t.

7.3.5.7.4.2 Improved Cross Redundant Data Cleaning Algorithm — The
cross information flow tuples that come from a set of readers will set up a hash index via the attribute *Rid* as the keyword. The node of index is called tag tuple (TG); it includes all of the cross information tuples that have the same attribute *Rid*. Since the tuples of TG may come from different readers, building a second index via the attribute *Tid* as the keyword is considered. In the second index, the content of every node is certain tuple cache queues that come from the reader and have a time sequence.

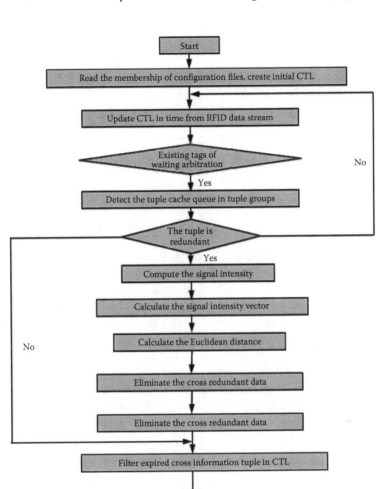

Figure 7.12   Flowchart of ICRDC algorithm.

With the constant incoming of tuples, the structure needs to be maintained incrementally. The steps involved in this algorithm are as follows, and they are further illustrated in Figure 7.12.

1. Read the membership of configuration files that bind with reference tags and readers; create an initial CTL.
2. Update CTL in time from the RFID data stream.
3. Check if the existing tags of waiting arbitration are nearby reference tags in CTL. If the answer is no, go to step 8; if there are some arbitration tags, go to step 4.

4. Detect the tuple cache queue in tuple groups and check whether or not the tuple is redundant.
5. Compute the affiliation $F(i, j, t)$.
6. Compute the signal intensity W_{ij}.
7. Calculate the signal intensity vector $G_j = (G_{j,1}, G_{j,2}, \dots, G_{j,n})$.
8. Calculate the Euclidean distance. Through an exclusive process, cross redundant data can be eliminated.
9. Filter the expired cross information tuple in CTL.

7.3.5.7.5 Numerical Analysis

To evaluate the performance of the algorithm, similar experimental setups (compared to other methods) have been prepared. The results of this algorithm have been compared with those obtained by the two state-of-the-art algorithms CRDC and SMURF. RFID networks with different numbers of sensor nodes have been tested, and experimental areas in which the locations of readers and tags are randomly generated have been set up. The compressibility of accuracy of the algorithm has been compared with those of CRDC and SMURF. With an increase in the experimental areas or the number of readers and tags, the rate of compressibility and accuracy has been quickly improved. All three algorithms are effective in improving the compressibility and accuracy of redundant data, but the improvement from improved cross redundant data cleaning algorithm (ICRDC) is more significant. The experiments show that ICRDC can clean redundant RFID data effectively and, more importantly, will not affect the integrity of RFID data.

7.3.5.7.6 Applications of Integrated Architecture

The developed integrated architecture and data cleaning algorithm can be applied in many different applications. In this section, its application in a relief supply management system is introduced. In rescue efforts after disasters occur, it is important to ship relief supplies in a timely manner and to manage the relief process efficiently. It is always difficult to monitor and manage the distribution of relief supplies in vast geographic areas. With RFID and WSN technologies, the implementation of intelligent logistic management and storage systems becomes possible, since such systems can play a significant role in acquiring information on resources, scheduling, and deploying relief supplies automatically. These systems can collect real-time data and then analyze, assess, and monitor the processes of distribution to ensure that relief supplies are distributed appropriately in a timely manner.

As shown in Figure 7.13, the integrated system consists of four components: relief supplies detection, relief supplies control, relief supplies warehousing, and relief supplies decision support.

The relief supplies detection module is used to detect all the sensors and collect all data. The source of the data includes various sensors, attributes stored in

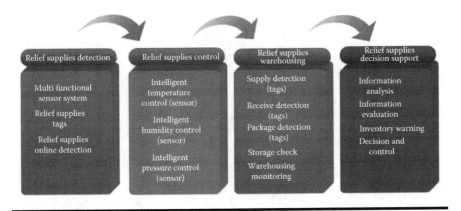

Figure 7.13 Components of a relief supplies management system.

RFID tags, the locations of the relief supplies, and many others. All of the data are transmitted to the host data distribution center through the ZigBee-based WSN. The relief supplies control module keeps the data stored in RFID tags along with the environmental data perceived by sensors regarding temperature, humidity, and pressure. This module also conducts comprehensive analysis and assessment; it is equipped with a software system for real-time monitoring and displaying. The software system can verify the attributes and can control the distributing process. The relief supplies warehousing module is used to manage the relief supplies to be stored into specified units of storage. It monitors the transportation of relief supplies from suppliers to storage. If errors occur during the process, this module recommends the solution.

The decision support module is used to analyze transmission data using business intelligence tools. It focuses on the basic attributes of relief supplies and is capable of rescheduling the distribution based on the latest needs of the various relief procedures. It verifies the locations of relief supplies to determine whether or not the supplies have been transported to destinations.

7.3.5.7.7 Implementation

The deployment of relief supplies management using RFID and ZigBee technology is shown in Figure 7.14. In order to manage relief supplies more effectively, a number of readers and terminals are set in the warehouse. The steps include the following:

1. For each piece of supply material, an RFID tag or another type of tag is attached. Readers are installed in the warehouse. This allows the reader to obtain required information from tags when the supplies pass through the entrance.

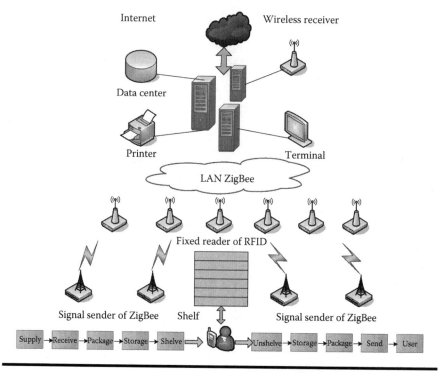

Figure 7.14 Deployment of relief supplies based on RFID and ZigBee.

2. In the warehouse, a certain number of RFID handheld/car terminals are set among shelves and exits. This is a convenient way to check inventory, track supplies, and retrieve accurate information on relief supplies. The network can ensure that the warehouse acquires accurate data.

3. In the storage areas, based on the size of area, the distance signal transceiver module is set to form a backbone wireless network. Especially on the main roads, a number of ZigBee nodes are installed. When a vehicle with an integrated label passes through the coverage area, the monitoring and controlling can determine the location of vehicle in a real-time basis.

With the consideration of dynamic, distributed, and decentralized environments, RFIDs and WSNs become the better choice for many applications over traditional wired networks. In this application example, the focus is on the integration of RFID and WSN, a five-layer system architecture has been developed to integrate WSNs with the RFID systems, and ZigBee has been selected as the communication protocol for the WSN, in order to meet the requirements of a large number of sensor nodes, vast areas, low cost, etc. To eliminate any redundant data in the integrated WSN, an improved cross redundant algorithm (ICRDC) has been developed. It has been verified via simulation and a comparison study that ICRDC can effectively improve data compressibility and accuracy. To validate the feasibility and effectiveness of the integrated architecture and the algorithm, they have been applied to real-world relief supplies management.

7.3.6 Challenges

Cooper and James (2009) discussed the challenges of data management in IoT. Databases are distributed that are different from traditional centralized ones; a very large number of nodes handle volumes that are vast, the speed is fast; and the data/information space is global. Currently, there are fast, reliable, inexpensive e-infrastructures that provide communication services. However, the required new network is far more complicated, since the web is growing not arithmetically but exponentially.

As far as the applications are concerned, many challenges exist in adapting IoT. Researchers have discussed the challenges of IoT in the aspects of data explosion, data interpretation, fault tolerance, interaction, power supply, scalability, security and privacy, software complexity, and interoperability. Atzori et al. (2010) agreed that it is challenging to make a full interoperability of interconnected devices possible, to enable the adaptation and automation for a high degree of smartness, and to assure security and privacy. Sperner et al. (2011) emphasized that a unified reference architecture is a key prerequisite for realizing interoperability with the IoT for integration with business processes. The concept called Business Process Model and Notation 2.0 was proposed. It focuses on activities and the implicated flow of process.

The key technological drivers, potential applications, and challenges in IoT have been studied recently. Researchers discussed ways in which the current existing "Intranet of things" can be evolved into an integrated and heterogeneous system, that is, IoT, in which the challenges are analyzed. IoT so far is a vast and mostly unexplored territory. IoT is seen as a pillar of the future Internet. It is believed that there is a clear need for developing a reference architectural model that will allow interoperability between different systems. The main enabling factor of IoT is the integration of several technologies and communication solutions: identification and tracking technologies, networks, enhanced communication protocols, and distributed intelligence for smart objects. From the perspective of business users, the most important considerations will be in automation, manufacturing, logistics, business process management, intelligent transportation, etc.

7.3.6.1 Future Work

IoT emphasizes the connections between things, which cover the reasoning, context-awareness, and transactions. The emerging technologies, including sensing, ubiquitous computing, cloud computing, wireless sensing, make IoT capable of connecting M2M networks, sensor networks, and even the ubiquitous networks. IoT will provide our daily lives with more connectivity and intelligence. The trend in IoT is the fusion of sensing and the Internet, in which all sensed things are able to provide services intelligently.

7.3.6.2 SOA for Internet of Things

The key idea of IoT addresses the fact that things are interconnected. A well-designed IoT framework is necessary to guarantee the operations of IoT; it helps to reduce the gap between the physical and virtual worlds. The architecture of IoT involves multiple key issues including architecture design, networking and communication, smart objects, services and applications, business models and corresponding processes, cooperative data processing, and security. Figure 7.15 summarizes the basic SOA of IoT, which comprises four layers, depending on the basic functionalities:

- *Sensing layer*, which is integrated with existing hardware to sense the information of things
- *Networking layer*, which is the basic networking support over a wireless or wired network

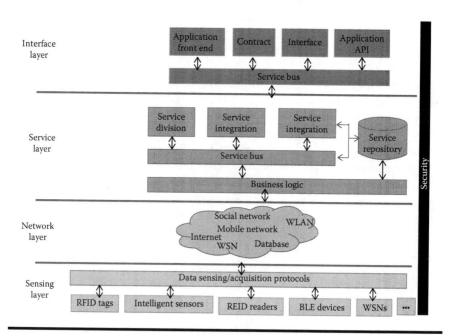

Figure 7.15 Service-oriented architecture for IoT.

- *Service layer*, which is used to create and manage services to users or applications
- *Interface layer*, which provides the interaction methods to users/applications

The SOAs can be imperative for the service providers and requestors in IoT and can help achieve interoperability between heterogeneous devices. In designing the architecture for IoT, the extensibility, scalability, and interoperability among heterogeneous devices and their business models should be taken into consideration. Since things move on a real-time basis and interact with their environment, an adaptive architecture is required in order to make devices interact dynamically with other things to support the unambiguous communication of events. The decentralized and heterogeneous nature of IoT requires an architecture that can provide IoT with efficient event-driven capability.

An SOA approach will allow the decomposing of the complex system into multiple simpler and well-defined subsystems in IoT; by doing this, the software and hardware in IoT can be effectively reused. SOA has been successfully applied in existing WSNs. For IoT, a commonly accepted SOA should provide IoT extensibility, scalability, modularity, and interoperability among heterogeneous things, where the functionalities and capabilities can be abstracted into a common set of services. An extension of the SOA vision of IoT paradigm is shown in Figure 7.15 (Roman and Lopez, 2009).

7.3.6.2.1 Sensing Layer

IoT can be considered as a truly interconnected network, in which things can be connected and controlled remotely. The ensemble of applications and services leveraging such technologies opens a plethora of new business and market opportunities in many areas, including healthcare and industrial processes. In the sensing layer, the wireless smart systems on tags or sensors are able to automatically sense and exchange information between devices.

In the past few years, technological advances in sensing and wireless communication capabilities have made devices equipped with RFID or intelligent sensors more accessible and more versatile. The capability of IoT to sense and identify things or environments has been significantly improved, and therefore its usability has been enhanced for more and more applications. In IoT, the identification capability and retrievability make a thing hold a digital identity that can be easily specified in the digital domain. In some industry sectors, an intelligent services deployment scheme has been developed in which each service is assigned a UUID that can be recognized by other things that need that service. Through this, a device can be easily used to look up and retrieve the appropriate information. Such UUIDs are critical for the deployment of successful services in a huge network like IoT.

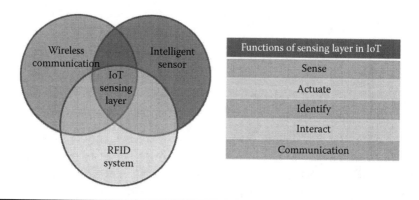

Figure 7.16 Functions of sensing layer in IoT.

In designing the sensing layer of an IoT (see Figure 7.16), attentions have been paid to the following issues:

■ Cost, size, resource, and energy consumption: The things might be equipped with sensing devices such as RFID tags, sensor nodes, or intelligent devices that should be designed with limited resources and should be cost effective.
■ Deployment: The sensing things (RFID tags, sensors, etc.) in IoT might be deployed on a one-time, incremental, or random basis, depending upon the requirements of applications.
■ Heterogeneity: The variety of things with different properties makes the IoT very heterogeneous.
■ Communicability: All things should be accessible, retrievable, and communicable.
■ Networked: Things are organized as multihop, mesh, or ad hoc networks.

IoT is a very complex network in which a large number of hardware and software platforms are involved. IoT may consist of a number of heterogeneous systems, for which the following issues need to be addressed:

■ Energy-efficient communications between things
■ Cognitive radios of things
■ Communication spectrum and frequency allocation
■ Wireless coexistence of WLAN, ZigBee, Bluetooth, etc.

In the sensing layer, the hardware issues include wireless identifiable systems design, ultra-low-cost tags, and smart/mobile sensors.

7.3.6.2.2 Networking Layer

The role of the networking layer in the IoT is to connect all things together. In the networking layer, things can share the sensed information (that is very important for intelligent event processing and management related to IoT) with the connected things. Even more, the networking layer might be capable of aggregating information from existing IT infrastructures (e.g., business, transportation, power grids, healthcare, and ICT systems). In SOA-IoT, services are always provided by things that are deployed in the heterogeneous network. This process might involve a QoS guarantee for the services that is very different according to the requirements of its users/applications. On the other hand, it is essential for a dynamically changing network to automatically discover and map things in networks. Things need to be assigned roles automatically to deploy, manage, and schedule the behaviors of things and to be able to switch to any roles at any time, as required. This enables devices to collaboratively perform tasks. In the networking layer, following issues required to be addressed:

■ Network management technologies for heterogonous networks, including fixed, wireless, and mobile
■ Energy efficiency in networks
■ QoS requirements
■ Discovery and search engine technologies
■ Data and signal processing
■ Security and privacy

Information confidentiality and human privacy security are two challenging issues in IoT. For information confidentiality, the encryption technology can get help from the WSN or from other networks. In IoT, as many daily-life things are connected, the issue of privacy becomes serious. It should be noted that many issues are still need to be studied such as the privacy and legal aspects. Existing network security technologies can provide a basis for privacy and security in IoT, but more work still needs to be done.

7.3.6.2.3 Service Layer

The service layer relies on the middleware technology, which is a key enabler of services and applications in IoT. Middleware technology can provide the IoT with a cost-efficient platform in which the hardware and software platforms can be reused.

The service layer in IoT involves activities in the field of middle service specifications; most of them are undertaken by various standards that are developed by different organizations. Therefore, a commonly accepted service layer is important for IoT. A well-designed service layer will be able to identify a minimum set of common requirements of applications, application programming interfaces (APIs), and protocols to support required applications and services.

This layer processes all service-oriented activities, including information exchange and storage, the management of data, an ontologies' database, search engines, and communication. This layer includes the following components:

- Service discovery: Finding objects that can provide the required service and information in an efficient way.
- Service composition: Enabling interaction among connected things. The discovery phase exploits the relationships of things to find the desired service, and the service composition component is able to schedule or re-create more suitable services in order to obtain the most reliable services for the request.
- Trustworthiness management: Aiming at understanding how the information provided by other services has to be processed.
- Service APIs: Providing the interactions between services required in IoT.

7.3.6.2.4 Interface Layer

In IoT, a large number of devices, which have been made by different manufacturers and hence do not always comply with the same standards, are involved. These heterogeneous things might inevitably cause interaction problems between things, such as information exchanging, communication, and cooperative processing of dynamical events. The increasing number of things in an IoT makes it difficult to dynamically connect, communicate, disconnect, and operate. There is a strong need for an effective interface mechanism to simplify the management and interconnectedness among and between things.

An interface profile (IFP) can be seen as a subset of service standards that allows minimal interaction with applications running on application layers. One illustration of an interface layer is the implementation of Universal Plug and Play (UPnP), which specifies a protocol for seamless interaction among services that are offered by different things. IFPs are used to describe the specifications between applications and services. The services on the service layer run directly on limited network infrastructures in order to effectively find new services for an application as it connects to the network, to dynamically retrieve metadata about it and the services it hosts. A SOCRADES Integration Architecture has been proposed that can be used to effectively interact between applications and the service layer. A representational state transfer is defined to increase interoperability for a loose coupling between services and distributed applications, which will make the interaction between services and applications more effective. Traditionally, the service layer provides a universal API for applications. Recent research on SOA-IoT shows that a service provisioning process can effectively provide interaction between the applications and services.

7.3.7 Open Problems and Future Directions

7.3.7.1 Technical Challenges

In addition to the challenges introduced earlier, although a lot of IoT research efforts have been made, many other challenges still exist:

1. In designing a service-oriented architecture, service-based things related performance and is a challenge. In addition, the automated service composition according to the requirements of applications is still a challenge.
2. From the viewpoint of the network, IoT is a very complicated heterogeneous network, which includes the connection between various types of networks based on various communication technologies. Developing devices for addressing and optimizing things management is a challenge.
3. From the viewpoint of service, the lack of a commonly accepted service description language makes the services development incompatible in different implementation environments. A powerful services discovery and search engine should be helpful to spread the IoT technology.
4. IoT is taking place in an ICT environment, and it is affected by everything connected. It is a challenge to integrate IoT with the current IT systems.

7.3.7.2 Standardization

The rapid growth of IoT makes standardization difficult. The standardization in IoT aims at lowering the entry barriers to the new service providers and users, which can improve the interoperability and can allow products or services to better compete at a higher level. The open standards in IoT (such as security standards, communication standards, and identification standards) might be the key enablers for the spread of IoT technologies and will embrace some cutting-edge technologies. The specific issues regarding IoT standardization include interoperability, radio access level issues, semantic interoperability, and security and privacy issues.

7.3.7.3 Information Security and Privacy Protection

Social acceptance of new IoT technologies and services will strongly rely on the trustworthiness of the information and the protection of private data. Although a number of projects have been launched in the area of security and privacy, a reliable security protection mechanism for IoT still needs to be researched:

Technically, the following issues need to be solved: (1) the definition of security and privacy from the viewpoint of social, legal, and cultural aspects; (2) a trust mechanism; (3) communication security; (4) the privacy of communication and user data; and (5) security on services and applications.

7.3.7.4 Innovation in IoT Environment

IoT is a complex network that might be managed by a number of stakeholders, wherein services should be provided openly. Therefore, open new services/applications development should be supported without excessive market entry or operation barriers. The cross-domain systems-supported innovation is a challenge for IoT.

7.3.7.5 Development Approaches

Currently, depending on the area or country, IoT has been developed following three approaches:

- Opportunity investment approach: In the United States, the short- or mid-term return on investment drives the development of smart energy and others.
- Stakeholder approach: In the EU, the IoT are developed using a "stakeholder approach" in which a number of short-term (4–5 year) IoT projects are launched by public–private partnership investments. This approach is cost-efficient and convenient and has been widely used in some IoT applications such as healthcare, automotive, and home appliances.
- Integrated approach: In China, the IoT infrastructure, software, and services/applications are integrated. A number of state-supported projects have been launched (such as "sensing China"), which are trying to integrate IoT fully into the IT architecture.

Although each approach has its unique merits, all of them can promote the fundamental development of IoT.

In the past few years, IoT has developed very quickly, and a large number of fundamental technologies have been proposed. This section introduces the recent research on IoT from the viewpoint of technologies. We introduce the service-oriented architecture models of IoT and the fundamental technologies that can be used in IoT. We introduce the applications of IoT from several aspects: business, social networks, healthcare, infrastructure, security, and surveillance. We list the open research problems in IoT. The IoT is the trend of next Internet. Relating to the main theme of this book, IoT will greatly impact on future-generation ESs in the foreseeable future.

References

Alam, S., Chowdhury, M., and Noll, J. 2011. Interoperability of security-enabled Internet of Things. *Wireless Personal Communications*, 61(3), 567–586.

Atzori, L., Iera, A., and Morabito, G. 2010. The Internet of Things: A survey. *Journal of Computer Networks*, 54(15), 2787–2805.

Bonivento, A., Fischione, C., Necchi, L., Pianegiani, F., and Sangiovanni-Vincentelli, A. 2007. System level design for clustered wireless sensor networks. *IEEE Transactions on Industrial Informatics*, 3(3), 202–214.

Cooper, J. and James, A. 2009. Challenges for database management in the Internet of Things. *IETE Technical Review*, 26, pp. 320–329. Available at http://tr.ietejournals. org/text.asp?2009/26/5/320/55275.

Fielding, R. and Taylor, R. 2012. Principled design of the modern web architecture. *ACM Transactions on Internet Technology*, 2(2), 115–150.

Fox, M., Chionglo, J., and Barbuceanu, M. 1993. *The Integrated Supply Chain Management System*. Toronto, Ontario, Canada: University of Toronto.

Gama, K., Touseau, L., and Donsez, D. 2012. Combining heterogeneous service technologies for building an Internet of Things middleware. *Computer Communications*, 35(4), 405–417.

Giachetti, R. 2004. A framework to review the information integration of the enterprise. *International Journal of Production Research*, 42, 1147–1166.

Guinard, D. et al., 2010. Interacting with the SOA-based Internet of Things: Discovery, query, selection, and on-demand provisioning of web services. *IEEE Transactions on Service Computing*, 3(3), 223–235.

Haller, S., Kanouskos, S., and Schroth, C. 2009. The Internet of Things in an enterprise context. Domingue, J., Fensel, D., Tranerso, P. (Eds.), In *Future Internet Systems (FIS)*, LNCS, Vol. 5468, Heidelberg, Germany: Springer, pp. 14–28.

Hasselbring, W. 2000. Information system integration. *Communications of ACM*, 43, 32–38.

Heer, T. et al. 2011. Security challenges in the IP-based Internet of Things. *Wireless Personal Communications*, 61(3), 527–542.

Hernandez-Castro, J., Tapiador, J., Peris-Lopez, P., and Quisquater, J. 2008. Cryptanalysis of the SASI ultralightweight RFID authentication protocol with modular rotations. CoRR 2008.

Hsu, C. and Wallace, W. 2007. An industrial network flow information integration model for supply chain management and intelligent transportation. *Enterprise Information Systems*, 1(3), 327–351.

ITU. 2005. The internet of Things, International Telecommunication Union (ITU) Internet Report, Geneva, Switzerland.

Juels, A. 2006. RFID security and privacy: A research survey. *IEEE Journal on Selected Areas in Communications*, 24(2), 381–394.

Kumar, S. and Budin, E. 2006. Prevention and management of product recalls in the processed food industry: A case study based on a porter's perspective. *Technovation*, 26, 739–750.

Logeais, G. 2013. The Internet of things in the context of manufacturing. SAP Research, http://docbox.etsi.org/workshop/2008/200812_WIRELESSFACTORY/SAP_LOGEAIS.pdf. Accessed on October 2013.

Mangina, E. and Vlachos, I. 2005. The changing role of information technology in food and beverage logistics management. *Journal of Food Engineering*, 70(3), 403–420.

Miorandi, D., Sicari, S., De Pellegrini, F., and Chlamtac, I. 2012. Survey Internet of Things: Vision, applications and research challenges. *Ad Hoc Networks*, 10(7), 1497–1516.

Mousavi, A. et al. 2005. Tracking and traceability solution using a novel material handling system. *Innovative Food Science and Emerging Technologies*, 6, 91–105.

Pang, Z. 2013. Technologies and Architectures of the Internet-of-Things (IoT) for Health and Well-being. PhD Thesis, Royal Institute of Technology.

Pretz, K. 2013. The next evolution of the internet. http://theinstitute.ieee.org/, online available. Accessed on October 2013.

Puschmann, T. and Alt, R. 2004. Enterprise application integration systems and architecture-the case of the Robert Bosch group. *Journal of Enterprise Information Management*, 17, 105–116.

Roman, R. and Lopez, J. 2009. Integrating wireless sensor networks and the internet: A security analysis. *Internet Research*, 19(2), 246–259.

Sperner, K., Meyer, S., and Magerkurth, C. 2011. Introducing entity-based concepts to business process modeling. In *Lecture Notes in Business Information Processing*, Springer, New York, Vol. 95, pp. 166–171.

SRI International. 2008. Disruptive civil technologies: Six technologies with potential impacts on US interests out to 2025, National Intelligence Council.

Sun, E., and Zhang, X. 2012. The tailings dam monitoring and pre-alarm system (TDMPAS) and its applications in mines. *Proceedings of SME Annual Meeting*, February 2012, Seattle, WA, pp. 232–236.

Uckelmann, D., Harrison, M., and Michahelles, F. 2011. An architectural approach towards the future Internet of Things. In Uckelmann, D. et al. (Eds.), *Architecting the Internet of Things*, Berlin, Germany; Heidelberg, Germany: Springer-Verlag, pp. 1–24.

van Kranenburg, R. and Bassi, A. 2012. IoT Challenges. *Communications in Mobile Computing* 2012, 1, 9, doi:10.1186/2192-1121-1-9.

Vermesan, O. 2013. Internet of Things vision and the technology behind connecting the real, virtual and digital worlds. www.grifs-project.eu/data/File/CERP-IoT%20SRA_IoT_v11. Accessed on October 2013.

Wang, L., Lin, Y., and Lin, P. 2007. Dynamic mobile RFID-based supply chain control and management system in construction. *Advanced Engineering Informatics*, 21, 377–390.

Welbourne, E., Battle, L., Cole, G., Gould, K., Rector, K., Raymer, S., Balazinska, M., and Borriello, G. 2009. Building the Internet of Things using RFID: The RFID ecosystem experience. *IEEE Internet Computer*, 13(3), 48–55.

Wolfert, J. et al. 2010. Organizing information integration in agri-food-a method based on a service-oriented architecture and living lab approach. *Computers and Electronics in Agriculture*, 70, 389–405.

Zang, C. and Fan, Y. 2007. Complex event processing in enterprise information systems based on RFID. *Enterprise Information Systems*, 1(1), 3–23.

Chapter 8

Industrial Information Integration

8.1 Enterprise Application

As enterprises realize that the nonintegrated nature of existing information infra-structures prohibits them from complete success, the efficient coordination of intra- and/or interorganizational business processes can be achieved only by inte-gration, and the importance of a seamless integration of both intra- and interorga-nizational business processes, a streamlining enterprise integration for enterprises and extended enterprises has been a trend since the mid-1990s. Enterprises realize that they need to efficiently align, integrate, and manage all of the relevent aspects of business processes in order to accomplish the strategies and objectives of their enterprise. Enterprise integration has now become a key issue for many enterprises, in order to facilitate business processes through integrating and by streamlining processes both intra- and interorganizationally.

Enterprise application (EA) is the term that is used to describe the applications that an enterprise or an extended enterprise would use to share data, functionality, and processes for selective applications. The scope of EA is determined by needs. EA involves enterprise integration and assists an enterprise or an extended enter-prise in business operations, since it is clear that one of the major concerns today is the integration of applications within the enterprise or the integration of applica-tions in extended enterprises.

EA involves conceptual frameworks, methods, technologies, platforms, and tools. In an intraorganizational context, EA is supposed to provide infrastruc-tures for the integration of all or partial aspects of an organization's operations and processes. Theoretically, it is expected to integrate all relevant functional units within the organization by using systems such as intraorganizational ESs. It may involve interconnecting heterogeneous applications. In an interorganizational context, EA is supposed to provide infrastructures for integration across organizations in an extended enterprise context. Based on the definition mentioned earlier, an EA is an application that applies to an enterprise or an extended enterprise, that is conceptualized, designed, and implemented to integrate all or only the relevant applications used within or beyond the organization, and that creates an environment for efficient operations. An EA can be an application with various types of resources integrated. Its platforms can include BPM, CRM, BI, and ESs. For many enterprises, some existing applications have already been developed, which assist individual business operations. As such, the integration of existing applications with selected new applications is an economical approach to achieving process improvement and intra- and interorganizational collaboration. From the technological point of view, structures that support both existing application and new application integration should correspondingly be developed.

The term "application" can have several meanings. An EA first is a business application, then it is an IT application. In an IT application, a coherent set of components are used to conceptualize and computerize business processes for the involving domains in an organization. Research indicates that EAs can be characterized by heterogeneity, autonomy, dynamics, distributivity, and complexity.

Heterogeneity means that an EA may have its own data and process model and that it involves the heterogeneities in a variety of aspects, including semantic, technical, and other heterogeneities. The technical issues related to the heterogeneity are mainly caused by different data formats, devices, operating systems, networking technologies, middleware, programming languages, services, and interface technologies. Heterogeneity is one of the main research topics for enterprise integration, and many methods and techniques for it have been proposed in the literature. For example, for the multilingual semantic interoperations used in interorganizational ESs, semantic consistency is a research issue that has not been well resolved. Recently, methods for using multilingual semantic interoperation in interorganizational ESs have been proposed, which contribute to improving multilingual semantic interoperation by proposing a concept-connected near-synonym framework. Although standardized solutions are adopted in enterprise integration, technologies may still remain heterogeneous. As such, enterprise integration technology has difficulty in providing a common solution. One of the objectives of building industrial information infrastructures is to reconcile the heterogeneity of applications as well as other relevant resources. Autonomy means that an EA may be independent of other EAs. Autonomy is a property that is contradictory to the desirable property of interoperability. One of the challenges of enterprise architectures including SOA is to reconcile such contradictions. In EAs, one of the goals is to achieve the coexistence of autonomy and interoperability through technological means. "Dynamics" means that EAs may be required to rapidly adapt to the changes that take place in the environment. "Complexity" means that, just as in other large software systems, an EA may be treated as a black box. This means either that detailed codes and/or interfaces are not available for certain purposes or that a gray system (i.e., interfaces) may be available but not the detailed codes, due to the complexity of the system. Due to its complex characteristics, the integration task of EA is challenging. As such, before the concept of EA with enterprise integration emerged and its significance was well recognized, the phenomenon called "island of information" or "islands of automation" was not uncommon at all.

8.1.1 Intraorganizational EA

The purpose of intraorganizational EA is to manage a complete set of enterprise resources as described by the concept of entire resource planning proposed by Xu (Wu et al., 2009; Xu, 2011). Before the concept of entire resource planning emerged, intraorganizational EAs were mainly seen as ES applications as exemplified by ERP. An intraorganizational EA, within the context of ESs, includes applications such as PDM, MES, CAD, and all other related ES components, especially those supporting generic ES processes. In the oil industry, as an EA, RMES has been developed to fill the gap left from traditional ERP and to both enable consistent information flow to the enterprises and improve the agility of enterprises. In the energy and environmental protection sectors, in the framework of IIIE, EA efforts have been made to integrate the processes of the reverse logistics of used batteries and to contemplate the integration of related information flows in reverse logistics, as well as to examine the contents, sources, and collection of the reverse logistics of used batteries. Through such work, an EA in reverse logistics of used battery was developed. Such EAs are typical examples of intraorganizational EAs.

8.1.2 Interorganizational EA

The purpose of interorganizational EA is to manage a complete set of enterprise resources for multiorganizations as described by entire resource planning proposed by Xu (Wu et al., 2009; Xu, 2011). Before the concept of entire resource planning emerged, interorganizational EAs were mainly seen as ES applications, as exemplified by ERP. Applications that support interorganizational operations work with the concepts of extended enterprises and SCM. Due to the interorganizational nature of material flows, the reverse logistics system of used batteries mentioned earlier is an excellent example of how an EA initialized intraorganizationally can eventually evolve into an interorganizational one.

EAs have certain requirements to meet. For example, an EA is expected to enhance the efficiency and productivity of an organization with certain requirements involved. Such requirements include the integration requirement, the flexibility requirement, and the new technology requirement.

8.1.3 Integration Requirements

EAs require integration, which facilitates cooperative operations in a heterogeneous computing environment. Erl (2004) specified the integration requirements: (1) Application x can access Application y for data retrieval or exchange, (2) Application x can access Application y for its processing logic, (3) Application x and Application y exchange data and business processing logic and collaborate for certain business processes, and (4) Applications x and y coordinate for an *ad hoc* application to support a certain process.

"Flexibility requirements" mean the agility of the application to quickly react to constantly changing business requirements and technological evolution.

8.1.4 New Technology Requirements

One of the trends in EAs is to move toward cloud computing, in which an enterprise moves part of an infrastructure or its entire infrastructure to the cloud platform, a type of Internet-based computing, where services are delivered to an organization through the Internet as an on-demand service.

8.2 Integration Approaches

Enterprise integration has been intensively researched since the late 1980s. Enterprise integration aims at achieving a collaborative working environment either or both intra- and interorganizationally involving data flow, process flow, and resource flow with relevant techniques and methods, making the enterprise behave as a "whole." The integration methodology provides the architecture and the components required for an enterprise integrated in the enterprise integration context. For the framework of enterprise integration, the focus can be either or both intra- and interorganizational. If an intraorganizational focus is emphasized, enterprise integration is assumed to cover the entire or partial enterprise. If an interorganizational focus is emphasized, enterprise integration is expected to cover the entire or partial extended enterprise or supply chain. Enterprise integration can happen at the organizational level, the business process level, the IT level, the system level, the data level, etc. As enterprise integration has been applied to many sectors in various manufacturing and service industries, the integration of enterprises in different industrial sectors has attracted considerable research attention. It is generally agreed that enterprise integration consists of a set of methods, techniques, and tools that aim at facilitating coordination and synergistic cooperation within the organizational context for a purpose such as EA. The objective, then, is to achieve full-scale coordination of all relevant systems and subsystems so that they will work together in an application or applications that target the achievement of the strategic plan of the organization.

It is helpful to define the roles played by enterprise integration and integrated information infrastructures. Enterprise integration, as mentioned earlier, is an effort, first, to integrate the organizational structures and functionalities of an enterprise or an extended enterprise for strategic competition. To achieve such goals, an appropriate information infrastructure for the required integration is necessary. As any enterprise moves toward enterprise integration, it is necessary to create an integrated information infrastructure facilitated by methods in IIIE and by systems such as ESs. From the standpoint of enterprise integration, ESs seem to provide a partial technological means.

An enterprise may have existing legacy systems and may expect them to continue to serve while adding or migrating to a new set of applications. To address this issue, an EA solution with integration methodological support that can help achieve quality integration is referred to as EAI. EA is usually addressed with selective individual applications, and it is closely related to enterprise integration. The relationship between them is complementary. The implementation of EA with enterprise integration can be facilitated by ESs. In other words, EAI is the integration that generates interoperability between dissimilar applications with integration techniques and technologies and allows them to share data and business processes through connected systems such as ESs. EAI may use various software to integrate EAs and may not invariably involve the discarding of legacy systems or current applications. This illustrates that EA is usually application oriented and occurs individually or discretely. In the process of EA, enterprise integration is generally required. The enterprise integration that applies to EA is called EAI.

EAI is characterized by involving and integrating data and processes, reusing relevant and applicable data and processes, and generally requiring no discarding of or few changes to the existing infrastructures (although there can be technical integration issues with the existing legacy systems). Just as in EA, EAI is generally not approached as a single one-time effort. It is a continuous effort of individual integration tasks that individually contribute to a specific EA as it migrates from existing applications to a new application. This process neither happens instantly nor can an EA last forever. Most enterprises have a significant number of EAs, and they may not necessarily expect to integrate every single application simultaneously. They are more interested in having an integration mechanism in place, which will allow them to integrate certain applications at a certain time as needed. As such, many enterprises find it beneficial to develop an initial EAI platform and then to implement integration among selected applications gradually. Practical experience shows that the integration of existing applications with potential new applications is an economical approach that can be used to gain competitive advantages. Once the initial effort is successful, more EAI tasks can be initiated, building upon the existing efforts. From a business point of view, EAI makes it possible to adopt new applications in a flexible and economical way, making it possible to quickly develop new manufacturing or service platforms to improve an organization's business competitiveness.

Originally, EAI was focused only on integrating ESs with intraorganizational applications, but now it has been expanded to cover aspects of interorganizational integration such as extended enterprise or supply chains. It was in the intraorganizational context that much of the value of adopting EAI was first found. As long as a legacy system has much data and is based on dated technology, and as long as there is a need to develop integration links with disparate systems or new applications, EAI is required. As EAI enables the facilitation of the integration of both intra- and interorganizational systems, EAI can link interorganizational applications together to computerize and facilitate business processes to the greatest extent possible, even as it helps to avoid the making of a variety of changes to the existing applications, including data and systems. As mentioned earlier, EAI can seamlessly integrate heterogeneous systems so that isolated business applications can be interoperable and so that data, functionalities, and processes can be shared within an enterprise or an extended enterprise. In simple words, EAI makes information sharing between different existing systems in the enterprise or the extended enterprise possible. EAI is a method as well as a technology for integrating the heterogeneous applications that were originally created with different methods and were based on different platforms. From a technical perspective, EAI addresses integration problems at all integration levels including the data level and the process level throughout an infrastructure. Through creating a sharable structure, EAI connects the heterogeneous data sources, applications, and systems together, either or both intra- and/or interorganizationally (He et al., 2009); thus, the existing systems can be integrated, ensuring that organizational units can be functional while using different systems. Based on this idea, ESs and supply chains can exchange and share data, functionalities, and processes; more importantly, with EAI, an enterprise can have opportunities to redesign core processes and applications for the purposes of operation improvement.

Major EAI-enabling technologies range from EDI to web services and XML-based process integration and provide adaptable, flexible, and scalable EAI frameworks. Solutions comprise the efficient integration of diverse data and business processes across enterprises both intra- and interorganizationally, that is, conversion of varied data representations among involving systems, the integration and interoperation of intra- and interorganizational EAs, and the connection of proprietary/legacy data sources, processes, workflows, applications, and ESs for achieving integration intra- or interorganizationally (Qureshi, 2005). There are many additional advantages of EAI, such as a reduction in data redundancy and function overlapping, the ensuring of a greater degree of data consistency and integrity, and the ability to offer more functions and better services than individual systems (Chen et al., 2007).

The main integration methods can be summarized in terms of integration scopes, integration perspectives, integration layers, and integration levels.

Integration scopes: These mainly distinguish the integration approaches into intra- and interorganizational, as mentioned earlier. Intraorganizational integration is further divided into two subcategories: horizontal integration and vertical integration. "Horizontal integration" intends to motivate the cooperation of units within the enterprise for operations effectiveness while "vertical integration" puts its focus on the effective linkage between upstream and downstream structures. In other words, horizontal integration mainly puts emphasis on enabling different units to cooperate on the same task simultaneously, and vertical integration focuses on coordinating hierarchies to effectively carry out tasks for which they are jointly responsible.

In an intraorganizational scope, horizontal integration refers to cross-functional or horizontal linkage integration within the organization. Its purpose is to make cross-functional cooperation possible through the integration of data, functionalities, and processes based on the core business process of the organization, in order to reduce any low efficiency caused by inappropriately designed functionalities and processes and to improve functional effectiveness and efficiency. In this process, data integration is an essential step in horizontal integration as it makes integrated TP, which can reduce data inconsistency and improve TP efficiency, possible. A higher degree of horizontal linkage integration was the purpose of the early generations of ESs. Horizontal integration targets the integration of those functions and components that are generally considered the main tasks of ESs, including related applications that have already been developed. A typical example of horizontal application is ESs exemplified by ERP, in which an organization is trying to optimize the entire operation by different departments horizontally within the organization. Vertical integration aims at integration to be implemented at different levels of an organization, in order to seek more efficient and effective business processes through different levels in the organizational hierarchy. In other words, vertical integration attempts to integrate data, functionalities, and processes implemented at different levels of an organization in an innovative way. One of the purposes of vertical integration is to develop operational alternatives through a redesigning of business functions and processes at different levels. From a systems perspective, in general, it is not desirable to consider horizontal integration and vertical integration separately. For better operational efficiency, horizontal integration and vertical integration cannot be separated and should be approached systematically and holistically in order to achieve better performance, as horizontal and vertical integrations are complementary to each other from a systems point of view. Systematically approaching the relative positions of horizontal and vertical integrations will encourage an enterprise to achieve better performance, due to the system complexity that is involved.

"Interorganizational integration" mainly represents the integration of applications from different organizations. It involves applications in e-business environment including B2B and B2C. It relates to the third layer in Figure 6.1. The concept of the supply chain is closely related to interorganizational integration. Supply chain integration not only extends the scope of traditional intraorganizational integration but also extends the scope of interorganizational integration to form the concept of the integrated supply chain. B2B models are examples of integrated supply chains, as B2B facilitates the collaboration in the extended enterprises through enabling the integration of systems beyond the corporate walls of a single enterprise. It is predicted that with the development of interorganizational EAI technology, ESs, and supply chains, a supply chain as a "whole" can eventually be integrated together to form a new-generation collaborative e-business platform.

Integration perspectives: As the integration focuses on the modeling and automation of complex processes in heterogeneous environments, and as effective integration relies on quality integration methods, over the last decade, significant efforts have been made in the development of integration methods. Such efforts have been made based upon different modeling perspectives; in fact, some emphasize one view over others. Each aspect of integration can be modeled from different perspectives. These are a few perspectives from which integration can be approached:

1. *Conceptual perspective*: This perspective emphasizes the designers' view in modeling integration.
2. *External perspective*: This perspective focuses on the users' view of the integration (which is closely related to conceptual perspective).
3. *Programming and implementation perspective*: This perspective focuses on programming and implementation aspects.
4. *Systems perspective*: This perspective considers EAI as a complex system (Chen and Cheng, 2009; Qian et al., 1993).

Integration layers: As mentioned earlier, integration can be viewed in terms of layers: the data integration layer, the business process integration layer, the application integration layer, the platform integration layer, as well as the standards integration layer. Linthicum defines two main layers: the business model layer and the data model layer. The business model layer can be further decomposed into the application interface layer, the method layer, and the user interface layer (Linthicum, 1999). Lublinsky (2001) classified integration layers as the process layer, data layer, and message layer. He et al. (2009) classified the integration layers as the process layer, the application layer, the interface layer, and the presentation layer. Although there have been different classifications of the integration layers, research indicates that the taxonomies related to the integration layer share commonalities. Efforts have been made to reconcile the differences between each of the taxonomies, since the commonality does exist in all available taxonomies. The integration can occur at three or more layers. The first layer pertains to the "business integration," which represents the business process coordination and orchestration and link activities among the many different units within an organization. The representative methodologies include CIMOSA and others. The original purpose of CIMOSA is to provide a conceptual framework for enterprise modeling in which "generic architecture" is applicable to all manufacturing enterprises, "partial architecture" applies to a specific industrial sector, and "particular architecture" models the structure of a particular enterprise. The second layer corresponds to "data and application integration," which focuses on the interoperability of applications on heterogeneous platforms including data sharing and accessing remote application services (Izza, 2009). It involves the concept of unified data modeling. The tools used in the process of integration and orchestration include languages such as HTML and XML, and platforms such as CORBA and OMG, in addition to the WfMS. The third layer concerns the "physical system application" such as the integration of operating and communication systems. The representative methods and tools include relevant ISO/OSI standards and specialized protocols such as SOAP.

Integration levels: Integration is supposed to occur at multiple levels, such as business process integration, semantic application integration, syntactical application integration, physical integration, and also intra- and interorganizational integrations. As such, to build an integrated system such as ESs that is capable of providing required functionalities, those incompatibilities (the incompatibilities between different communication networks and protocols that exist at a physical level, and the incompatibilities between data and software, etc., that exist at various levels) must be overcome. As an example, although some existing applications are using XML, incompatibilities exist in both data models and schemas. Another classification of the integration approach is based on the integration of semantic, syntactical, hardware, and platform levels (Izza, 2009). The "semantic level" refers to the interchange of data where there is an explicit representation of the meaning of the data. The "syntactical level" refers to the interchange of data without regard to an explicit representation of its meaning. This includes parameter passing, external data access, and timing mechanisms. Semantic integration may help integrating applications that are more robust and of better quality. The hardware level involves differences in hardware, networks, etc. The platform level involves the differences in database platforms, operating systems, etc.

Integration is of a multidimensional nature. In terms of the scope of integration, horizontal and vertical integrations (as introduced earlier) are considered the broad scope of integration, in which much more complex multidimensional integration is involved. Izza (2009) proposes a 4D framework of the integration: level, layer, scope, and view. This framework constitutes recent efforts in developing a methodological framework for integration techniques, although it has been realized that this framework is still not complete. From a multidimensional perspective, an nD framework is a more appropriate proposal, in which efforts toward multidimensional modeling are desirable. Chapter 9 discusses the law of requisite variety that requires a match between the dimensionality implicitly represented and the dimensionality encompassed. In this law, an integration implicitly represents an integer dimensionality K_s, and an integration method is assumed to define an integer dimensionality K_m. In a 4D framework, the potential risks are any underspecified dimensions due to a currently limited capability in conceptualization and mathematical representation, that is, $K_m < K_s$. A desirable integration method should exhibit requisite variety, that is, $K_m > K_s$. The concept of dimensionality analysis in information systems research, as proposed in 2000, can be applied to build the nD integration methodology (Xu, 2000). The concepts of dimensions and subdimensions, as well as all involving interactions, are particularly useful in this context. With this as a theoretical framework, some interesting results have been obtained (Izza, 2009). For example, different integration approaches have emerged that have resulted from an observation of the interactions that exist at the level of level and scope: (1) a syntactic integration approach that includes syntactic intraenterprise integration and syntactic interenterprise integration approaches and (2) a semantic integration approach that includes semantic intraenterprise integration and semantic interenterprise integration approaches (Izza, 2009). At the level of level and layer, there will be at least three different integration approaches that result from the interactions that exist: (1) a data integration approach that includes both syntactic data and semantic data integration approaches, (2) a service integration approach that includes both syntactic service and semantic service integration approaches, and (3) a process integration approach that includes both syntactic process and semantic process integration approaches. The third instance is layer and scope, in which the different integration approaches resulting from the interactions include (1) a data integration approach that includes both intra- and interorganization data integration approaches, (2) a service integration approach that includes both intra- and interorganization service integration approaches, and (3) a process integration approach that includes both intra- and interorganizational process integration approaches.

An RA is a framework proposed to guide the development of an integrated system such as ESs through the formalization of its methods. RA targets the integration of the entire organization (not only the technological aspects but also other aspects, including the social and economic aspects). Although RA is not as complete as what entire resource planning has proposed, it was an early endeavor to integrate the multiple aspects of an organization. Among the most well-known architectures include CIMOSA, GIM, and PERA. A significant part of this effort in the early 1990s resulted in the establishment of the IFAC/IFIP Task Force on Architectures for Enterprise Integration. The ARDIN Research Project, initiated in the 1990s, is an effort to move RA for enterprise integration forward. The ARDIN RA is organized into five interrelated dimensions: the first dimension guides the development of an ES with business process perspectives in consideration of the capability for dynamic evolution. The second dimension focuses on the conceptual design. The third dimension emphasizes the other relevant enterprise structures. The fourth dimension includes a set of tools that can be applied to the conceptualization, design, and execution phases. The fifth dimension focuses on how an enterprise can better perform for continuous improvement purposes and is capable of adapting to dynamic evolution. Methodologically speaking, ARDIN RA recognizes the importance of those complex interactions among the different dimensions (Chalmeta et al., 2001).

The concept of dimensionality is also important in developing methods for the interoperability of EA, since developing the interoperability of EAs requires input

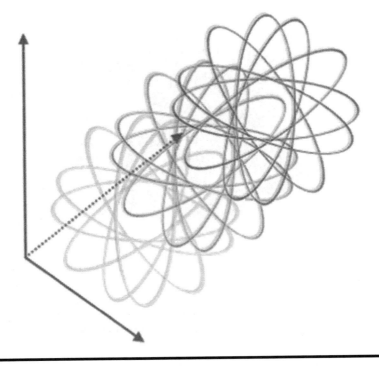

Figure 8.1   An illustration of *n*D framework.

from multiple domains, including enterprise modeling, ontologies, architecture, and platform. The benefit of developing a multidimensional framework is to provide conceptual relationship considerations between different methods, which can give rise to consistency between different models.

The *n*D framework mentioned earlier is interesting and constitutes a framework for developing appropriate integration methods and techniques. Based on the initial concept of dimensionality in information systems research (Xu, 2000), in Figure 8.1, an *n*D framework is proposed, which is more comprehensive and which shows that more effective integration approaches can be developed from the interactions that exist among multidimensions for both intra- and interorganizational integrations. Such a framework may guide researchers in the development of more new and practical integration techniques that will meet the challenges of the complexity of the integration task, especially as various current techniques have their strengths as well as their weaknesses. For example, due to the lack of formal service semantics, dynamic service mediation, and orchestrational support in SOA and web services, SOA and web services provide neither data nor behavior mediation. This weakness reduces the efficiency of current service-oriented integration approaches in the context of industrial information integration (Izza, 2009). CIMOSA does focus on a multiple perspective enterprise modeling

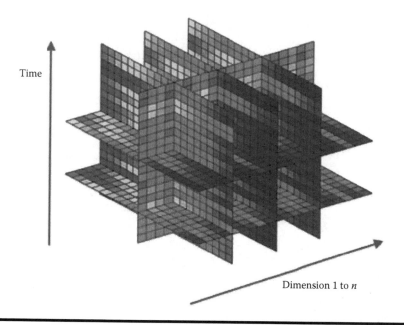

Time

Dimension 1 to n

Figure 8.2 Time and space model of information flows within an organization.

framework; however, the CIMOSA framework does not support interorganizational modeling. A methodological framework may help in developing integration techniques that can minimize weaknesses. Another interesting concept in the nD framework is the time and space model (see Figure 8.2). It is important to integrate and coordinate all ongoing activities in organizations in terms of the measurement of both time and space, in order to realize the best flow of synchronized information.

8.2.1 Syntactic Integration Approaches

In integration approaches, the following main methods or techniques can be listed: SOA, BPM, middleware-based techniques, and other methods, such as standard-based techniques.

8.2.2 SOA

EAI has evolved from point-to-point integration to process integration, and now it is evolving toward SOA (Zhang et al., 2009). SOA is a recent trend in integrating heterogeneous systems and different middleware systems. SOA is an enterprise level service-oriented system architecture; one of the objectives in developing its methodology was to integrate applications. The key advantage of using SOA in EA is that it can accommodate functions across various platforms despite different systems that are used; that is, it can perform across multiple platforms. The advantages of SOA include that any architecture based on SOA will start from the specific needs of an enterprise and that it will provide the needed flexibility. This flexibility refers to the ability to quickly respond to the changing and competitive business environment. Due to its flexibility, SOA enterprise architecture is capable of meeting these possible future challenges.

SOA offers the possibility of realizing the important goals of an organization, including meeting the need for integrating systems, meeting the need for integrating different application systems across different platforms, and meeting the need for integrating systems for better business performance that will change existing business models. SOA provides guidelines about how services are described and used. In SOA, developers organize and package different software applications as services. Each service includes an interface that lists the operations it provides and the set of messages that it can accept and send. Services can be reconstructed and reused to create new applications. In industry, SOA has been successfully applied to manufacturing systems, MESs, power systems, production systems, healthcare systems, transportation systems, SCM systems, natural disaster warning systems, etc. SOA promises to provide flexibility and extendability to legacy systems. It is assumed to be one of the best architectures for EAI.

8.2.3 BPM

BPM has been proposed to standardize the management of business processes, both intra- and interorganizationally. In BPM, a tool called the process broker, an extension of the message broker, is used for integration purposes. Process broker can encapsulate process logic in integrating applications. Examples of BPM technologies include *WfMS*. The Workflow Management Coalition is an organization that defines standards and conventions for workflow management systems. The Workflow Management Coalition develops standardized models for workflows. As introduced in Chapter 6, *WfMS* can support integration based upon workflow and business process methods and techniques.

8.2.4 Middleware-Based Techniques

Middleware-based techniques use middleware to integrate applications. Middleware is reusable software that resides between the applications and the underlying operating systems and hardware. Intraorganizational integration can be implemented through standardized middleware, while interorganizational integration is focusing more on the aspects of collaborative frameworks. In general, middleware can be classified into the following categories: (1) data-oriented middleware, (2) component-oriented middleware, (3) message-oriented middleware (MOM), (4) application servers and transaction monitors, and (5) DEC-RPC. Linthicum (1999) and Ruh et al. (2000) classify integration technologies into these categories: (1) database-oriented middleware such as Open Database Connectivity (ODBC) or JDBC; (2) message-oriented technologies such as XML, Remote Procedure Call (RPC), and MOM; (3) object-oriented technologies such as CORBA and Distributed Common Object Model (DCOM); and (4) transaction-based technologies such as application servers and transaction process monitors; and (5) interface-oriented technologies such as APIs.

Data-oriented middleware: Data-oriented middleware is used to integrate the databases or proprietary APIs. It allows connecting heterogeneous databases such as ODBC or JDBC. ODBC was originally developed by Microsoft during the early 1990s. It is a standard C programming language middleware API for accessing DBMS.

Component-oriented middleware: Component-oriented middleware provides integration between distributed components. The main techniques in this category are CORBA and COM/DCOM, etc. DCOM by Microsoft can transfer primitive data and some specific type of data such as VARIANT defined by Microsoft in COM/DCOM (Ni et al., 2006).

Message-oriented middleware: MOM can be used to integrate independent and autonomous applications into a single integrated system using a message exchange mechanism. Because of this strength, it is considered to be one of ES integration's main technologies.

8.2.5 Object-Oriented Technology

8.2.5.1 CORBA

CORBA is considered to be an industry standard technology infrastructure for integration purposes. It has applications in many domains including manufacturing, healthcare, and telecommunications. CORBA is commonly used at the component level. Unlike DCOM, CORBA is platform and language independent. CORBA supports a variety of systems and programming languages and has the ability to integrate legacy software applications. The Object Request Broker (ORB) is a middleware component that implements the CORBA bus and is responsible for delivering requests from the client side to the server applications. The ORB is responsible for a CORBA object's implementation including request receiving and response preparing. The CORBA specification provides a uniform framework across the entire distributed environment and makes applications built using an ORB portable across diverse platforms. However, CORBA can be complicated and difficult to properly use. As a result, interest in CORBA has declined. Currently, CORBA has virtually lost its leading position in the middleware marketplace; however, some of CORBA's strengths have been incorporated into technologies such as J2EE and web services.

8.2.5.2 DCOM and COM

DCOM is a Microsoft technology for communication among software components distributed across networked computers. DCOM is the distributed extension to COM, which provides a set of interfaces that allow clients and servers to communicate within the same computer. Using DCOM, two objects on two separate computers are able to call each other's methods. DCOM supports object-oriented languages such as C++ and Java and is well supported by Windows platforms. However, DCOM supports only a few non-Windows operating systems, which limits the use of DCOM for heterogeneous networks. DCOM is one of the distributing computing technologies (including CORBA, Java RMT, .Net, and Jini) used in distribution system components.

8.2.6 Transaction-Based Technology

Application servers and transaction monitors: Application servers provide services to the client nodes. Commercial application servers mainly include J2EE servers and proprietary servers.

DCE-RPC: DCE is a software infrastructure for developing distributed systems based on RPC. DCE was designed and developed to support distributed applications in heterogeneous environments in the early 1990s. DCE consists of many components such as RPC, cell and global directory services, security service, distributed time service, and distributed file service. DCE is supported by many different platforms and legacy operating systems. A main advantage of DCE is that the DCE RPC facility provides a way of communicating between software modules running on different systems. Compared with traditional networking programming methods, RPC is relatively simpler to code. However, DCE does not have strong support mechanisms for object-oriented languages because RPC is inherently procedural. As a result, DCE has lost much of its popularity in recent years.

Standard-based techniques are classified into main categories: (1) techniques that are based on standard exchange formats and (2) techniques that are based on standard data and process models (Izza, 2009). For (1), the techniques mainly include XML, ebXML, and RosettaNet, using standardized syntax of messages to facilitate the integration process; for (2), the techniques mainly include Business Process Modeling Language (BPML), Integrated DEFinition for Process Description Capture Method (IDEF3), and UML. The lack of standards deserves attention. For example, standards can be very important in describing and orchestrating process flows across applications and systems. Standardization plays an important role in achieving integration. However, currently, there are still only a limited number of standards that directly relate to enterprise integration.

XML, ebXML, and the RosettaNet standard: XML is a text-based markup language specified by the World Wide Web Consortium (W3C). A data interchange mechanism based on XML paves the way for high-level data integration.

As described by Maheshwari (2003), integration of the web with business applications such that each component shares a common data format is the motivation for developing the data integration language. XML was developed for this purpose. Specifically, it was developed to allow for the publishing of documents using a medium that provides structure, extensibility, and validity. XML has quickly found a place in EAI due to its capacity of providing a common data exchange format. XML has now become the mainstream in data exchange and is considered as the *de facto* standard in data exchange. By using XML-based message processing as the data communication format, web services can eliminate the differences among systems that use different component models, operating systems, and programming languages and can make heterogeneous systems operate (Ji, 2009). In general, XML is helpful in achieving data and application interoperability in EAs. The advantages of XML in B2B data exchange include flexibility, heterogeneity, and readability. XML and web agent have been proposed to replace CORBA and traditional EDI for achieving interorganizational data exchange. In concurrent engineering, CORBA and distributed multiagents have been used for extended distributed collaborative design tasks, and XML has been used for setting the Manufacturability Markup Language in data exchange. The application of XML in manufacturing is becoming increasingly popular. In MES, an XML data exchange mechanism can meet the internal EA needs as well as the external ones in which the RosettaNet standard has been adopted by interorganizational partners. In EAI, XML has been used for the seamless integration of EA, using available technologies. XML plays an important role for integration purposes, both intra- and interorganizationally.

The major advantage that XML can offer is the XML-based data exchange, which can facilitate data exchange in different ESs. However, XML has limitations in some aspects, such as degrading system performance and security. However, it is clear that the advantages of XML far outweigh its disadvantages, thus making it a preferred tool for use in EAI.

XML's standard can be classified into basic standard, vertical standard, and horizontal standard. RosettaNet represents the vertical XML standard, while the counterpart in the horizontal standard includes ebXML. RosettaNet standards can work effectively in a system-to-system exchange, but may not be as concise and/or efficient as possible, due to the different requirements of the architecture partners. One of the objectives of RosettaNet is to develop an SCM global industry standard. It focuses on developing frameworks, dictionaries, PIPs, and e-business processes for system-to-system exchanges. The RosettaNet PIP provides interfaces for system-to-system exchange by decomposing the core business process of a typical supply chain into eight clusters. As the eight XML-based B2B frameworks have been classified into three categories (process-centric, industry-specific, and cross-industry) in the industry-specific framework, RosettaNet provides a comprehensive description of business processes in a particular industry. ebXML focuses on business processes in the process-centric framework aiming at a cross-industry standard (that can also be developed for use within industry-specific standards). Practically, the industry-specific RosettaNet can be plugged into ebXML. The current effort is to converge standards between ebXML and RosettaNet. In such cases, partners with different standards such as ebXML and RosettaNet can be integrated.

UML: UML is a conceptual methodology for describing business processes. UML is able to provide different architectural views of a system, in which the most relevant for enterprise integration is the process view. As an activity-oriented process language, UML includes sublanguages for capturing activities; as such, UML provides a high-level intuitive approach to express relevant aspects in EAI. However, it may not allow for precise specifications, unless a specific semantic interpretation is available. The other relevant view is the design view. The design view is able to address both the static and dynamic properties of a system, which can be quite useful in modeling enterprise integration. Although UML captures modeling details well, it requires a certain level of knowledge to be able to employ those details. The other process models such as IDEF are relatively easier to understand than the UML models.

IDEF3 is a business process modeling method complementary to IDEF0. IDEF3 is characterized by scenario-driven process flow description. This method is part of the IDEF family of modeling languages.

BPML: There are many formalisms for describing a business process. BPML is one of the languages used in business process modeling. BPML builds on the foundation of WSCI and shares its roots in web services. BPML supports the modeling of business processes through its support for semantics.

Ad hoc techniques are characterized by implementing point-to-point integration; applications can possibly be integrated with $o(n^2)$ number of interfaces.

8.2.7 Semantic Integration Approaches

Many integration methods are semantics based. The main idea of these methods is the representation of semantics in the integration process.

Ontology-based integration methods: Ontology is generally defined as a model for conceptualization and computation purposes. An ontology is a kind of knowledge representation that provides conceptualizations of some domains. An ontology specifies a vocabulary including the key terms, their semantic interconnections, and some inference rules. Ontology approaches may use specific ontology architecture. These approaches can be classified as a single ontology approach, a multiple ontology approach, or a hybrid ontology approach. However, some ontology architectures use several ontologies. This can cause some ontology heterogeneity problems. For solving such problems, the main approaches include (1) ontology mapping, (2) ontology alignment, (3) ontology transformation, and (4) ontology fusion.

Several languages such as Web Ontology Language (OWL) and several methodologies have emerged. OWL is considered to be suited to business-oriented notions and constructs and is attracting applications due to its support for semantics.

Ontology-based approaches have been proposed for use in data integration, as well as for process integration. OWL Services (OWL-S) is an OWL ontology for services. OWL-S provides an OWL ontology for describing web services. An OWL-S description for a service consists of components such as service profile, service model, and service grounding. Web Service Modeling Ontology (WSMO) is a conceptual model for aspects related to semantic web services. It provides an ontology-based framework that supports the deployment and interoperability of semantic web services. WSMO has, as its main components, goals, ontologies, mediators, web services, etc.

The methods, techniques, and technologies mentioned earlier can be used to integrate different systems. Although these methods, techniques, and technologies are not fully mature yet, they represent the progress made in methodological and technological discovery. In combination, they can serve as a starting point toward a comprehensive integration method in the framework of IIIE and can eventually lead IIIE to its full potential.

8.3 Enterprise Application Integration

In the past, many industrial enterprises have acquired disparate systems and applications over the years; as such, the phenomenon of the "information island" is not unusual. An information island can result from a lack of integration, in which many EAs within the same enterprise are unable to communicate with each other. When this occurs, data and business processes are not sharable. In an interorganizational environment such as the e-supply chain, many EAs within the same chain are unable to communicate with each other. The need to integrate such diverse applications and systems is vital in order to meet business requirements and challenges.

Work on addressing the challenges of integrating different applications and systems is of great significance. This body of work is now becoming a part of the new framework termed IIIE. EAI represents the techniques and technologies that enable the integration of EAs both intra- and interorganizationally. The objective of EAI is to facilitate information exchange among business enterprises in a timely, accurate, and consistent manner and to support business operations targeting for seamless integration. For this objective, EAI entails integrating enterprise data sources and applications so that data and business processes can be easily shared. EAI is not a single technique or technology. Rather, it is a collection of methods, techniques, tools, and technologies from multiple disciplines that enable EAs to be interoperable. EAI involves the integration of multiple applications, existing or new, that were independently developed, and (to a certain extent) that integration of data and applications is expected to be accomplished without requiring significant changes to multiple applications. As such, EAI must be able to integrate heterogeneous applications that have been created with different methods and on different platforms. EAI is expected to be powerful enough to connect both existing and new applications in order to enable collaborative operation, both intra- and interorganizationally. In other words, EAI provides the infrastructure and mechanisms for integrating both legacy and new application systems. From the perspective of developing ESs, EAI can help overcome the limitation of the ESs by providing a more applicable infrastructure. From the perspective of EA, EAI is able to integrate discrete applications into a seamless whole to form an applicable, flexible, and manageable infrastructure for applications.

The integration of EAs includes the integration of data, business process, applications, and platforms as well as integration standards. Through creating an integrated structure, EAI connects the heterogeneous data sources, systems, and applications intra- or interorganizationally. EAI aims not only to connect the current system processes but also to provide a flexible and convenient process integration mechanism. With EAI, intra- or interorganizational application is expected to be integrated seamlessly and to ensure that different units within an enterprise or even within an extended enterprise can cooperate with each other, even if they use different systems. A complete EAI offers functions such as data and business process integration, since the goal of the EAI is to create integrated BPM. By coordinating the business processes of multiple EAs and by combining software, hardware, and standards together, ESs such as ERP can be used to share and exchange data seamlessly in a supply chain environment.

In general, those EAs that were not designed as interoperable can be integrated on an intraorganizational and/or interorganizational basis. Through such a process, legacy and newer systems can be integrated in order to provide greater competitive advantages. The constantly changing business requirements and the need for adapting to the rapid changes in the supply chain may require the help from SOA for architectural development. A more innovative enterprise architecture will not only provide new functionality but also leverage investments in the legacy systems running the enterprise's key applications.

EAI can facilitate the integration of both intra- and interorganizational systems. EAI can link applications within a single organization or supply chain in order to simplify and automate business processes to the greatest extent possible, while at the same time avoiding make sweeping changes to the existing data structures or applications. In other words, EAI is the sharing of data and business processes among any connected application or data sources in the enterprise or supply chain. EAI can streamline the processes within the enterprise, integrating diverse applications such as ESs and legacy applications. However, although EAI may be applicable both intra- and interorganizationally, it does not necessarily translate across organizations. The main reason is that although an enterprise or an extended enterprise might choose to integrate applications, it is not realistic to expect its partners to choose to integrate their applications.

EAI is a key research area in IIIE. As introduced earlier, IIIE is a set of concepts and techniques that are used to facilitate the industrial information integration process. One of the main purposes for establishing IIIE is to respond to the challenging demand from the industry. In 1996, the International Electronics Manufacturing Initiative revealed that many existing industry information systems were not interoperable. For example, a factory system may involve an order processing application, a production scheduling application, and a production control application, which might reside on three separate servers. The three distributed applications might be required to cooperate with each other for operations purposes. After 1996, substantial efforts were made to integrate nonintegrated applications and systems in industry sectors. Today, many industrial sectors (such as manufacturing, logistics, energy, and telecommunications) have integrated distributed EAs in factory automation, production systems, e-manufacturing systems, and many other sectors. The evidence shows that integrated distributed EAs can perform well with fewer resources than can centralize systems in industrial applications due to the increasing demand for agility, flexibility, and scalability. As an interesting topic in industrial information integration, in the following, we will introduce the architectures and technologies for integrating distributed EAs, illustrate their strengths and weaknesses, and identify research trends and opportunities in this increasingly important area in IIIE.

In the past 20 years, distributed EAI has received much attention. Much research has been conducted on enterprise integration, either about stand-alone application or about collaborative applications. A distributed EA is defined as an application with software components residing in more than one system in a network. Often, the network is heterogeneous and is composed of diverse devices, systems, and operating systems. In an industrial environment, many enterprises have numerous applications, devices, technologies, and protocols that are distributed across a network. Distributed EAs typically require their distributed components to interact with one another through certain remote communication mechanisms such as message passing and remote invocations in their networking environments. However, distributed EAs often have problems when trying to communicate with one another, due to different formats and protocols. As distributed applications in an enterprise continue to grow, integrating distributed EAs has become a challenging task. For example, many industry enterprises have trouble when attempting to integrate industry applications effectively. Such applications include PDM systems, engineering management systems, MESs, and CAD systems running on different hosts. As an example, according to the IEC61499 (the International Electrotechnical Commission's Function Block specification), a distributed control system consists of a number of applications that may be distributed among multiple devices. Often, a control processing application resides in a device, but its output conversion application resides in another device. Sometimes, a function of an application may be distributed among several devices and may require the cooperation of different subsystems in order to function properly. Additionally, the devices and applications may have been developed or provided by different vendors with different programming languages, formats, and/or protocols. In these cases, significant integration efforts are required in order to realize and enhance the interoperability of these devices and applications.

For many industry enterprises, it is imperative for their industry systems to cooperate in order to achieve certain business objectives. As the environment in the industry becomes increasingly distributed and heterogeneous across multiple organizational and geographical boundaries, there are strong demands to integrate various distributed applications in order to help enterprises' operations. As such, integrating distributed industrial applications has attracted much attention in enterprise computing. To date, many enterprises have invested heavily to integrate distributed EAs due to infrastructure upgrades, corporate restructuring, mergers and acquisitions, joint ventures, outsourcing, and the adoption of new technology, as integrating various applications owned by an enterprise can increase or enhance an enterprise's competitive advantages. Research indicates that EAI can reduce the cost of implementing and maintaining distributed systems interorganizationally. As EAI encompasses technologies that enable distributed and heterogeneous applications to interact with one another across networks, it helps to integrate many individual applications into a whole. It consists of the methods and tools used to coordinate various applications and to support the integration of both intraorganizational and interorganizational systems. EAI solutions comprise the efficient integration of diverse data and business processes across the enterprises, as well as the integration of intraorganizational and interorganizational EAs. With EAI, intra- or interorganizational distributed EA systems can be integrated effectively and can ensure that different divisions, different units, and even different enterprises can cooperate with each other.

8.3.1 Distributed EA Architectures

Distributed EA architectures have undergone an extensive evolution. Early-generation EAs were built on centralized mainframes. As the capacity of computers has increased, many applications and tasks have been moved to the user's computers in order to better satisfy business processing needs. The first-generation distributed EAs (those in the 1980s) were mainly developed using a two-tier client and server architecture. In this architecture, the client is responsible for presenting the application to the user, and the server is in charge of storage and management. As the complexity of transactions and the amount of data continued to increase in the mid to late 1990s, a three-tier architecture became popular in EAs. In a three-tier architecture, software components are divided into three layers such as the data layer, the application layer, and the presentation layer. The client tier focuses on the user interface and interacts with the middle tier. The middle tier focuses on business process and interacts with the data tier. Middleware such as CORBA is often deployed in the middle tier to integrate distributed EAs (including independently developed applications). In addition, TP monitors often run at the middle tier for resource allocation, workload balancing, and scalability.

As web applications become widespread, the three-tier architecture has been extended to the web-centric architecture by adding web clients (mainly browsers) and web servers (e.g., Apache). In a web-centric architecture, the web client sends HTTP requests to the web server for contents. The web server may choose to process the request directly or to forward the request to an application server (if the request cannot be handled directly). Next, the application server will interact with the back-end database and will send responses back to the client (mainly browsers). Furthermore, the three-tier architecture continues to evolve, and subsequently, the multitier architecture appears. For example, additional tiers are often introduced between the client and other layers (such as the data layer) to ensure security, workload and resource balancing, and performance monitoring.

8.3.2 Integration of Distributed EAs

Enterprise integration includes business integration, application integration, and physical system integration. EAI focuses on integrating intraorganizational applications but now has been expanded to deal with interorganizational integration. EAI provides ways to integrate heterogeneous applications on different systems and platforms. So far, many efforts have been made with EAI solutions to help achieve quality application integration.

Integration can be studied through a variety of dimensions, including integration scopes, integration perspectives, integration layers, and integration levels. As mentioned earlier, enterprise integration can be divided into horizontal and vertical integrations. Integration levels can be at the business process integration, syntactical integration, semantic integration, physical integration, and also hardware and platform levels. For integration layers, researchers found it useful to study the integration in terms of data layer, business process layer, application layer, as well as systems layer. The following is a brief overview of key integration technologies for selected layers.

8.3.3 Business Process Layer Integration

Integration at this layer includes integration in the sublayers such as basic coordination, functional interfaces, business protocol and policies, and nonfunctional properties. A traditional method of application integration involves complex low-level network and operating system programming. Thus, maintaining the traditional integrated ES is quite challenging. To make application integration at the business process level easier, the main focus has been on developing middleware technologies. A number of middleware technologies have been developed in the past two decades to integrate distributed EAs.

The use of middleware can make distributed systems integration less challenging. By adopting middleware, applications can be isolated from the variety of ever-changing hardware platforms, operating systems, networks, and protocols that make the ESs integrated. Due to the advantages brought by the middleware, there has been an extensive use of middleware technologies. Middleware can help ensure reliability by providing scalability and by contributing to ESs for better performance. As important integration technologies, middleware technologies are often used by industrial enterprises to develop new applications through integrating legacy applications with the emerging technologies. Typically, a middleware for communication comprises two types of remote communications: message passing and/or remote invocation. Specifically, "message passing" includes synchronous and asynchronous messaging. "Remote invocation" includes synchronous, client-side asynchronous, and server-side asynchronous remote invocations. Middleware has become popular for exchanging data across EAs and in support of heterogeneous applications. However, middleware may be limited by the applicable scope as it is required to respond to the specification changes of the supported applications. Figure 8.3 illustrates the use of middleware in distributed applications.

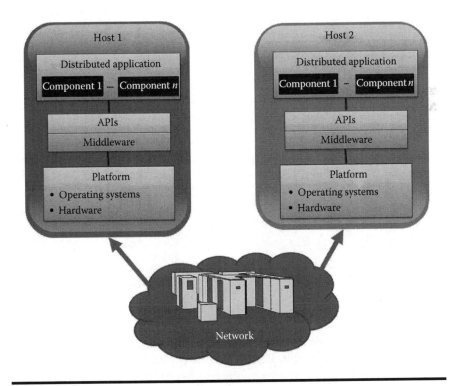

Figure 8.3   Use of middleware in distributed applications.

There are many types of middleware such as RPC-based middleware, MOM, event-based middleware, database middleware, TP monitors, security middleware, agent-based middleware, and service-oriented middleware. Some large companies such as SAP implement their own custom middleware as part of their application solutions. Each of these technologies and approaches has its advantages and disadvantages. For example, middleware is generally technology and application dependent and provides little or no visibility of business processes. However, EAI focuses on the holistic needs of the enterprise process. The most commonly used general-purpose middleware technologies at the business process layer for integrating distributed EAs include RPC-based (remote invocation) middleware and MOM. Some of the major middleware technologies were introduced in Section 8.2. These include CORBA, DCOM, and DCE, in addition to remote method invocation (RMI) and MOM, which will be briefly introduced later.

8.3.3.1 Remote Method Invocation

Java RMI was released by Sun around 1997. It provides a distributed computing environment for developing client and server applications using Java language. Java RMI offers developers a way to implement distributed Java applications. For example, a Java application on a computer can invoke the methods of other remote Java objects on different computers. As Java is platform independent, Java RMI-based applications can be executed on many different computing platforms. However, RMI heavily relies on Java language and does not have direct support mechanisms for other common languages such as C or C++.

8.3.3.2 Message-Oriented Middleware

MOM relies on messages to enable communication between separate systems. Using the message passing and queuing mechanism, MOM is able to carry information and action requests between distributed applications. In fact, MOM is a loosely coupled and asynchronous technology. MOM provides strong support for asynchronous communications. The main disadvantages of MOM include limited support mechanisms for scalability, lack of industrial standards, and poor portability. However, MOM has been used successfully in some industrial systems such as integrated manufacturing systems.

8.3.4 Data Layer Integration

In EAI, data integration means the sharing of relevant data between two or more applications. Data integration is also good for the standardization of data. The main objectives in any data integration task include data accuracy improvement, productivity improvement, greater agility and flexibility, and facilitating system replacement/organizational mergers (Gleghorn, 2005). Data integration requires tasks of data conversion and mapping that involve source schema, target schema, and the mapping between them. Understanding and maintaining the underlying schemas to regularly address changes is required. The data integration process mainly consists of operations such as extraction, transportation, transformation, and insertion. The extraction process is to prepare data to meet the requirements of the target application in a format that is acceptable for transportation. During the data transportation, some important factors (e.g., reliability and security) need particular attention. In the transformation step, data from the source application are formatted to meet the target application's requirements. In the insertion step, the inbound data are imported into the target system. A wide variety of integration products that provide infrastructures for performing the tasks mentioned earlier are available. Point-to-point, spoke-and-hub, messaging bus, and ESB with BPM are among the common integration approaches that have been used. In "the point-to-point approach," an integration program is used to connect each source and each target application. In other words, the corresponding interface objects are developed between any two of the required data exchange points. The main operation is to extract corresponding target data from the data source system through interface objects. "The spoke-and-hub approach" leverages reusable components to perform generic integration functions; since this approach is configurable, extensible, and reusable, it makes rapid integration with various systems possible. Research on data integration mainly deals with moving data between heterogeneous data sources under different operating systems.

8.3.5 *Communication Layer Integration*

Integrating distributed applications requires those separate applications to be able to communicate with one another and to exchange data and information. For example, an application may need to know the current status and operations of a remote application in order to perform certain tasks (such as scheduling). Typically, protocols such as HTTP and Internet Inter-Orb Protocol (IIOP) can be used to transport and exchange data and information between two different applications.

8.3.6 *Presentation Layer Integration*

The integration in the presentation layer mainly focuses on user interface integration. Presenting an integrated and dynamic view for users is the main goal of user interface integration. So far, there has been much research conducting for the presentation level. A recent example of user interface integration is web mashup (such as integrating Google Map with other applications). Portlet is another user interface integration technology that has the technical potential to help produce more customizable and flexible portal applications. More research is needed regarding the presentation level for effective user interface integration to take place.

8.3.7 Other Integration Technologies

In this section, we briefly review the integration of distributed EAs using technologies such as J2EE, .Net and web services, SOA (see Section 8.2), and ESB. Some authors have considered that the two major approaches to EAI are based on J2EE and .Net. .Net focuses on web services, as does J2EE.

8.3.7.1 J2EE

J2EE (Java 2 Enterprise Edition) has emerged as a leading platform for developing and integrating EAs. J2EE has three main layers: business objects, presentation, and back-end layers. J2EE contains a number of technologies such as EJB, RMI, Java Server Pages, and XML language. As the main layer in J2EE, the business object layer consists mainly of EJB. EJB provides a simplified method to develop component-based distributed applications over heterogeneous environments. J2EE has been extensively used in industrial systems. For example, J2EE was used as a system framework to integrate systems for supply chain alliance. The back-end layer in J2EE is associated with ESs, since this layer handles interactions with ESs. J2EE has the features of being platform independent, secure, stable, and scalable. Due to these important features, J2EE has been considered the industrial standard and is a preferred platform for EAI.

8.3.7.2 Net Framework

The .Net refers to Microsoft's framework for web services software. The .Net framework allows objects developed by different programming languages to interact with each other. It allows programmers to create programs in different languages and allows the programs' execution on different runtime systems and environments. The .NET framework also provides remoting infrastructure that allows an object on one computer to invoke the methods of an object on another computer. Using .NET remoting, objects can communicate with each other even though they reside on different computers.

8.3.7.3 Web Services, SOA, and ESB

As traditional middleware such as CORBA and DCOM is typically used for Intranet applications and often encounters difficulties in crossing firewall boundaries, web services have been developed to support the integration of Internet applications. A web service usually contains a number of functions that can be organized as a single entity and can be published to the network as a service to be used by other programs. Web services use the HTTP protocol transport information and thus can easily pass requests through firewalls.

Web services are building blocks for constructing web-based distributed applications and can be viewed as suitable middleware for use in Internet-related application integration. Both J2EE and .Net can be used as programming tools by developers to create web services. Web services play an important role in integrating different middleware systems, and can provide middleware for the middleware abstraction layer for integration purposes. Since different middleware has different advantages, many enterprises have used a variety of middleware over the years. As a result, these enterprises face the "middleware islands" issue caused by the use or adoption of several middleware approaches, since middleware technologies and products from different vendors often do not easily interoperate. As a result, enterprises have had to find other ways to integrate these different middleware systems. In some cases, *ad hoc* techniques such as adapters can be used or developed to support the integration needed by enterprises. But often, integration is not easy due to its complexity, the technical expertise required, and the costs involved. Web services and its underlying principle named SOA are considered good solutions for such middleware-to-middleware interactions.

8.3.7.4 Enterprise Service Bus

ESB and its application in SOA are relatively recent. Service-oriented integration depends on ESB. ESB offers capabilities such as data format transformation, message routing, mediation, monitoring, and service management. ESB is able to work across different middleware products and standards to implement enterprise-wide SOA. ESB can shield from different protocols (e.g., CORBA IIOP, RMI) to realize a smooth data flow and exchange among different application systems, thus improving and enhancing interoperability. As such, ESB is considered an infrastructure to facilitate SOA, and SOA-based ESB is often viewed by researchers and practitioners as a new middleware technology. Chen et al. consider that in order to fully support the variety of interaction patterns that are required in SOA, ESB is particularly good at supporting the three main styles of enterprise integration—SOA, message-driven architecture, and event-driven architecture—in one infrastructure (Chen et al., 2007). Figure 8.4 illustrates an SOA-oriented integration environment using ESB. In summary, web services, SOA, and ESB provide a promising and valuable framework for interenterprise integration.

8.3.8 Future Perspectives

8.3.8.1 Trends

The integration of various EAs is an ongoing task for industry enterprises, especially those that are adopting new technologies. Some new trends in this area include the following:

1. As web services, SOA, and ESA are being increasingly used to integrate legacy and existing applications, ensuring QoS for effective integration is also becoming increasingly important. As different web service applications

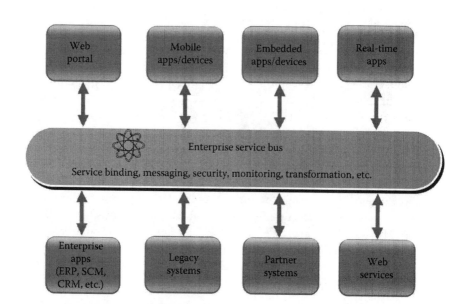

Figure 8.4   An SOA-oriented integration environment using ESB.

often have different QoS requirements and may have conflicts between each other for resources (such as bandwidth and processing time), it is desirable to develop empirically tested QoS integration models to check QoS parameters, to examine the service levels and quality, and to implement mechanisms for dynamically selecting QoS-aware web services. Models such as Trusted and the Autonomic Service Cooperation model to be deployed in industrial enterprises are expected to appear.

2. As the amount of data in industrial applications increases exponentially, there is an increasing need to integrate data sources for decision support, data mining, knowledge discovery, and OLAP. Application-specific middleware such as data mining middleware will be increasingly developed and deployed for industrial information processing purposes.

3. Although the semantic web and social networking technologies are still in their infancy in terms of their use in industrial applications, the integration of semantic web and social networking technologies with sensor data is expected to grow in industrial applications and to add more value to customers and partners in industrial settings. Various ontology approaches for semantic integration and interoperability will become mature and more applicable in industrial environments.

4. Mobile applications, smart embedded devices, and embedded systems have been increasingly deployed in industrial applications. This creates new opportunities and challenges to build communication and interoperability between industrial systems and embedded devices or systems. Due to its strengths in

both autonomy and interoperability, SOA approaches have the capability to implement data and communication exchanges between embedded devices/systems and various applications. We expect to see that industry applications will become increasingly integrated with services running on networked, resource-limited mobile devices, and smart embedded devices using SOA approaches. On the other hand, techniques such as embedded web servers, gateway implementation, XML, and TCP/IP-based protocol application programming will continue to be used to exchange data with a variety of heterogeneous Networked Control Systems and with the commercial programmable logic controllers deployed in the industry over the years.

8.3.8.2 Some Research Challenges

1. As more and more industrial enterprises are adopting web services to integrate various devices and applications, security challenges are becoming more prominent. For example, security risks may arise from the processes of command transferring, remote diagnosis, and maintenance, because information exchange and device access currently have to go through multiple networks.
2. Complexity of distributed software. Writing distributed software can be complicated by synchronization, deadlock, fault tolerance, heterogeneous operating systems, and other complexities of concurrent systems.
3. Many industrial enterprises have distributed real-time control systems (DRCSs), which interact with sensors and actuators over the network. Generally, DRCSs need to be more robust, efficient, and reliable than data-centric applications. Particularly, industrial enterprises (such as power plants) have strict response time requirements and performance constraints on their industrial systems and applications. For example, many industrial automation and control systems collect data from heterogeneous sensors and require real-time analyses for the data collected from a variety of heterogeneous sensors. Since DRCS is playing an increasingly crucial role in critical industrial operations, new integration techniques (such as real-time distribution middleware, distributed real-time Java, and real-time SOA solutions) are needed to address performance concerns, to mitigate integration risks, and to ensure data quality and security.
4. Integrated enterprise devices and applications may collapse due to unanticipated events. Thus, research on integration resilience and reliability (including data and information reliability) is highly valued in EAI. New guidelines and methods for fault-tolerant task load balancing, message scheduling, and transaction mechanisms need to be further developed in order to ensure the robustness, reliability, and maintainability, and rapid diagnosing of integrated EAs across various environments.
5. As new technologies are constantly being introduced into industrial systems, user interface integration for industrial systems poses many new challenges

such as various interface types, service interface definitions, and the creation of standards. Interface integration requires a solid understanding of various applications, devices, and enterprise-wide integration requirements. Currently, there is a lack of conceptual modeling techniques that can effectively elicit, represent, and analyze enterprise-wide integration requirements.

6. In recent years, new technologies such as WSNs, RFID, and IoT have been deployed to industrial systems such as logistics systems, material flow systems, and SCM systems. These new technologies make the integration of EAs and various interfaces yet more challenging. There is a lack of services, standard architecture, quality assurance, and guidelines for handling the interactions of heterogeneous devices, sensors, aggregators, actuators, and diverse applications while reducing security and privacy risks. For example, the lack of services for connecting users to the appropriate sensor networks becomes very apparent as the number of data sources in a sensor network increases. The ability to include built-in and dynamically deployed user services in a physical device still needs research to ensure the expected functionality and performance. Various middleware solutions and architectures (e.g., publisher and subscriber architecture, message bus architecture, ontology architecture) for integrating industrial applications, WSN, RFID, and IoT have been proposed. However, these solutions and architectures are typically designed for respective domains. The development of an ontology that precisely defines the concepts and properties of an enterprise architecture domain and integrates these new technologies is considered to be challenging. Developing these middleware solutions and architectures is a challenging issue.

7. Complexity and systems approach

Research indicates that the tasks involved with enterprise integration are of a complex nature and partially involve unstructured elements. Enterprise integration is an extremely complex process that involves different organizational and technological factors. Its complexity can be further reinforced by the multidimensionality involved in the integration process (see Figure 8.1). Experiences show that application integration often results in highly unstructured and complex models—not only the process models but also the data models. Enterprise integration modelers are relatively frequently used to create well-structured process models in application integration. As models for enterprise integration tend to become large and complex, it is possible to overlook some of the details that are needed to maintain completeness, consistency, and validity. In such model reduction processes, reducing the complexity level has been seen as a scientific approach represented by scientific inquiring systems (also see Chapter 9). However, how the complexity level can be reduced and how such complexity can be represented by formal methods is a research question. Due to the fact that the enterprise integration is characterized by complexity, a systems approach is always an appropriate method.

According to Themistocleous and Irani (2003), EAI incorporates multiple intra- and interorganizational applications with a central integration infrastructure, and this infrastructure is responsible for the coordination of all integration tasks. Meanwhile, no integration technology addresses all of the integration problems, as each was designed to address a broad category of integration issues (Maheshwari, 2003; Themistocleous and Irani, 2003). Themistocleous and Irani also note that EAI providers realize the capabilities of integration technologies and configure EAI using a subset of integration technologies only. The practice of using a subset and not a complete set of relevant techniques may result in incomplete functionality. Nonetheless, it is necessary to understand and evaluate the capabilities of each EAI implementation as a whole, within a systems framework. The complexity of integration technologies may encourage us to seek a solution to the integration problem through thinking in a systems perspective.

Puschmann and Alt (2001) indicate that EAI systems presuppose an effective information system architecture planning. Khoumbati et al. (2005) note that the aim of EAI is to integrate individual applications into a seamless whole. Both the concept of planning and the concept of "whole" are systems research issues. EAI has been implicitly considered as a systems research issue since it was described as a system-of-systems issue by Fernandes et al. (2010). In the following, at least three angles of systems perspectives can be helpful and supportive.

8. Combination of methods

Combining methods means to synthesize the complementarities among existing methods, to improve the existing methods through incorporating new methods, and to reorganize knowledge. Themistocleous and Irani (2003) indicate that in addressing enterprise integration problems, a diversity of techniques such as XML and CORBA have been proposed; however, it appears that no single method solves all integration problems. Their research reveals that there is no single EAI package that is capable of addressing all integration problems, nor do any of the EAI packages evaluated in their study meet all of the evaluation criteria. Although WSDL, SOAP, UDDI, and Business Process Execution Language (BPEL) have brought new potential to SOA, Zhang and Zhou (2009) indicate that WSDL, SOAP, UDDI, and BPEL provide only partial solutions to the issue of interoperability. Ontology-Driven XSLT Generation (ODXG) can support the characterization of XML messages, in both format and interpretation. The implementation tool for ODXG supports the development of both ontology models using IDEF5 and process models using IDEF3. According to Fernandes et al. (2010), the combination of these two modeling methods (along with their extensions) allows for the design-time and runtime support of semantic EAI. According to Themistocleous and Irani (2003), EAI incorporates multiple intra- and interorganizational applications and provides a flexible integrated architecture, since EAI is based on a set of more than 16 integration technologies used to piece applications together. As such, an effective integration solution largely depends on the right combination of technologies (which is also discussed in Chapter 9). The evolution of EAI itself provides evidence of such point of view, since EAI addresses integration problems through combining a diversity of integration technologies.

9. Dynamics

Maheshwari (2003) indicates the importance of building a technology infrastructure that can adapt to changes as integration architecture evolves. Dynamics is also characterized by the dynamic allocation of resources involved in the integration process. As Chapter 9 describes, the solution to the integration problem is to think in a systems perspective.

10. The concept of system, subsystem, and dimensionality

Enterprise integration is multidimensional (see Figure 8.1). The dimensions involved in enterprise integration have been introduced and discussed in this chapter. A typical application requires the combination of the processes of different enterprises to yield one integrated system. In this process, differences between different data formats and computational platforms have to be integrated; typically, the applications used have to be developed using different models and methodologies, as well as different tools. A major technical issue is to achieve consistency between different subsystems. Khoumbati et al. (2005), Zhang (2009), and other authors indicate that the integration problem needs to seek a multidimensional integrated application solution. In both intra- and interorganizational application integration efforts, an interesting question is this: Should we model all concepts and processes relevant to the subsystems and dimensions of the enterprise and achieve consistency between different systems, subsystems, and dimensions? In addition to the examples regarding the dimensionality involved in enterprise integration that were discussed in this chapter, the concept of dimensions of abstraction was proposed. Dimensions of abstraction involve dimensions such as the data dimension, the process dimension, the structure dimension, and the policy dimension. Ever since the concept of dimensionality in informatics was proposed in Xu (2000), research on dimensionality has attracted attention, since a theory of dimensionality will guide us to model how all relevant systems, subsystems, and dimensions can be related to one another in an effective manner. It is generally agreed that the solution to the integration problem is to think in a systems perspective and focus on the holistic view of the involved enterprise.

8.4 Summary

Figure 8.5 shows a classification of enterprise modeling framework based on the existing work and the recent development in the area. At the top level are the enterprise modeling frameworks such as CIMOSA, GIM-GRAI, and PERA,

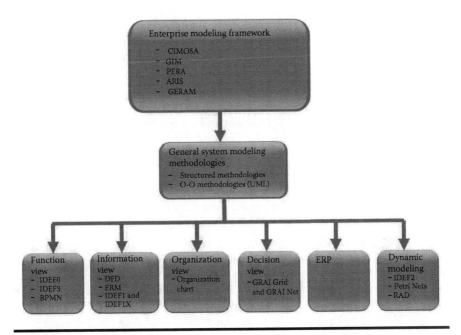

Figure 8.5   A classification of modeling framework for enterprise modeling. (Adapted from de le Fuente, M.V. et al., *Comput. Ind.*, 61, 702, 2010.)

as they provide generalized RAs and methodologies. Under this level, there are methods under functional view, information view, dynamic modeling view, and the newly developed ERP (entire resource planning) view. Current efforts are moving toward enhancing the existing methods or integrating or unifying these methods.

Since the 1990s, enterprise integration has received much attention, and much research has been conducted on enterprise integration either by integrating existing applications or by developing new applications with integrated existing applications. Since EAI concerns itself with the interoperability of various applications on heterogeneous platforms, it is the key to support the business processes among different application systems within an enterprise, since it integrates various heterogeneous systems. So far, many techniques have been developed to integrate EAs. We have mainly introduced the current status of the integration of distributed EAs and discussed future directions for research. As enterprises continue to add and/or deploy more applications, integrating distributed EAs becomes inevitable for enterprises that need to achieve a competitive advantage. As industry enterprises are increasingly adopting multitier client–server, Internet, and service-oriented architectures for their EAs and industrial devices, we have to address the need and the requirement for interoperability in the industrial enterprise environment.

EA is currently a rather new field. From the standpoint of enterprise integration, ESs seem to be able to provide the solution. Currently, it still offers many research challenges, such as the integration of hybrid wireless networks in ESs, user interface integration, reliability, performance management and security risk management, data mining for distributed applications, architecture design for real-time sensor networks, and control architecture for the sensor, actuator, and control service implementation. These issues need to be addressed in order for industrial systems to become more effective.

In the past, EAI has been focused on integrating existing intraorganizational systems. Now EAI has evolved into a variety of interorganizational applications, and consequently, the success of more and more enterprises is heavily reliant upon the integration of intra- and interorganizational systems. As such, EAI incorporates interorganizational ESs and SCM. As more organizations struggle with integration challenges and are increasingly striving to improve their operations, the adoption of even more integrated ESs becomes an imperative. Puschmann and Alt (2001) gave an example of the success of EAI at the Robert Bosch Group, a manufacturing company whose intra- and interorganizational business processes are supported by a homogenous information system architecture. Integration of existing systems, enterprise-wide or extended enterprise-wide, represents one of the most urgent priorities of many organizations. EAI helps an organization to maximize its ability to achieve its business objectives while remaining competitive. EAI contributes to this success through facilitating information sharing, facilitating workflow, and reducing the presence of information islands. EAI is considered to be an effective method for realizing information sharing, both intra- and interorganizationally.

Integrating existing systems or integrating existing systems with new applications is becoming one of the most urgent priorities of many organizations, in order to meet increasing challenges. Over the last decade, the architectures and technologies used for EAI have been improved. More new ideas and methodologies are expected to bring substantial improvements, both theoretically and practically, in order to achieve real-time enterprise-wide and/or extended enterprise-wide systems integration. EAI has emerged to overcome the limitations of the early generations of ESs and to extend the scope and capacity toward a new-generation ES. It can be predicted that the technological capacity provided by EAI today will become part of the technological capacity provided by new-generation ESs. The objective is to develop an architectural framework in which the synergies between EAI technologies can be achieved at their best levels and then used to solve the integration problem. Given the bright outlook of its future, EAI is poised to continue to grow and to adapt to new requirements. The ultimate goal of EAI is to integrate all relevant applications or systems in such a way that any application can access any other relevant applications within the right time frame, to materialize the idea of a zero-latency enterprise. The objective is the integration of EAs to yield one integrated system in which the boundaries between different applications have been bridged.

References

Chalmeta, R., Campos, C., and Grangel, R. 2001. References architectures for enterprise integration. *Journal of Systems and Software*, 57, 175–191.

Chen, H., Yin, J., Jin, L., Li, Y., and Dong, J. 2007. JTang synergy: A service oriented architecture for enterprise application integration. *Proceedings of the 2007 11th International Conference on Computer Supported Cooperative Work in Design*, Melbourne, Australia, pp. 502–507.

Chen, Z. and Cheng, H. 2009. Research on enterprise application integration categories and strategies. *Proceedings of 2009 International Forum on Computer Science Technology and Application*, Chongqing, China, pp. 372–375.

de la Fuente, M. V., Ros, L., and Ortiz, A. 2010. Enterprise modeling methodology for forward and reverse supply chain flows integration. *Computers in Industry*, 61, 702–710.

Erl, T. 2004. *Service-Oriented Architecture-A Field Guide to Integrating XML and Web Services*. New York: Prentice Hall.

Fernandes, R., Li, B., Benjamin, P., and Mayer, R. 2010. Applying semantic technologies for enterprise application integration. *Proceedings of 2010 International Symposium on Collaborative Technologies and Systems*, Chicago, IL, pp. 416–422.

Gleghorn, R. (November/December) 2005. Enterprise application integration: A manager's perspective. *IT Professional*, 7(6), 17–23.

He, X., Li, H., Ding, Q., and Wu, Z. 2009. The SOA-based solution for distributed enterprise application integration. *Proceedings of 2009 International Forum on Computer Science Technology and Application*, Chongqing, China, pp. 330–336.

Izza, S. 2009. Integration of industrial information systems: From syntactic to semantic integration approaches. *Enterprise Information Systems*, 3(1), 1–57.

Ji, X. 2009. A web-based enterprise application integration solution. *Proceedings of Second IEEE International Conference on Computer Science and Information Technology*, Beijing, China, August 8–11, 2009, pp. 135–138.

Khoumbati, K., Themistocleous, M., and Irani, Z. 2005. Integration technology adoption in healthcare organisations: A case for enterprise application integration. *Proceedings of 38th Hawaii International Conference on Systems Sciences*, Waikoloa, HI, pp. 1–9.

Linthicum, D. 1999. *Enterprise Application Integration*. Harlow, U.K.: Addison-Wesley.

Lublinsky, B. 2001. Achieving the ultimate EAI implementation. *eAI Journal*, 3(2), 26–31.

Maheshwari, P. 2003. Enterprise application integration using a component-based architecture. *Proceedings of 27th Annual International Computer Software and Applications Conference*, Washington, DC, pp. 557–562.

Ni, Q., Lu, W., Yarlagadda, P., and Ming, X. 2006. A collaborative engine for enterprise application integration. *Computers in Industry*, 57, 640–652.

Puschmann, T. and Alt, R. 2001. Enterprise application integration-the case of the Robert Bosch Group. *Proceedings of 34th Hawaii International Conference on Systems Sciences*, Maui, Hawaii, USA, pp. 1–10.

Qian, X., Yu, J., and Dai, R. 1993. A new discipline of science: The study of open complex giant system and its methodology. *Journal of Systems Engineering and Electronics*, 4(2), 2–12.

Qureshi, K. 2005. Enterprise application integration. *Proceedings of IEEE 2005 International Conference on Emerging Technologies*, Islamabad, Pakistan, pp. 340–345.

Ruh, W., Maginnis, F., and Brown, W. 2000. *Enterprise Application Integration: A Wiley Tech Brief*. New York: Wiley.

Themistocleous, M. and Irani, Z. 2003. Towards a novel framework for the assessment of enterprise application integration packages. *Proceedings of 36th Annual Hawaii International Conference on Systems Sciences*, Waikoloa, HI, p. 234, DOI: 10.1109/HICSS.2003.1174608.

Wu, S., Xu, L., and He, W. 2009. Industry-oriented enterprise resource planning. *Enterprise Information Systems*, 3(4), 409–424.

Xu, L. 2000. The contribution of systems science to information systems research. *Systems Research and Behavioral Science*, 17, 105–116.

Xu, L. 2011. Enterprise systems: State-of-the-art and future trends. *IEEE Transactions on Industrial Informatics*, 7(4), 630–640.

Zhang, J. 2009. Apparel enterprise application integration model based on service-oriented architecture. *Proceedings of IEEE International Conference on Automation and Logistics*, Shenyang, China, pp. 1374–1377.

Zhang, L., Zhou, S., and Zhu, M. 2009. A semantic service oriented architecture for enterprise application integration. *Proceedings of 2009 Second International Symposium on Electronic Commerce and Security*, Nanchang, China, pp. 102–106.

Chapter 9

Systems Approach to Industrial Information Integration

9.1 Complexity

Complexity is a complex multiplicity of elements/relations among objects. There are two perspectives on studying complexity. One school considers that there currently exist no unified definitions of complexity; as such, complexity has various meanings. Despite this, however, in several scientific fields, complexity has a relatively defined meaning. Some of these disciplines include mathematics, information theory, systems science and engineering (SSE), and software engineering. The other school considers that complexity science has already been established as it attempts to study the nature of complex systems, including complex adaptive systems. Complexity exists in many natural and societal systems. Enterprises, extended enterprises, and supply chains are societal systems with complexities that have grown dramatically. Along with that complexity, significant challenges have been posed that need further scientific research. From a systems science perspective, inquiry into complexity is somewhat like the study of an ill-structured system. Instead of applying pure scientific methods, the advantage of applying a systems approach is that it can acquire solutions that will potentially be more applicable in real-world settings, as systems methods are not limited to algorithms or to models that are relevant only to limited dimensionalities.

A complex system is defined as system $S = (M, R)$ with complexity if it satisfies the following conditions:

1. The system S has a large number of subsystems $S_i = (M_i, R_i)$ and a large number of levels.
2. There is at least one nonlinear relation between the levels.
3. The system S possesses the wholeness property P, which cannot be derived by using any analytical model of the subsystems or components from any level.

In this system, S is an ordered pair of sets, M is the set of all objects of S, and R is a set of relations defined on M. The sets M and R are called the object set and the relation set, respectively. For any relation $r \in R$, r is defined as there exists an ordinal number $n = n(r)$, which is a function of r, called the length of the relation r, such that $r \subseteq Mn$ (Lin et al., 2013).

In the context of enterprises and their environment, the following types of complexities deserve attention (Lin et al., 2013):

1. Physical complexity: This is the complexity that exists in a physical system. In the process of evolution, physical systems exhibit complexities at the physical level.
2. Functional complexity: This can be created by complex enterprise structures.
3. Structural complexity: This can be created and enhanced by enterprise structures. A complex enterprise structure creates a process in which both the quantity and characteristics of complex interconnections are increasing.
4. Organizational complexity: With the increase in complexity (such as functional complexity and structural complexity), a corresponding organizational complexity develops.
5. Social complexity: Social systems are complex open systems with the complexity that is characterized by complicated structures, multilevels, and uncertainty.

Organizational complexity means that the state of the organization of concern is in complex motion, which is represented by complex latitudinal, longitudinal, and spatial patterns. It is not difficult to show that as the latitudinal, longitudinal, and spatial complexities increase, the complexity of the organization increases. Enterprises or supply chains are considered to be social systems. The degree of complexity of such systems is much greater than that of physical systems. In the process of evolution, influenced by uncertain internal and external factors, social systems are subject to constant and dynamic changes in both structures and functions.

In the context of ESs and enterprise integration, the following types of complexities deserve attention (Lin et al., 2013):

1. Algorithmic complexity: Algorithmic complexity exists in ESs (Sommerville et al., 2012). Algorithmic complexity is addressed by theories such as the complexity theory in mathematics, information theory, and computer science.
2. Integration complexity: This mainly addresses the effect of the system's components on its emerging behaviors and their interconnections, in addition to addressing the complexity that exists in the process of integration. Integration complexity mainly consists of the complexity of the internal structure, the complexity of the external environment, the complexity of interactions, and the use of complex adapting mechanisms.

In ESs, as we study the phenomena of complex systems, it is important to focus on the complexities mentioned earlier and to thoroughly consider their interactions and evolution in a complex environment.

As an SSE approach for resolving complexity, the Warfield version of systems science (WSS) proposes the equation DOU = DON ∪ DOC for resolving complexity, in which DON, DOC, and DOU are defined as DON = domain of normality, DOC = domain of complexity, and DOU = domain of the universe of learnability. The former contains what is known, and the latter contains what is unknown. The key point of the equation is the progression from DOC to DON through DOU when confronting major challenges of complexities. This equation represents complexity and an approximate means of resolving complexity. WSS measurement of complexity departs from the traditional measurement, as it moves toward a philosophical and pragmatic representation while retaining the coverage of uncertainty as seen in the form of problems or interrelationships among problems (Warfield, 2007).

The two major purposes of the equation are (1) migrating content from DOC, the domain of complexity into DON, and the domain of normality and (2) resolving complexity in problematic situations to the benefit of organized human activity. These two purposes can be served at the same time by carrying out the work program of complexity (WPOC). The means of activating the syllogism in the WPOC is Harary's Reachability Matrix (HRM). Two processes can be applied in a variety of ways while the concept of HRM is employed. These two are the nominal group technique (NGT) and interpretive structural modeling (ISM). NGT and ISM were proposed in the early 1970s and have been widely applied since then. The WPOC typically yields a portfolio of products, tangible or intangible, including problem set, problem field, problem categories, problematique Type 1, problematique Type 2, complexity metrics, options field, optionatiques, design alternatives, and DELTA charts. All of these are described in great detail with good application examples in references (Warfield, 2007). Table 9.1 shows the relationship between SSE, ISM, and WPOC.

ESs are highly complex. They involve a variety of interactions and are highly interdisciplinary, with science, engineering, social, and ecological factors

Table 9.1   Relationship between SSE, ISM, and WPOC

Time Period	Action Component	Comments
1968–1974	System Engineering	Work as structured around A.D. Hall III activity Matrix. The definitive book amplifying this is Warfield, J. N. and Hill, J. 1972. *A Unified Systems Engineering Concept*, Columbus, OH: Battelle.
1974–1981	Interpretive Structural Modeling (ISM)	Work was structured around ISM software. The definitive books on this subject are Warfield, J. N. 1974. *Structuring Complex Systems*, Columbus, OH: Battelle Memorial Inst. Monograph No. 4. Warfield, J. N. 1976. *Societal Systems: Planning, Policy, and Complexity*, New York: Wiley Interscience. Warfield, J. N. 2003. *The Mathematics of Structure*, Palm Harbor, FL: AJAR.
1981–2001	Interactive Management (IM)	Work was expanded to incorporate NGT followed by ISM and to incorporate fields, profiles, and DELTA charts. The definitive book on this subject is Warfield, J. N. and Cardenas, A. 1994. *A Handbook of Interactive Management*, Ames, IA: Iowa State University Press.
2000–2009	Work Program of Complexity (WPOC) or Enhanced Interactive Management (EIM)	Renaming recognized the contributions coming from experience with three major programs: (1) Ford Motor Company large-scale design project, (2) Defense Systems Management College projects, and (3) ITESM teaching and design projects. The definitive book on this subject is Warfield, J. N. 2006. *An Introduction to Systems Science*, Singapore: World Science
2009	Logilectic	Renaming recognizes the importance of linguistic simplicity and enhanced descriptions of processes

Source: Warfield, J., Generic systems science: Logilectic in action, Working paper, 2009.

intertwined (see Figure 5.27). Some ESs problems cannot be easily solved by only applying pure science and engineering methods; for ESs, the complex interactions among components and elements need to be considered when attempting to solve certain problems. The quantity of interdisciplinary research is growing, and new methodologies that are beyond the traditional mode of thinking (such as reductionism) are emerging. In ESs, one of the main tasks in studying complexity is to investigate the complexity of ESs in the framework of SSE to explore the complex emergent behavior that appears in the evolution with the interactions between systems, subsystems, and their universes.

9.2 Design Science

The term "design science" was introduced in the 1960s (or even much earlier). Design science was defined as a science for systematically studying design in different fields. In general, design science means the scientific study of design, often called "science of design."

As an independent discipline, design science has been developing its formal methodologies. In different areas such as architecture, engineering, planning, and defense, in addition to their discipline-oriented core concepts and methods, as long as the task of design is involved, a set of design methodologies have been proposed.

In IIIE, there is a growing need to plan, design, and develop ESs on the basis of a systematic body of knowledge contributed to by design science, in addition to more discipline-oriented core concepts and methods. Hevner et al. (2004) propose a set of seven guidelines in designing information systems. The seven guidelines address design as an artifact and include problem relevance, design evaluation, research contributions, research rigor, design as a search process, and research communication (Hevner et al., 2004).

The guideline states that "design-science research must produce a viable artifact in the form of a construct, a model, a method, or an instantiation" (Hevner et al., 2004). The second guideline notes that "the objective of design-science research is to develop technology-based solutions to important and relevant business problems" (Hevner et al., 2004). The third guideline states that "The utility, quality, and efficacy of a design artifact must be rigorously demonstrated via well-executed evaluation methods" (Hevner et al., 2004). The fourth guideline notes that "Effective design-science research must provide clear and verifiable contributions in the areas of the design artifact, design foundations, and/or design methodologies" (Hevner et al., 2004). The fifth guideline states that "Design-science research relies upon the application of rigorous methods in both the construction and evaluation of the design artifact" (Hevner et al., 2004). The sixth guideline states that the search for an effective artifact requires utilizing available means to reach desired ends while satisfying laws in the problem environment (Hevner et al., 2004).

Warfield has set forth seven challenges for information systems designers. They are second-order thought, behavioral pathologies, discursivity, quality control principles, metrics of complexity, physical infrastructure, and synergy (Warfield, 2008). All of them are incorporated in systems science as defined in his tutorial paper (Warfield, 2007), and all of them are illustrated in a case study of the Ford Motor Company (Staley and Warfield, 2007).

Warfield's design science is characterized by the four main components as a science of description, a science of design, a science of complexity, and a science of action. Warfield's science of generic design and its component "interactive management" has been successfully applied to the research, development, and implementation of ESs and illustrates how ESs can benefit from applying systems science. In the 1980s, automobile manufacturing required significant ESs integration efforts to be made. Components of ESs such as CAD, CAE, and CAM might have been separately supplied, as such, integration quickly became a necessity. Integrating the product system that put the three Cs together as "the C3P system" was the challenge facing the Ford Motor Company. At Ford, the theory of science of design as well as a theory in systems science was applied by a cross-functional team to design and implement Ford's C3P system, beginning in 1995. It has continued to serve the company since its implementation. The integrated system has been used in design, body engineering, interior styling, interior layout and test development, manufacturing plant layout, work cell design, tool design, and assembly line development.

In Warfield's design science method, the design method consists of two sequential tasks: discovery and resolution. The successful design of the C3P system reflects the essence of the method. In the discovery stage, six substages are involved. They are generating a problem set, filtering the problem set, structuring the problematique, computing complexity metrics with data from the problematique, placing problems in dimensions using ISM and discussion, and delivering results. After a clear understanding of complexity is achieved with the problem set, the filtered problem set, the problem field, the dimensions, the complexity metrics, and the interpretation, the resolution stage begins. This stage includes substages such as generating options, determining the interdependency of option categories, forming options profiles, comparing and choosing an alternative, arraying the options in dimensions, and developing the work breakdown plan. Based on WSS and using WPOC, patterns of problems and patterns of options can be developed, both with a substantial number.

The law of requisite variety was introduced by W. Ross Ashby. Ashby's law of requisite variety provides an essential guideline for designing the corrective measures that can resolve a situation. At Ford, Ashby's law of requisite variety was successfully applied to design alternatives, and at least one option was chosen from each design dimension to formulate a design alternative. Ford's early effort in integrating ESs is an excellent example that shows how design science, relying on systematic approaches, can handle the challenge of complex systems integration.

In WSS, the characteristics of design science include the front-end science, the neutral science, the multiple-observer science, the four-component science, the learning-architecture-dependent science, the two-methodology science, and the prose-graphics-oriented science. The details are (Warfield, 2007) as follows:

- The front-end science: This is the science to be applied when a problematic situation is to be examined, in order to organize it for further detailed study.
- The neutral science: This is the science that is independent of all specific sciences to the largest possible extent.
- The multiple-observer science: This is the science that depends on the empirical verification of theories.
- The four-component science: This is the science, hierarchically formed from a science of description, to a science of design, to a science of complexity, to a science of action.
- The learning-architecture-dependent science: This is the science that requires access to a wide space for wide-angle vision.
- The two-methodology science: This is the science that depends on two methodologies: nominal group technique and interpretive structural modeling, in order to handle complexity.
- The prose-graphics-oriented science: This is the science that relies on extensive prose-graphics communication.

WSS methodology can be summarized in terms of correlates, proposition-based descriptions, functional descriptions, and well-defined attributes. Some very useful terms include structure of scientific character, discursivity, discursivity structure, discursivity structure quantitative, global insight, and relational synthesis. Warfield's theory of generic design is described in great detail with good application examples in his work (Warfield, 1990).

9.3 Systems Approach

The second half of the twentieth century ushered in the age of systems movement. Systems movement began during the 1940s and 1950s. Many systems terms have been developed since the 1940s, such as cybernetics, systems theory, systems analysis, systems methodology, systems science, systems engineering, systems approach, and systems thinking. Today we can consider these terms under the common umbrella provided by SSE. For more than half a century, SSE and the ideas behind it have penetrated many disciplines of natural science and social sciences, including IIIE.

SSE can be considered as a basis for industry information integration and can be applied to IIIE. Traditional methods focus on reductionism and decomposition, with an emphasis on studying the components of a system. SSE focuses on complex interconnections and emergent behaviors. Pragmatically, SSE concepts have been adopted by the field of industry information integration; however, the significance and impact of SSE on industry information integration has not been well demonstrated in the literature until recently. In 2012, Sommerville et al. indicated that while there is increasing awareness of related issues in large-scale complex IT systems, the most relevant background work comes from SSE (Sommerville et al., 2012).

This section aims to introduce the contribution of SSE to industrial information integration research and to discuss how concepts and findings in SSE can be applied to industrial information integration. In addition to Sections 9.1 and 9.2 where we introduce the systems concepts of complexity and design science, this section will further point out that SSE is a necessity in order to deal with the overwhelming systems complexity in industrial information integration (Sommerville et al., 2012).

The diverse literature included in this section has important relationships to industry information integration, as well as to their home fields. In the following, we will briefly introduce selected systems concepts with ESs application examples. We will also introduce a recommended inquiry system for ESs research and development. We believe that, in long run, SSE will profoundly impact and change industry information integration research.

9.3.1 Information Integration: An SSE Perspective

Industry information integration can be defined in terms of a number of perspectives; for example, one includes relating to its functions or to its structure. From a functional perspective, industry information integration is a technologically implemented system with the purpose of coordinating all the elements including the processes, resources, and technology of an enterprise working toward the realization of industrial integration purpose. From a structural perspective, information integration consists of a collection of data, algorithms, processes, technology, people, and organizations, all forming a cohesive structure that serves the purpose of enterprise integration. In an information integration process, it is necessary to view information integration systematically as a sociotechnical system and to develop ESs using a wide spectrum of methodologies including SSE, which is superimposed over the complex sociotechnical interactions. SSE provides the basis for taking a broader view of information integration.

9.3.2 Systems Concepts

Information integration, to some extent, has been developed as a science in the image of the established natural sciences and of engineering, with the belief that the success of information engineering in the natural sciences and engineering can be repeated if the methodology emulates the methods of the natural sciences and engineering. This view identifies three major paradigms. The first of these is theory; the second is abstraction; and the third is design. Proponents of this view assume that information integration problems can largely be resolved by sophisticated methods and techniques. SSE indicates that information integration involves a variety of well-structured subsystems that are clearly found in the natural sciences and engineering, as well as ill-structured subsystems that involve complexity. The traditional model of information integration may be criticized because it leads to a narrow view of information integration in organizations. From an SSE point of view, information integration can be treated as a technical process as well as a complex social process.

9.3.2.1 Complex Systems

Section 9.1 has given the definition of a complex system in which R is the main factor in making a system complex. Complex systems interactions have been studied by many researchers. In ESs, some components can operate as independent systems. However, when they are incorporated into a system, their behavior depends on their interactions with other systems components. Systems interdependence is essential in integrated information system functioning; complex contexts must be made explicit. Staley and Warfield provided an example with the experience at Ford. Numerous interactions of different systems/subsystems/dimensionalities may be involved, such as interactions within a system or a subsystem, interactions between systems and/or subsystems, and/or interactions between a system and its environment. Such interrelated subsystems may result in dynamic changes. Interactions can occur between interactions as well, and some interactions emerge only when the system is viewed as a whole. ESs technologies, due to their complexity, are amplifying such higher-order interactions. If the information integration is to be implemented successfully, these interactions must be understood and the methods for designing and implementing ESs must take the higher-order relationships into account (Richards and Gupta, 1985). However, these highly complex, higher-order interactions are not being heavily investigated. A partial reason is that existing research methods are not capable enough of handling higher-order relationships, either conceptually, mathematically, or practically.

Sommerville et al. provided an interesting example of complex interactions. "On the afternoon of May 6, 2010, the U.S. equity markets experienced an extraordinary upheaval. Over approximately 10 min, the Dow Jones Industrial Average dropped more than 600 points, representing the disappearance of approximately $800 billion of market value" (Sommerville et al., 2012). The trigger event was identified as a particular sale that began a complex pattern of interactions between the high-frequency algorithmic trading systems. It was caused by the interactions of independently managed software systems (Sommerville et al., 2012).

Based on system $S = (M, R)$, "complexity stems from the number and type of relationships between the system's components and between the system and its environment. If a relatively small number of relationships exist between system components and they change relatively slow over time, then engineers can develop deterministic models of the system and make predictions concerning its properties. However, when the elements in a system involve many dynamic relationships, complexity is inevitable. Complex systems are nondeterministic, and system characteristics cannot be predicted by analyzing the system's constituents. Such characteristics emerge when the whole system is put to use and changes over time, depending upon how it is used and on the state of its external environment" (Sommerville et al., 2012). "Complexity stemming from the dynamic relationships between elements in a system depends on the existence and nature of these relationships. Engineers cannot analyze this inherent complexity during system development, as it depends on the system's dynamic operating environment. Coalitions of systems in which elements are large software systems are always inherently complex. The relationships between the elements of the coalition change because they are not independent of how the systems are used or of the nature of their operating environments. Consequently, the nonfunctional (often even the functional) behavior of coalitions of systems is emergent and impossible to predict completely" (Sommerville et al., 2012).

Although some methods and techniques have been developed to focus attention on the complex multiplicity of subsystems/dimensions, however, these methods are either unable to sufficiently describe the complex relationship or lack of means for considering connected partial areas as a whole. SSE has the potential to help conceptualize higher-order complexities. As indicated by Sommerville et al., to help address complexity, we must adopt systems perspectives (Sommerville et al., 2012).

9.3.2.2 Dimensionality

Dimensionality is a fundamental concept in physical science (Figure 9.1). It has recently been introduced into feasibility studies, information retrieval, information systems planning, software development, and the information systems development process. Ashby's law of requisite variety is based on the following: (1) There exists a system that is in change; (2) there exist criteria for exerting control over the system in change, and the controller will satisfy the criteria; and (3) the system can be characterized by number n of dimensions, and the "variety" of the system is measured by n. The key idea of the Ashby's law of requisite variety is this: For successful control purposes, the variety available to the control mechanism must be n, the same as the "variety" of the system. According to Warfield, "if we assume that a system includes identifiable subsystems, there is reason to suppose that there might be collections of dimensions, some of which would apply to particular subsystems, some of which would apply and have meaning only when the total system is assessed" (Warfield and Christakis, 1987). The concept of dimensionality provides an essential basis for applying Ashby's law of requisite variety.

In ESs, a variety of dimensions are involved, and they interact with one another. ESs success can be considered as a multidimensional system. The dimensions

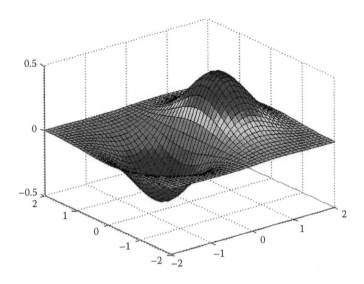

Figure 9.1   The concept of dimensionality in physics.

include the vendor/consultant quality, system quality, information quality, individual impact, workgroup impact, and organizational impact (Ifinedo and Nahar, 2007). The dimensions involved in ESs may not be independent; they have effects on each other. An important question is this: How can multidimensional systems such as sociotechnical factors be integrated into systems and methods?

Most current methods support the development of one dimension, such as the technical dimension, but not the others, such as the social or organizational dimensions (Sommerville et al., 2012). However, such nontechnical dimensions significantly affect the integration of systems (Sommerville et al., 2012). There are plenty of examples of missing dimensions that can be found in past information integration practice. In current information integration practice, some nontechnical dimensions are beginning to be recognized and addressed; however, the study of such nontechnical dimensions has lagged behind technical studies. This has led to information integration, which might have been technically elegant but was not ideal from a social standpoint. Recognizing that the information integration includes technical systems with social dimensions, the study of multidimensional information integration has emerged. However, much more research is required before multidimensional analyses can be used routinely for complex industrial information integration.

9.3.2.2.1 Temporal Dimension

In an information integration process, each step in the process has an effect on all of the other steps; as the design process advances, information integration may become complicated, and the process of designing and maintaining integrated systems may become more of a dynamic process. The typical methodological mismatch in this aspect is that some existing formal methods are insufficient for modeling the time dimension.

Modeling passing value indeterminacy and describing batch processing functions provide examples. Interorganizational cooperative systems are typical ESs. In real-time cooperative systems, heterogeneous purchasing processes and electronic contracts are not uncommon (Liu et al., 2012), and some important actions are subject to timing constraints. As such, the importance of the time dimension cannot be underestimated. To model such systems with time constraints, LTWNs and ILTWNs were proposed (Du and Jiang, 2003). ILTWNs can be used to model passing value indeterminacy and to describe the batch processing functions of real-time cooperative systems. A basic property of ILTWNs and LTWNs is soundness. Soundness means the correctness of ILTWNs and LTWNs. It is used to verify whether the structure and dynamic behavior of ILTWNs and LTWNs are consistent with the requirement specifications of the modeled systems and to test whether the systems terminate at an acceptable state. Although soundness can be determined, however, it is EXPSPACE hard. As such, soundness has been discussed from its subclasses such as T-fair ILTWNs (Liu et al., 2012). The significance of this method is to verify the soundness preservation of ILTWNs composed of multiple T-fair LTWNs. It helps us to understand and to model the cooperative workflows, as the method can reduce the complexity of the analysis, based on the static structure of ILTWNs.

The important implications here include the following: (1) the criticality of time dimension; for example, in modeling passing value indeterminacy and in the batch processing functions of real-time cooperative systems, the correct behavior of such systems depends not only on the logical correctness of the results obtained through running workflows but also on the time that it takes to produce them before critical deadlines; (2) SSE theories and related mathematical tools may be valuable as they can reduce the complexity of the analysis.

9.3.3 ESs' Subsystem Integration: Workflow Management

Workflow management technology combines techniques and tools to facilitate the process of developing intra- and interorganizational information integration. Under the heading of workflow management, there are numerous variants of the technique used. In the workflow and BPM field, there is a lack of confidence that information integration according to traditional methods will sufficiently serve their users and bring competitive advantages, since currently no single method covers all aspects required, as discussed in Chapter 6. The lack of a suitable systems framework concerning the workflow process as a whole helps to explain many of the shortcomings in the existing methods, and SSE can eventually provide the basis for workflow management. It is inevitable that information integration will try to buttress traditional methods with new SSE concepts. For example, SSE concepts and practice including WSS, the concept of dimensionality, and macromathematics can be applied. These systems theories can be applied to workflow and process management through contributing systems ideas and methods.

9.3.4 ESs' Subsystem Integration: Manufacturing Systems

During the last decade, the manufacturing environment has changed dramatically. Worldwide competition among manufacturers and the development of new manufacturing technologies have contributed to today's competitive environment. SSE has profoundly changed today's manufacturing industry as well as its manufacturing information systems. Research indicates that modern manufacturing technology is a hybrid generated by knowledge from manufacturing, computer science and engineering, SSE, management, and marketing. SSE is a necessity for facilitating the large-scale introduction of information technology into manufacturing, since integrating systems concepts into manufacturing systems enable manufacturers to take more and more aspects of the manufacturing process into account. Evidence shows that cybernetics, control theory, and SSE concepts (including WSS) have been successfully applied to manufacturing systems integration, in the context of ESs such as the C3P system at Ford.

9.3.5 Inquiring Systems

Sommerville et al. (2012) have indicated: "Engineers are concerned primarily with building technical systems from hardware and software components and assume system requirements reflect the organizational needs for integration with other systems, compliance, and business processes. Yet systems in use are not simply technical systems but 'sociotechnical systems.' To reflect the fact they involve evolving and interacting communities that include technical, human, and organizational elements, they are sometimes also called 'sociotechnical ecosystems,' though the term sociotechnical systems is more common. Sociotechnical systems include people and processes, as well as technological systems. Defining technical software-intensive systems intended to support an organization's work in isolation is an oversimplification that hinders software engineering. So-called system requirements represent the interface between the technical system and the wider sociotechnical system, yet requirements are inevitably incomplete, incorrect, and/or out of date. Coalitions of systems cannot operate on this basis." In addition, "software engineering has focused on reducing and managing epistemic complexity, so, where inherent complexity is relatively limited and a single organization controls all system elements, software engineering is highly effective. However, for coalitions of systems with a high degree of inherent complexity, today's software engineering techniques are inadequate" (Sommerville et al., 2012). In essence, "we need to examine the essential divide-and-conquer reductionist assumption that is the basis of all modern engineering" (Sommerville et al., 2012). In addition, The SEI ULSS report has argued that current engineering methods strictly based on reductionism are inadequate, saying: "For 40 years, we have embraced the traditional engineering perspective. The basic premise underlying the research agenda presented in this document is that beyond certain complexity thresholds, a traditional centralized engineering perspective is no longer adequate nor can it be the primary means by which ultra-complex systems are made real" (Paige et al., 2008).

There are two distinct types of paradigms described in the literature: the science paradigm and the system paradigm. Each paradigm features a number of inquiring systems. The science paradigm acknowledges analytical thinking in the form of the reductionist nature of the inquiring process. Its representative inquiring systems are Leibnitzian, Lockean, and Kantian inquirers. The system paradigm acknowledges that systems comprise parts, emphasizing a whole system with interrelated parts. Its representative inquiring system is the Singerian inquirer. The Leibnitzian inquiring system takes a system and constructs a mathematical representation of it; the Lockean inquirer is an empirically based inquiring system; and the Kantian inquirer is a partially analytical (Leibnitzian) and partially empirical (Lockean) inquiring system, wherein the basis is the degree of match between the formal model (Leibnitzian) and the data collected (Lockean), with an emphasis on truth being synthetic, that is, it puts forth that any analytical models based on reduction and data are inseparable. In addition, the Hegelian inquirer is an inquiring system that determines relevant information for a system by presenting two or more conflicting views, and the Singerian inquirer is a meta-inquiring system that includes Leibnitzian, Kantian, Lockean, and Hegelian inquiring systems (Churchman, 1971). Churchman's well-known book titled *The Design of Inquiring Systems Basic Concepts of Systems and Organizations* relates the principles of SSE to the philosophical systems of Leibniz, Locke, Kant, Hegel, and Singer. Among these inquiring systems, Leibnitzian (analytically based), Lockean (empirically based), and Kantian (analytically and empirically based) inquirers are suited to problems that generally are well structured and for which an analytic solution exists. A Hegelian inquiring system is appropriate for ill-structured problems such as organizational perspectives on ESs, and a Singerian inquiring system is of special value since holistic yet analytical modeling is appropriate for many of the interrelated well-structured and ill-structured subsystems involved in ESs.

Although the literature in information integration research seldom refers to the role played by inquiring systems explicitly, researchers generally adopt Leibnitzian (analytically based), Lockean (empirically based), and Kantian (analytically and empirically based) inquirers, as introduced in this book. The key idea here is reductionism. "Reductionism is a philosophical position that a complex system is no more than the sum of its parts, and that an account of the overall system can be reduced to accounts of individual constituents. From an engineering perspective, this means systems engineers must be able to design a system so it is composed of discrete smaller parts and interfaces allowing the parts to work together"

(Sommerville et al., 2012). This assumption assumes that an integrated system can be built with the system elements and can integrate them to create the desired overall system without considering complex system characteristics such as social aspects. As indicated by Sommerville, for many years, researchers generally adopted this reductionist assumption, and their work concerned finding better ways to decompose systems and better ways to complete system integration. Underlying all of the software-engineering methods and techniques are reductionist assumptions (Sommerville et al., 2012). Some software project failures are a consequence of adherence to the reductionist view (Sommerville et al., 2012). SSE recognizes that systems often exhibit behaviors more than what those individual elements exhibit in decomposition.

Inquiring systems are of considerable importance to integrated information systems. Past experience has shown that the science paradigm and its inquiring systems have been overemphasized in studying some ill-structured subsystems in integrated information systems that actually cannot be analyzed effectively by existing formal models. For example, human factors and social processes that contribute to software process dynamics have been modeled and represented by pure analytical models. It is obvious that some existing analytical methods cannot adequately address the behavioral aspects of social processes.

More authors have recently realized the importance of choosing appropriate methods to represent various subsystems/dimensions in information integration (Sommerville et al., 2012). The information integration process must not only be aware of the existence of subsystems and interactions but also be able to identify the types and characteristics of subsystems and interactions in order to define the level of abstraction at which an analysis can be carried out. Then, it must select appropriate inquiring systems. Studies have demonstrated that the type and characteristics of the system to be represented determine the inquiring systems to be used. When inquiring systems are incorrect, research will be correspondingly weak or even erroneous. Research indicates that current information integration engineering method based on reductionism is not sufficient as expected (Sommerville et al., 2012). Inquiring systems are of considerable importance in addressing the urgent need for comprehensive approaches to help construct large-scale complex ESs.

9.3.6 *Methodological Development*

Sommerville et al. (2012) revealed the challenges posed by complex systems, noting that "Even when the relationships between system elements are simpler, relatively static, and, in principle, understandable, there may be so many elements and relationships that understanding them is practically impossible. Such complexity is called 'epistemic complexity' due to our lack of knowledge of the system rather than some inherent system characteristics. For example, it may be possible in principle to deduce the traceability relationships between requirements and design, but, if appropriate tools are not available, doing so may be practically impossible."

Warfield (1990) presented his version of the law of requisite variety, which requires a match between the dimensionality implicitly represented. In his perspective, a system implicitly represents an integer dimensionality K_s, and a system designer defines an integer dimensionality K_m. In many practices, systems are underspecified in dimension due to poor system conceptualization, that is, $K_m < K_s$. In fact, a design process must exhibit requisite variety, that is, $K_m = K_s$. Regulation is possible only when system designers have access to sufficient information.

The concept of dimensionality in information integration provides an essential basis for applying Ashby's law of requisite variety in IIIE. Xu (1995) proposed that complex system design should start with a systems concept phase to develop a complete understanding of the system and to define an overall design approach. The suggested systems concept phase has the following important tasks: (1) identify and define systems/subsystems/dimensions and interactions, (2) examine systems/subsystems/dimensions separately and collectively, (3) apply the law of requisite variety, and (4) tackle systems/subsystems/dimensions and interactions of different types/characteristics with appropriate inquiring systems and tools with theoretical developments.

Warfield (1986) pointed out that micromathematics is the primary tool of traditional methods in science and engineering, and Bellman (1977) also indicated that new mathematical approaches are needed. Macromathematics potentially is the primary tool of synthesis and overview (Warfield, 1986). Traditional methods rely primarily on micromathematics, and the importance of macromathematics has not been paid sufficient attention. Research indicates that complex systems cannot be properly modeled with micromathematics such as calculus, differential equations, and optimization methods.

A number of systems methodologies for information integration have appeared in the literature over the years, addressing the various phases of information integration. These appear to signal that significant developments have occurred in practicing SSE in the area of industrial information integration. Within these developments, there are efforts being made, which attempt to formulate distinctive and comprehensive methodologies, while some others are focusing on improving existing methods. The observation can be made that various methodologies are viewed as complementing one another and that the diverse efforts present an active and substantial trend in applying SSE to information integration.

9.3.6.1 Interdisciplinary Study

SSE is attempting to integrate all relevant disciplines in an effort to form a structured problem-solving approach. IIIE is attempting to develop a structural framework that will integrate relevant disciplines for industrial information integration purposes in order to engage those challenges faced when applying methodologies in industrial information integration (see Figure 5.27). It is recognized that information integration will require an interdisciplinary approach and the appropriate taxonomies to adequately reflect the complex subject covered. IIIE is moving toward the development of a structural framework to address the challenge of constructing ultra-large-scale systems such as ESs by applying SSE. There is a need to explore and envision a framework within which all relevant disciplines and methods can experiment with different systems, subsystems, dimensions, and the interactions involved.

In 2006, Northrop et al. identified seven research areas for ultra-large-scale systems: human interaction, computational emergence, design, computational engineering, adaptive systems infrastructure, adaptable and predictable system quality and policy, and acquisition and management (Northrop et al., 2006). These seven areas apparently acquire expertise from a range of disciplines—from science, engineering, and SSE to social science. Researchers support the idea that research required is interdisciplinary.

9.3.6.2 Selection of Methods

Some authors suggest that integrated information system developers should select the methods that best fit a specific task. In other words, designers should select the combination of methods most useful to the specific systems/subsystems/dimensions and interactions with which they are dealing (Northrop et al., 2006). Some methods may coexist. Many examples are available: (1) large-scale systems that use strategies borrowed from architecture and other engineering disciplines are being successfully developed (Hess, 1990); (2) in designing a complex system, system designers use throwaway prototyping for some parts and evolutionary prototyping for others (Davis, 1992); and (3) in designing large-scale systems, the traditional process model of software development has been complemented by the layered behavioral model (Curtis et al., 1988). As pointed out by Jackson, the use of diverse methods, models, tools, and techniques from different paradigms in combination with appropriate methodological and theoretical development is encouraged. The objective, however, is not to build a toolkit based on the criterion of what works in practice (Jackson, 1997).

9.3.6.3 Theory for Integrated Information Systems as a Whole

The development of a complete theory about integrated information systems as a whole is one of the most important goals in industrial information integration engineering and SSE research. It requires the existence of well-founded concepts and methods for modeling the knowledge of different systems, subsystems, and dimensions. SSE and mathematics can provide the appropriate concepts and related methods that will allow a high degree of knowledge integration. Meta synthetic engineering (MSE) has emerged as a methodology for dealing with open complex giant systems (OCGSs). Each OCGS features a variety of subsystems with complex structures and interrelations. The goal of MSE is to integrate the large amount of dispersed knowledge from different disciplines in a whole structure (Qian et al., 1993).

In terms of the relationship between SSE and mathematics, research indicates that systems theory works well for most types of mathematical models. It has been used in graph theory, combinatorial optimization, simulation modeling, and systems dynamics (Muller-Merbach, 1983). The systematic procedure used in systems theory helps modelers move toward a specific structure of thought and toward a certain precision of analysis and synthesis. In terms of the compositions of mathematical theories, traditional mathematical system theory subsumes almost all the classical models of physics and engineering. Research suggests that those mathematical branches that provide overview capabilities such as algebraic geometry, combinatorial theory, graph theory, logic, set theory, and topology should be included (Warfield, 1986). The term "macromathematics" was coined by Warfield (1986). Macromathematics includes algebraic topology, algebraic theory, category theory, combinatorial theory, general mathematical systems, general topology, graph theory, homological algebra, logic, ordered algebraic structures, probability, set theory, and statistics. The main characteristics of macromathematics are (1) the ability to convey frameworks, (2) the ability to symbolize, (3) the capacity to represent any relations involving parts of the system graphically, (4) the ability to portray both aggregation and disaggregation, (5) the capacity to represent ideas in ways that are translatable from prose to graphics to mathematics, and (5) the capacity to represent quantitative and qualitative information in the same framework with equal facility (Warfield, 1986).

It is believed that more research will emerge, which will direct us in the development of a complete theory of integrated information systems, with the help of SSE.

9.3.7 Summary

This section traces the impact of SSE on industrial information integration as well as the contributions made by systems science to industrial information integration. It demonstrates that industrial information integration has deep roots in systems science. The section indicates that SSE has been used to pursue various intellectual activities in industrial information integration and notes that much of the ongoing ESs research shows that there is an awareness of the importance of SSE research. It is clear that there are many opportunities for significant research in applying SSE to industrial information integration. It is also clear both (1) that in designing ESs, the application of SSE becomes a necessity and (2) that much remains to be done to create a complete systems theory that is both powerful and applicable to industrial information integration. We can expect that the wealth of research in industrial information integration in the framework of SSE will produce both an astonishing array of theoretical results and empirical insights and a large suite of tools and methods. SSE promises to be an important foundation for industrial information integration in the years to come (see Figure 5.27).

9.4 CMFT: A New Theory in Systems Perspectives and Its Implication to IIIE

In 2008, a new theory on MFs called the CMFT revealed the essence of MF objects and phenomena. In this theory, MFs are specified in terms of the MF in the economic dimension, the social dimension, and the natural dimension, as well as the interrelationships among them. It was pointed out that the MF is not only an economic phenomenon but also a social and a natural one, as Figure 9.2 shows. In other words, there exists not only an economic MF but also social and natural flows. The economic MF is the core of the MF, whereas the social and natural flows are the basis of the MF (Xu, 2008).

CMFT theory is a successful example of the application of SSE, as well as a typical example involving both macromathematics and micromathematics. In CMFT, the concepts of systems, subsystems, dimensions, and their interactions have been comprehensively considered. As such, this theory provides extremely important insights into future enterprise architecture and integration, especially in the consideration of sustainable economic and societal development and growth. Based upon this, the concept of next-generation ESs has been proposed. This next-generation ES is called entire resource planning (ERP) or CRP; its design has been extended to comprehensively encompass the resources used and produced by enterprises in different industrial sectors, within the macro context of both economic

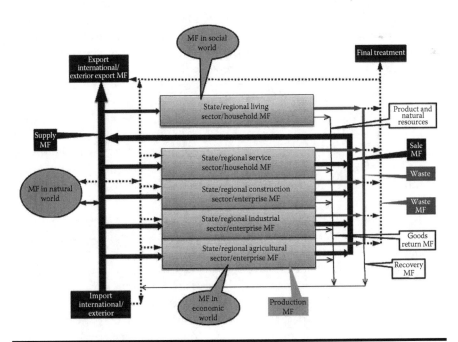

Figure 9.2   CMFT theory.

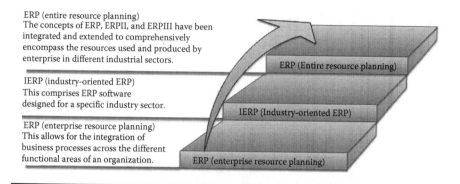

ERP (entire resource planning)
The concepts of ERP, ERPII, and ERPIII have been integrated and extended to comprehensively encompass the resources used and produced by enterprise in different industrial sectors.

IERP (industry-oriented ERP)
This comprises ERP software designed for a specific industry sector.

ERP (enterprise resource planning)
This allows for the integration of business processes across the different functional areas of an organization.

ERP (Entire resource planning)

IERP (Industry-oriented ERP)

ERP (enterprise resource planning)

Figure 9.3   The evolution of enterprise systems in terms of three stages of development efforts.

and societal developments. In ERP, not only is economic MF included, but social and natural MFs are included as well. ERP is considered a significant step forward in the evolution of ESs. Figure 9.3 shows the evolution of ESs in terms of three stages of development efforts.

The major factors that have contributed to the birth of ERP are obvious. The past decade has brought fundamental changes both to the global economy and to global business operations. The main challenges facing the enterprises are that (1) both the globalization of operations and the implementation of SCM are moving toward a deeper level and that (2) the expectations for sustainable economic growth and global environmental protection are rising. The limitations of the existing systems restrain their flexibility to cope with either the challenges described earlier (i.e., global operations at a deeper level) or the increasing requirements of sustainable economic growth and global environmental protection. The existing systems are just a part of the solution, and mere modest modification of the existing ESs will not meet the need for change. Only a new type of system will be capable of coping with the incoming challenges. The key characteristic of the new system is the provision of comprehensive coverage for all relevant types of resource planning. Its main objective is to close the gap between the existing systems and the proposed systems.

From the macro aspect, CMFT provides a general systems framework of future ESs, especially addressing the consideration of sustainable economic and societal development and growth. From the micro aspect, the theory is advancing toward a higher level of methodological development that will include the application of micromathematics, macromathematics, physics, SSE, operations research, and other micromathematical tools. XPMF is a new type of MF in the framework of CMFT. The concept of XPMF was proposed from the perspective of bioevolution complexity including factor-type primitive reconstruction driving type XPMF, enterprise primitive fractal synergetic type XPMF, and the adaptation of primitive swarm intelligence XPMF. XPMF, as revealed in its name, has obviously considered the complexity of dimensionality. Research has been conducted on fractal synergetic XPMF formation and control. Synergetics itself is an interdisciplinary science that studies the formation and self-organization of patterns and structures appearing in open systems that are far from thermodynamic equilibrium. Essential in synergetics is the concept of order parameters, which was originally introduced in the Ginzburg–Landau theory in order to describe phase transition in thermodynamics. The concept of order parameters is useful for describing the self-organization behavior of the system (Lin et al., 2013). Based on the theory of synergetics, an XPMF synergetic operation model based on an OIR order parameter 3D synergetic space was developed (see Figure 9.4). OIR stands for Objective Relational Grade, Information Share Grade, and Resource Complementary Grade. The model includes a PMF coordination mechanism, a control mechanism, and a problem-solving mechanism. In OIR 3D space, multi-PMF generates a synergetic effect in processing complex MFs in enterprises or extended enterprises.

9.5 Microscopic Perspectives: BI in ESs

Not only are macroscopic perspectives necessary in IIIE, but microscopic perspectives and micromathematics are required in the activities of IIIE. BI provides an excellent example of the micro aspect.

BI is the process of gathering correct information in the correct format at the correct time and delivering results for decision-making purposes. As such, BI has a positive impact on business operations, tactics, and strategy in enterprises. BI focuses not only on real-time data but also on real-time analysis that can instantaneously change the parameters of business processes. BI does not provide the same functionality as traditional information systems, but instead operates on data that are extracted from operational data sources exist in ESs. It provides an effective means to propagate actions back into business processes and operations (Zeng et al., 2012). This newly emerging area not only will improve applications in ESs but will also play a very important role both in ESs and in industrial information integration.

The traditional ESs focus on how to provide efficient and productive operations. However, as enterprises continue to grow, they can no longer stay competitive by merely providing productive operations. They face the challenge of processing and analyzing huge amounts of data and turning it into timely decisions for strategic competitiveness. In 2000, Langenwalter defined the executive direction and support function of ERP. Langenwalter considered that ERP is the means of communicating executive direction throughout the enterprise (2000). ERP's contributions are not only those individually available components that have been successfully employed for operations purposes. One of the main powers of ERP is in creating the function of executive decision support. In the IIIE framework, the integration

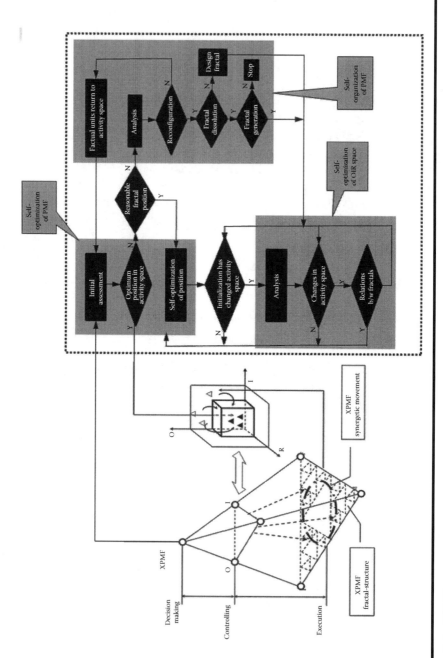

Figure 9.4 XPMF.

of executive decision support component has been well recognized. As one of the main tools for executive decision support, BI is regarded as a key approach to increasing the value of the enterprise based upon the BI component in ESs within the framework of IIIE. We can see the significant boost in BI research in the area of ESs and IIIE in recent years.

As mentioned earlier, in ESs, it is best to select methods that are most suitable to the systems/subsystems/dimensions with which they are dealing. It is true that abundant applications of BI in ESs are available, showing that micromathematics is considered to be a useful tool. BI has been used to extract useful patterns through methods such as process mining, clustering, and outlier detection. In last decade, BI methods have been applied to quality control functions in ESs. This component of ESs is called a quality management system (Langenwalter, 2000).

BI is increasingly important as enterprises strive to stay competitive, as it provides many opportunities to enterprises to sustain and grow. In the following, we present a brief review of BI that is related to ESs.

9.5.1 Business Intelligence

It is widely accepted that the technological components of BI mainly encompass data warehousing, data mining, and OLAP. A data warehouse is a repository of data collected from multiple sources; data mining is the core component in BI that allows analyzing data, detecting trends, and identifying patterns, and OLAP is the front-end analyzing tool. Data mining uses analytical methods to generate a hypothesis about the patterns found in the data and then to predict future behavior. Its objective is to discover patterns, trends, and rules to evaluate business operations, tactics, or strategies, which in turn will optimize the business processes and improve the competitiveness and profitability of enterprises (Zeng et al., 2012).

BI has been increasingly employed in ESs, especially for next-generation ESs. Researchers in ESs are increasingly interested in applying BI, since the knowledge discovery from data results from BI will affect the performance of ESs. We focus on data mining that allows for the discovery of interesting patterns in industrial processes that, in turn, provide high-quality information sources and directly affect the performance of ESs.

We categorize data mining techniques into supervised learning and unsupervised learning methods. Using supervised learning methods, the learning of the classifier is supervised; when using unsupervised learning, the set of classes to be learned may not be known in advance (Han and Kamber, 2006). The goal here is to introduce BI as micromathematical tool in ESs, rather than researching data mining itself, and the focus is on introducing selected state-of-art algorithms of supervised and unsupervised learning. Supervised learning methods include the decision tree, Bayesian statistics, neural networks, the support vector machine (SVM), regression, nearest neighbors, and unsupervised learning methods, including clustering and itemset mining.

9.5.2 Supervised Learning Methods

1. Decision tree: In a decision tree model, each inner node denotes a test on an attribute, branches represent conjunctions of features that lead to those classifications, and leaves represent classifications. Decision trees offer users the ability to predict the value of a categorical variable and also to directly use categorical variables as input or predictor variables. Decision trees are well suited for dealing with large quantities of input variables as well as a mixture of data types. A decision tree algorithm recursively selects an attribute to partition records into smaller subsets from each branch in a top-down manner. The most important step is to select an attribute for the splitting purpose. Three benchmark measures—information gain, gain ratio, and the Gini index—are used. Information gain is based on the pioneering work done by Claude Shannon on information theory (Lin et al., 2013). Information gain calculates the difference between the original entropy and the new entropy after the partition on a selected attribute, which explains to what degree the attribute can improve the purity of partitions. Gain ratio uses the split entropy to regulate information gain, and the Gini index measures the impurity of a data set.

2. Bayesian statistics: Bayesian classification is based on Bayes's theorem. Bayesian classifiers can handle both nominal and numeric attributes well, for nominal prediction. For data sets with many attributes, it is challenging to compute conditional probability. Therefore, the naive Bayesian classifier assumes that attributes are independent from each other. Since the assumption of independence is usually violated in any practical application, Bayesian belief networks are proposed to solve this problem. They allow dependencies between attributes and construct a directed acyclic network to calculate the joint conditional probability. The most challenging part is how to construct the network.

    Bayesian classifiers are supported by statistical theories and work best if the assumptions are satisfied. Before using naive Bayesian classifiers, we must make sure attributes are independent of each other.

3. Neural networks: A neural network is composed of connected artificial neurons with a weight associated with each edge. A neural network consists of three or more layers: an input layer, one or more hidden layers, and an output layer. Although there are many different types of neural network algorithms, the most popular one is back-propagation. This method iteratively adjusts the weight on each edge by comparing the prediction with the actual value. Since the modification is made in a backward direction, it is called back-propagation. Basically, neural networks generate the value for numeric prediction. However, by applying the same transforming function used by logistic regression, neural networks can also predict binary value. If the hidden layer is removed and the structure contains only input and output layers, the

neural network is equivalent to the linear regression model, which can easily be observed from the final prediction function. The neural network is equivalent to the polynomial regression model if the hidden layer has only one layer, and it can make any nonlinear separation if the hidden layer has two layers.

Compared with a decision tree and/or with Bayesian classifiers, it is hard to explain how results are produced. However, neural networks usually provide better results than decision tree and Bayesian classifiers. Neural networks are well suited for both numeric and nominal attributes, while decision trees handle nominal attributes better than numeric attributes.

4. Support vector machines: SVMs are usually not chosen for large-scale data mining problems because their training complexity is highly dependent on the size of the data set. SVMs search for the hyperplane that most clearly separates two classes. SVM was originally proposed for a linear separation. Later, a nonlinear kernel function was proposed that would map the original data into a higher linear separable dimension space. Three commonly used kernel functions are polynomial function, Gaussian radial basis function, and the Sigmoid function. SVM is generally considered to be a highly accurate classifier. SVM can only be applied to nominal prediction but is less prone to overfitting than other methods. If the mining purpose is to achieve highly accurate nominal prediction and to avoid the overfitting issues, SVM is a good choice.

5. Regression: The general linear model can handle the correlation between attributes. The logistic regression model is specialized for binary prediction, and multivariate adaptive regression splines deal with nonlinearity and interactions. Regression models can be applied to many applications because they are good at prediction and can handle different types of data. However, regression models are not specially designed just for the prediction purposes. The accuracy of their performance is generally not as good as neural networks and SVM.

6. Nearest neighbor: Nearest-neighbor classifiers are based on learning by analogy, that is, by comparing a given record with the historical records that are similar to it. There are two important issues related to the performance of nearest-neighbor classifiers: distance function and feature selection. If absolute value is important, Euclidean distance is recommended. If the absolute value is not important, Cosine distance is recommended. Any distance-based function intrinsically assigns equal weights to each attribute.

9.5.3 Unsupervised Learning

9.5.3.1 Clustering

Clustering analysis is a common technique. Clustering is a typical form of unsupervised learning. It classifies similar objects into different groups or, more precisely, partitions a data set into clusters so that the data in each subset ideally share some common traits. Clustering assigns records into groups. Numerous clustering algorithms have been proposed. No matter how the algorithms are designed, their goal can be summarized in terms of two objectives: (1) the objects in the same group are similar to each other or (2) the objects in different groups are dissimilar to each other. Different types of clustering algorithms strike a different balance between these two objectives. The most widely used clustering algorithms are classified into three categories: partitioning, hierarchical, and density-based clustering.

Partitioning clustering uses an iterative relocation procedure to move objects from one group to another. The most popular clustering algorithm is the k-means algorithm. It randomly selects the k of the objects and assigns each object to its closest center. When all the partitions are made, it recalculates the center of each partition and reassigns objects to the new centers. The algorithm stops when the k centers do not move any further. However, the k-means algorithm tends to partition the whole space to equal size in order to minimize the global variance, and the value of k is usually hard to determine. In order to solve this problem, a more robust expectation–maximization clustering algorithm was proposed. The goal is to estimate the parameters of possible mixing Gaussian distributions. It alternatively estimates the expected probability for each object as a member of each group and alters the parameters of each group to maximize those probabilities.

Although hierarchical clustering can find arbitrary shape clusters, it only considers cluster proximity while ignoring cluster interconnectivity, and an outlier is still assigned to the closest cluster. To discover clusters with arbitrary shape and outliers, density-based clustering methods have been developed. Density-Based Spatial Clustering of Applications with Noise (DBSCAN) finds core objects that have dense neighborhoods. This method relates core objects and their neighborhoods to form dense regions as clusters in terms of a density-based connectivity analysis. Another density-based algorithm, DENsity-based CLUstERing (DENCLUE), is a clustering method based on a set of density distribution functions. DENCLUE is regarded as generalization of DBSCAN. Both DBSCAN and DENCLUE use a global density parameter to find clusters. As a common property of many practical data sets is that their intrinsic cluster structures cannot be characterized by global density parameters, different local densities may be needed to reveal clusters in different regions of the data space. As such, Ordering Points to Identify the Clustering Structure (OPTICS) was proposed to extend DBSCAN to produce a cluster-ordering obtained from a wide range of parameter settings. The basic idea of the method is similar to DBSCAN, but it addresses one of DBSCAN's major weaknesses: the problem of detecting meaningful clusters in the data of varying density. OPTICS computes an ordering of all objects and, for each object, calculates the core distance and a reachability distance. Local Outlier Factor (LOF) is an algorithm proposed for finding anomalous data points by measuring the local deviation of a given data point with respect to its neighbors. LOF shares some concepts with DBSCAN and OPTICS such as the concepts of core distance and reachability distance. LDBSCAN method is able to detect not only arbitrary shapes and different density clusters, but also clusters that reside in other clusters.

9.5.3.2 Itemset Mining

The first research on itemset mining is the frequent itemset mining method, which searches sets of items that appear together frequently in a data set by the Apriori algorithm. Apriori is an algorithm to mine association rules. The Apriori algorithm makes use of an important downward-closed property to prune the exponential superset search space. It uses a breadth-first search strategy to count the support of itemsets and uses a candidate generation function that exploits the downward closure property of support. Numerous frequent itemset search algorithms were proposed later including FP-Growth and ECLAT. FP stands for frequent pattern. ECLAT is a depth-first search algorithm using set intersection. Progress has been made in formulating the problem as a constraint programming problem and then applying an existing solver to the constraint programming problem for speeding up the search. These algorithms will speed up the search procedure; however, a more important issue related to frequent itemsets is its effectiveness. Therefore, the concept of maximal frequent itemsets was proposed, in order to reduce redundant information.

9.5.4 Opportunities Provided by BI to ESs

Significant research progress has been made in BI. In this section, we identify some of the opportunities provided by BI to applications in ESs.

9.5.4.1 Process Mining

Process mining complements traditional BI and BPM. Process mining assumes a volatile environment and emphasizes the construction of process models to explain the real and partly hidden process flows of enterprises. Process mining is a growing research area, as process discovery accounts for about 40% of the costs of implementing BPM (Ingvaldsen and Gulla, 2012). By distilling a structured process description from system logs, process mining can answer many important industrial or business process questions. For example, how do the workflows proceed? Are there any system bottlenecks? How can the process be redesigned to improve its efficiency? An annotated transition system from historical data was constructed in the ProM framework (van Aalst et al., 2007, 2011), and its function has been extended to predict how long a given process will take.

9.5.4.2 Outlier Detection

Outlier detection refers to finding patterns that do not conform to what is expected. The executive direction and support component in ESs is interested in outlier detection because outlier mining can provide very useful information for many applications in ESs. It has been extensively applied in ESs, and numerous papers are related to this topic.

The analysis of data outlier can be performed by either supervised or unsupervised methods. The easiest way to identify outliers by a supervised method is to build a model for specified attributes and find the record with value that does not conform to the predicted value. However, it should be noted that unsupervised methods are more widely used. Three widely used types of unsupervised methods are statistical distribution–based, distance-based, and density-based methods. Statistical distribution–based methods use distributions to analyze a given data set. Distance-based methods assume that outliers are far away from most records or their nearest neighbors in the entire data set. Distance-based outlier detection methods may encounter difficulties when analyzing data with different density distributions. Therefore, density-based methods are proposed to handle this type of problems.

The most successful density-based outlier detection method is LOF. It calculates the ratio of deviation to its neighborhood density in order to indicate the abnormal degree. LOF can only be used for single-point outliers. Duan et al. (2009, 2011) combined LOF and DBSCAN for cluster outlier detection. Combining LOF and DBSCAN can successfully detect small clusters with objects numbering less than 10, which can be hard to achieve when using traditional clustering methods; this makes this algorithm especially suitable for cluster outlier detection. Just as in the high-dimensional issue exists in clustering, the feature selection is important for high-dimensional outlier detection; it is becoming a research area of increasing importance. A related concept for high-dimensional outlier detection is contextual outliers. Traditional techniques focus on single-point outliers and treat attributes independently. However, a record can be abnormal, due to its specific context. Such outliers can only be detected by specifying some attributes as contextual attributes in order to divide the whole data set into smaller partitions.

9.5.4.3 Graph Data

Traditional data mining methods treat objects independently in a data set. However, in many applications, objects are related to each other and can be more naturally represented in a graph model, wherein each object is a node and there is an edge between two nodes if they interact with each other. For example, devices communicate with each other to perform a certain function, just as industrial firms collaborate with each other to produce a product. The modern science of graphs helps us greatly as we attempt to understand complex systems. This type of graph data is gaining increasing attention from the BI research community. Linkage analysis, community detection, and graph summarization are interesting research topics.

Linkage analysis helps us find the potential partners in industrial groups or networks or to establish a link between the two procedures in order to improve the productivity. The linkage prediction can be achieved through either a supervised or an unsupervised method. Supervised learning deals with the imbalance and sparse connections, while unsupervised learning searches for an effective measure according to the characteristics of different networks.

Since nodes in the graph are not isolated from each other, detecting nodes that work together can help us understand the entire system. Two commonly used methods are spectral based and modularity based. Spectral-based methods use the eigenvectors of the adjacency matrix for community detection, since the change of the representation induced by eigenvectors makes the community structure more obvious. Modularity-based methods make use of the modularity function to search for better partition. Duan and Street (2011) found the connection between the modularity function and the correlation measure leverage. By using other correlation measures (such as the likelihood ratio) to change the objectives, this new method can better detect small communities.

9.5.4.4 Summary

BI plays an important role in ESs. Ning et al. (2009) published their research on predicting retailer demand and ways to correspondingly adjust inventory replenishment. This research significantly reduces operation costs and improves retailer satisfaction. It is related to the marketing operation function and other functions that are included in ESs (Langenwalter, 2000). However, it is not the only application connected to ESs. Take product design as an example: Enterprises can gather the demographic information of the customer who purchased the product from ESs. Itemset mining can help search the correlation between product features and customer demographic attributes. The extracted pattern can help designers customize the product for a target customer group. In addition, enterprises can also conduct opinion mining from the online customer reviews for their products. Opinion mining can detect market reaction and can search for the product features that customers like and dislike from the unstructured text reviews. Such information helps businesses promptly change the product design and the quantity produced. This is part of the functions performed by the customer integration component in the engineering subsystem in ESs (Langenwalter, 2000).

Industrial applications also challenge BI methods. For example, distributed and continuous data challenge the traditional BI methods. For instance, sensor networks are widely used in industry today. Each sensor has limited computing power, but sensors are widely distributed in different locations. As data mining algorithms are used in distributed environments, more research is required to coordinate the distributed data mining. When an industrial process is monitored, data are received continuously. The pattern search conducted on the data may not be able to provide timely results. Novel algorithms are needed for such a purpose.

Since BI can be applied to many potential applications in ESs, it is not possible to list all of the possible applications in this section. In summary, BI has a bright future and deserves research attention from ESs and IIIE.

9.6 Resilient ESs

Warfield pointed out that micromathematics is the primary tool of traditional methods in science and engineering. Macromathematics has the potential of becoming the primary tool of synthesis and overview (Warfield, 1986). Traditional methods primarily rely on micromathematics.

Sections 9.4 and 9.5 provide examples of the macro and micro perspectives required in IIIE. Resilient ESs show another example of how both macro and micro perspectives are applied in IIIE, in particular, how SSE and the ideas behind it have penetrated ESs and how systems theory has been a basis for ESs.

Resilience is a concept in systems science (Holling, 1973) and has been used as an attribute of a complex system. The term "resilience" was also used to describe the property of a material that can absorb external energy when it is forced to deform elastically and then can recover to its original form and release that energy (Liu et al., 2010). Holling defined resilience as a system's ability to absorb external stresses. It originally appears in the field of ecology, where it has been defined as the ability of an ecosystem to absorb and respond to disturbance (Holling, 1973). In complex systems research, resilience refers to "the capacity of a system to absorb disturbance and reorganize while undergoing change so as to still retain essentially the same function, structure, identity, and feedbacks" (Walker et al., 2004). Resilience was also defined as a property of the system that shows how the system can still function to a desired level when the system suffers from a partial fault. Since 2000, the resilience of human-made systems has attracted much research attention. The general question here is whether human-made systems such as ESs can have a high resilience property, as ES consists of human and machine system components. In the context of ESs, Guelfi et al. (2008) defined resilience as the capacity of a business process to recover and reinforce itself when facing changes.

In the extended enterprise structures in which integrated ESs are implemented, ESs often have high data rates and intensive processing requirements; as a result, the system can possibly lack sufficient system resources for processing to maintain high reliability. Due to the fact that the level of integration can be high, it can introduce an urgent need for fault tolerance, as the increased integration level will introduce more complexities that make enterprises vulnerable to unknown and unexpected disruptions. ESs are expected to be resilient, as the system and its environment are complex. For example, in a distributed system, failures anywhere may cause cascading failures of the entire system. Thus, resilience emerges as a highly desirable characteristic. Resilience is a system property that has been paid increasing attention in ESs and IIIE. The resilience property is critical to the functioning of ESs even from a single engineering dimension point of view.

Researchers are exploring the principles of the design of resilient systems within the framework of systems science. Zhang summarized four axioms related to the resilience of the engineered artifact system as follows (Zhang and Lin, 2010):

> "Axiom 1: A damaged component of the artefact system can never be fully recovered to its original one without external intervention.
>
> Axiom 2: A component or a part of the component of the artefact system for function-A may be trained to do function-B of another component.
>
> Axiom 3: There will always be a distinct identity of the embodiment of the artefact system that includes a control system and physical entity at one time in one place to perform one distinct function.
>
> Axiom 4: A system's failure is an emergent consequence of the system's internal vulnerability and system's external mishap."

From these four axioms, Zhang and Lin (2010) derive five design principles for resilient systems:

> "Principle I: A resilient system should be designed to have a certain degree of redundancy preferably functional redundancy. The more redundancy the system has, the higher the degree of resilience.
>
> Principle II: A resilient system should be designed to have a special controller which is responsible for (1) redundancy management and (2) function learning.
>
> Principle III: A resilient system should be designed to have a sensor which is responsible for both temporally and spatially (1) monitoring the system's function and performance, (2) monitoring the utilization of the system's capacity and (3) monitoring the system's demand.
>
> Principle IV: A resilient system should be designed to have a predictor which is responsible for (1) predicting potential threats to the system and (2) analyzing potential vulnerabilities of the system.
>
> Principle V: A resilient system should be designed to have an 'actuator' along with a 'physical' entity which is responsible for (1) implementation of changes of the system both in cognitive and physical domains and (2) implementation of trainings of one component or sub-system of the system to perform a new function."

In information systems research, resilience refers to a system's ability to provide and maintain an acceptable level of service in the face of various faults and challenges to normal operation (Liu et al., 2010). Researchers have proposed the architecture of resilient ESs. In the proposed architecture, ESs are viewed as a system consisting of subsystems including data and infrastructure subsystems. The infrastructure subsystem includes software, hardware, and human operations. The data subsystem includes data that are ESs specific.

9.6.1 Enterprise Resilience and Resilient Enterprise Systems

The complexity of resilience can be imagined. Erol et al. (2010) proposed the four attributes of extended enterprise resilience: decreased vulnerability, increased flexibility, adaptability, and agility. In other words, extended enterprise resilience is a function of an extended enterprise's level of vulnerability, flexibility, adaptability, and agility. Extended enterprises, due to their scope, scale, and complexity, can be more vulnerable to unforeseen disruptions. In today's competitive business environment, enterprises must have the ability to cope with unforeseen events, to quickly adapt to the turbulence, and to survive. As complex systems, ESs are more vulnerable to security issues, since systems are more independent of each other, thus having a higher risk of cascading failures (Zhang, 2008). ESs can be more sensitive to unforeseen disturbances, since there are more interfaces in such networked systems. As most of current ESs are connected with the Internet, the boundaries of ESs become more dynamic, and this makes enterprises more vulnerable to unforeseen events. Therefore, designing and developing resilient ESs have become a new and interesting topic. As humans develop the capacity to predict and cope with future disturbances to natural and societal systems, the resilience of both extended enterprises and ESs has attracted attention. The CMFT developed by Xu has pioneered the research in this direction (2008). One of the key questions is whether ESs can be designed with a high resilient capability. In ESs, particularly in the context of transaction processing, consistency, together with atomicity, isolation, and durability, is considered as the major property (Haerder and Reuter, 1983). Brewer (2000) made the following conjecture, known as the CAP theorem:

> *Theorem*: It is impossible to have a network-based system that has all three properties of consistency, availability, and tolerance to network partitions.

This conjecture was later formally proved by Gilbert and Lynch (2002). Liu et al. (2010) applied this theorem to the case of resilience and derived the following corollary:

> *Corollary*: It is impossible for a system to have all three properties of consistency, resilience, and partition tolerance.

As Liu et al. note, "this corollary implies that a tradeoff between consistency and resilience has to be made when a system is designed." However, research indicates that the kinds of ESs that we are interested in developing in the future are those that are resilient. Efforts have been made in this direction. From an SSE point of view, the concept of fault tolerance in ESs can be further concretized in a strategy to minimize the cost rather than the probability of occurrence. This is a "safe-fail" strategy (in contrast to "fail-safe" strategy of pure engineering design).

Although achieving resilience in ESs is challenging, multiple efforts in inter-disciplinary research will contribute to the development of a new generation of resilient ESs.

9.7 SSE Serves ESs

ESs are complex, especially for large-scale industrial organizations. ESs are now considered a significant challenge of these times. The need for ESs is obvious. Research indicates that ESs are a systems challenge. The question is this: Is SSE available to meet this challenge?

Warfield's version of systems science (WSS) has major implications for enter-prise integration involving complexity. WSS was applied by a large cross-functional team of Ford engineers and system developers in the mid-1990s to create the ES known as the C3P system (CAD/CAE/CAM/PIMS) that is applied to automobile design and manufacturing in the entire supply chain. The theoretical foundation of the C3P system is SSE and WSS.

SSE targets the integration of all relevant disciplines in an effort forming a sys-tematic problem-solving approach. IIIE works to develop a structural framework to integrate relevant disciplines for industrial information integration purposes in order to face the challenge by applying SSE methodologies (see Figure 5.27). So far, numerous information integration methods/techniques represent different perspec-tives, and different directions have been proposed. IIIE is moving toward the direc-tion of developing a structural framework to address the challenge of constructing ultra-large-scale systems such as ESs through applying SSE. The current urgent need is to set out a roadmap outlining how IIIE, in combination with SSE, can lead us to succeed in facing the research and development challenges of ESs.

In IIIE, efforts have already been made to reduce the three principal sources for poor intellectual productivity:

1. Unproductive emulation: This refers to the practice of developing ESs with-out the discipline of underlying science. IIIE seeks to build a discipline cover-ing all underlying science and engineering (see Figure 5.27).
2. Cultural lag: This refers to the failure to apply what is known to be effective in ES research. IIIE looks to integrate existing methodologies in related dis-ciplines in order to provide a scientific foundation (see Figure 5.27).
3. Spurious saliency: This refers to proceeding without any scientific basis for assigning priorities among the very large number of possible action compo-nents in developing ESs. IIIE seeks to emphasize SSE, which in turn enables the establishment of scientific underpinnings.

References

Bellman, R. 1977. Large systems. In Linstone, H. and Simmonds, W. (Eds.), *Future Research New Directions*. Reading, MA: Addison-Wesley, pp. 100–103.

Brewer, E. 2000. Towards robust distributed systems. *Proceedings of the 19th Annual ACM Symposium on Principles of Distributed Computing*, Portland, OR, p. 7.

Churchman, C. 1971. *The Design of Inquiring Systems Basic Concepts of Systems and Organizations*. New York: Basic Books.

Curtis, B., Krasner, H., and Iscoe, N. 1988. A field study of the software design process for large systems. *Communications of the ACM*, 31(11), 1268–1287.

Davis, A. 1992. Operational prototyping: A new development approach. *IEEE Software*, 9(5), 70–78.

Du, Y. and Jiang, C. 2003. Towards a workflow model of real-time cooperative systems. *Proceedings of Fifth International Conference on Formal Engineering Methods*, Singapore, pp. 452–470.

Duan, L. and Street, W. 2011. Community detection through correlation. *Proceedings of INFORMS Annual Meeting*, Seattle, WA, Charlotte, North Carolina.

Duan, L., Xu, L., Liu, Y., and Lee, J. 2009. Cluster-based outlier detection. *Annals of Operations Research*, 168(1), 151–168.

Erol, O., Sauser, B., and Mansouri, M. 2010. A framework for investigation into extended enterprise resilience. *Enterprise Information Systems*, 4(2), 111–136.

Gilbert, S. and Lynch, N. 2002. Brewer's conjecture and the feasibility of consistent, available, partition-tolerant web services. *ACM SIGACT News*, 33(2), 51–59.

Guelfi, N., Muccini, H., Pelliccione, P., and Romanovsky, A. (Eds.) 2008. *Proceedings of the 2008 RISE/EFTS Joint International Workshop on Software Engineering for Resilient Systems*, Newcastle Upon Tyne, U.K.

Haerder, T. and Reuter, A. 1983. Principles of transaction-oriented database recovery. *ACM Computing Surveys*, 15(4), 287–317.

Han, J. and Kamber, M. 2006. *Data Mining: Concepts and Techniques*. Amsterdam, the Netherlands; Boston, MA: Elsevier.

Hess, M. 1990. Information systems design in industrial practice. In Sage, A. P. (Ed.), *Concise Encyclopedia of Information Processing in Systems and Organizations*. Oxford, U.K.: Pergamon Press, pp. 295–301.

Hevner, A., March, S., Park, J., and Ram, S. 2004. Design science in information systems research. *MIS Quarterly*, 28(1), 75–105.

Holling, C. 1973. Resilience and stability of ecological systems. *Annual Review of Ecology and Systematics*, 4, 1–23.

Ifinedo, P. and Nahar, N. 2007. ERP systems success: An empirical analysis of how two organizational stakeholder groups prioritize and evaluate relevant measures. *Enterprise Information Systems*, 1(1), 25–48.

Ingvaldsen, J. and Gulla, J. 2012. Industrial application of semantic process mining. *Enterprise Information Systems*, 6(2), 139–163.

Jackson, M. C. 1997. Towards coherent pluralism in management science. Working paper, Lincoln School of Management, University of Lincolnshire and Humberside, Lincoln, U.K.

Langenwalter, G. 2000. *Enterprise Resource Planning and Beyond*. Boca Raton, FL: St. Lucie Press.

Lin, Y., Duan, X., Zhao, C., and Xu, L. 2013. *Systems Science Methodological Approaches.* Boca Raton, FL: CRC Press.

Liu, D., Deters, R., and Zhang, W. 2010. Architectural design for resilience. *Enterprise Information Systems*, 4(2), 137–152.

Liu, W., Du, Y., and Yan, C. 2012. Soundness preservation in composed logical time workflow nets. *Enterprise Information Systems*, 6(1), 95–113.

Muller-Merbach, H. 1983. Model design based on the systems approach. *Journal of the Operational Research Society*, 34(8), 739–751.

Ning, A., Lau, H., Zhao, Y., and Wong, T. 2009. Fulfillment of retailer demand by using the MDL-optimal neural network prediction and decision policy. *IEEE Transactions on Industrial Informatics*, 5(4), 495–506.

Northrop, L. et al. 2006. Ultra-large-scale systems: The software challenge of the future. Technical Report, Carnegie Mellon University Software Engineering Institute, Pittsburgh, PA. http://www.sei.cmu.edu/library/abstracts/books/0978695607.cfm.

Paige, R., Charalambous, R., Ge, X., and Brooke, P. 2008. Towards agile development of high-integrity systems. *Proceedings of the 27th International Conference on Computer Safety, Reliability, and Security*, Newcastle, U.K., pp. 30–43.

Qian, X., Yu, J., and Dai, R. 1993. A new discipline of science: The study of open complex giant system and its methodology. *Journal of Systems Engineering and Electronics*, 4(2), 2–12.

Richards, L. and Gupta, S. 1985. The systems approach in an information society: Reconsideration. *Journal of the Operational Research Society*, 36(9), 833–843.

Sommerville, I., Cliff, D., Calinescu, R., Keen, J., Kelly, T., Kwiatkowska, M., Mcdermid, J., and Paige, R. 2012. Large-scale complex IT systems. *Communications of the ACM*, 55(7), 71–77.

Staley, S. and Warfield, J. 2007. Enterprise integration of product development data: Systems science in action. *Enterprise Information Systems*, 1(3), 269–285.

van Aalst, W., Schonenberg, M., and Song, M. 2011. Time prediction based on process mining. *Information Systems*, 36(2), 450–475.

van Aalst, W., van Dongen, B., Günther, C., Mans, R., Alves de Medeiros, A., Rozinat, A., Rubin, V., Song, M., Verbeek, H., and Weijters, A. 2007. ProM 4.0: Comprehensive support for real process analysis. In *Lecture Notes in Computer Science*, Berlin, Germany; New York: Springer-Verlag, Vol. 4546, pp. 484–494.

Walker, B., Holling, C., Carpenter, S., and Kinzig, A. 2004. Resilience, adaptability and transformability in social-ecological systems. *Ecology and Society*, 9(2), 5.

Warfield, J. 1986. Micromathematics and macromathematics. *Proceedings of the 1986 IEEE International Conference on Systems, Man and Cybernetics*, Atlanta, GA, pp. 1127–1131.

Warfield, J. 1990. *A Science of Generic Design.* Salinas, CA: Intersystems Publications.

Warfield, J. 2007. Systems science serves enterprise integration: A tutorial. *Enterprise Information Systems*, 1(2), 235–254.

Warfield, J. 2008. Seven challenges for information system designers. In Xu, L. (Ed.), *Frontiers in Enterprise Integration*, London, U.K.; New York: Taylor & Francis, pp. 1–3.

Warfield, J. 2009. Generic systems science: Logilectic in action. Working paper.

Warfield, J. and Christakis A. 1987. Dimensionality. *Systems Research*, 4, 127–137.

Xu, L. 1995. Systems thinking for information systems development. *Systems Practice*, 8, 577–589.

Xu, S. 2008. The concept and theory of material flow. *Information Systems Frontiers*, 10(5), 601–609.

Zeng, L., Li, L., and Duan, L. 2012. Business intelligence in enterprise computing environment. *Information Technology and Management*, 13(4), 297–310.

Zhang, W. 2008. Resilience engineering: Overview [online]. Presented at a seminar at the Chinese Natural Science Foundation. Available at http://homepage.usask.ca/*wjz485/Other%20Publication.htm [Accessed November 3, 2009].

Zhang, W. and Lin, Y. 2010. On the principle of design of resilient systems-application to enterprise information systems. *Enterprise Information Systems*, 4(2), 99–110.

Chapter 10

Future Evolution

10.1 Overview

The development and growth of the Internet and other related technologies have had tremendous impact on ESs. The Internet has evolved from a packet-switching backbone for computer interconnection supporting traditional applications to a powerful infrastructure for computing services and information sharing. With the emergence of the IoT, cloud computing, pervasive computing, ambient intelligence, Web 2.0, and social media, the Internet is moving into an unprecedented new era. In particular, the development of the IoT extends the idea of smart devices. New technology allows devices of much smaller scale and lower cost to be embedded in physical objects. These devices can perform specific functions and can be used to monitor and interact with physical objects in the environment. Objects thus can be identified and located through computing systems from a remote location via wireless networks. Such new technology opens the possibility of a plethora of new applications, especially when the interactions between the physical and digital world are expected in applications such as environmental monitoring, healthcare monitoring, logistics, and transportation. The new IoT applications can be offered through systems such as ESs.

As web services become increasingly prevalent, many service providers and end users have created a variety of web services and have made them available on the Internet. The rapid development of cloud computing, such as the new deployment and delivery of software as a service (SaaS), is making the offering and use of web services much easier and efficient via the cloud (Buyya et al., 2008). It is expected that the development of such web services will eventually result in a business services ecosystem, in which demand and supply will be able to be brought together at a global scale and partnerships will be very dynamic.

To make the business services ecosystem more efficient, a variety of new service intermediaries may also arise to support the matching of demand and supply, the management of dynamic partnerships, and the interoperability between the business partners. To implement a business services ecosystem, a new infrastructure coming from an enhanced Internet is expected. This new infrastructure is called the Internet of Services (Finzen et al., 2010). It is expected that the future Internet will be composed of both the IoT and the Internet of Services; it will offer a wide range of services for all aspects of business, social, and daily life. For example, the future Internet will not only have the capability to keep track of objects, processes, and their contexts but also offer real-time monitoring of the environment. More importantly, the future Internet will be able to quickly adapt to changes for sustaining operations, offering effective services, and managing collaborations. The potential emergence of the future Internet implies that the next-generation ESs will rely more upon sophisticated technologies than upon those that are available today (Li, 2007). For example, supply chain quality management systems can be designed with sophistication for the proper integration of existing and/or new technologies (Li, 2007). By adopting more advanced technologies such as the IoT, the future Internet, and cloud computing, enterprises can dramatically improve the quality of their SCM systems and can deliver better performance in business processes and enterprise operations.

Although the blend of the future Internet and the IoT has been identified as the key technology trend that will reshape ESs during the next decade, the future Internet and the IoT also pose significant challenges to enterprises. First, there are issues regarding security, privacy, and trust in this highly interconnected world. Users may be concerned about their privacy when such a huge amount of data can be collected from their daily activities. To mediate such concerns, the huge volume of data that is collected has to be managed well, with appropriate governance, security mechanism, rules, and processes. Second, as new technologies such as WSNs, RFID, and IoT are deployed in industrial systems (such as MF systems, logistics systems, and SCM systems), the integration of industrial applications, devices, and various interfaces becomes more challenging. There is a lack of guidelines, standard architecture, services, and quality assurance for facilitating interactions between heterogeneous devices, sensors, aggregators, actuators, and diverse applications (He and Xu, 2013). There is also a lack of services that can connect users effectively to the appropriate sensor networks as the amount of data sources in sensor networks increases (Gadea et al., 2010). Although various middleware solutions and architectures for integrating industrial applications, WSN, RFID, and the IoT have been proposed, these solutions and architectures are still in the prototype stage and are often lacking rigid testing in real industrial environments. Third, research on the integration of the IoT and the Internet of Services is still at its infancy stage. It is still not clear how to effectively use context information or how to determine the adaptations at both the application and business levels. These challenges will have to be addressed before enterprises will be able to exploit the future Internet and transform existing ESs to future-generation ESs for high-level enterprise operation and collaboration. In particular, systems architectural approaches will need to be developed, in order to deal with the increasing complexity at the application, business, and technical levels, as well as to mitigate alignment and compatibility among them.

So far, the rapid technological advances of new technologies including the Internet technology have greatly benefited ESs by adding numerous new features to ESs. However, many existing ESs have not yet employed these advanced technologies. For example, many ESs have not yet been deployed to the cloud, nor have most ESs been integrated with Web 2.0 or social media technologies. More ESs and applications that leverage the wireless sensors, RFID, and the IoT technologies are expected. Thus, more efforts are needed to improve existing ESs by integrating these new technologies into ESs. Specifically, researchers and practitioners will devote more effort into blending the capabilities of existing ESs and emerging technologies in order to allow ESs to become more effective in practice. This blending will have the potential to help industries harness the power of current and emerging technologies in order to dramatically improve the performance of enterprise processes and operations.

In summary, although future ESs based upon future Internet and the IoT are expected to play an increasingly important role in enterprises in the future, there are still many issues and challenges that need to be resolved. Designing ESs is a complex process due to their high dimensionality and ever-changing array of technologies. Implementing ESs that support interorganizational integration remains extremely challenging, although different approaches have been proposed. To meet the growing needs of information integration in the industry, there is a compelling reason to develop advanced methodologies (including both formal methods and systematic approaches) to sync ESs with the rapid technological development in order to fully realize the potential of ESs. As modern enterprises continue to demand ESs that support greater flexibility and interoperability, ESs need to constantly improve their existing architecture in order to embrace cutting-edge technology and techniques.

10.2 Major New Theories Impacting ESs

10.2.1 MF Theory

In 2008, a new theory on MFs called the CMFT was proposed by Xu (2008). This theory includes seven basic theories, as Figure 10.1 shows. They are MF theory, CMFT, MF element theory, MF nature theory, MF science and technology theory, MF engineering theory, and MF industry theory. A short description of the seven basic theories is presented here. According to Xu (2008), first, MF theory is

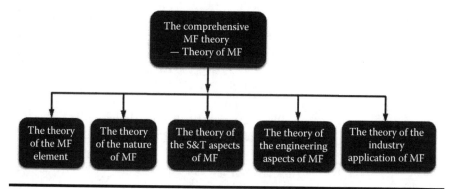

Figure 10.1 Seven basic theories supporting material flow theory.

put forward based on research regarding the scientific and technological aspects of MFs. Second, CMFT is put forward based on research on the MF objective matter and phenomena. Third, MF element theory is put forward based on research on the composition of the elements of MFs. Fourth, MF nature theory is put forward based on research on the nature of MFs. Fifth, MF science and technology theory is put forward based on research on the scientific and technological aspects of MFs. Sixth, material engineering theory is put forward based on research on the engineering aspects of MFs. Seventh, MF industry theory is put forward based on research on the industrial development related to MFs.

Xu (2008) further defines MF to be a comprehensive term for the flow at both the macroscopic and microcosmic levels. The MF theory reveals the essence of MF objects and phenomena. In this theory, MFs are specified in terms of the MF in the economic dimension, the MF in the social dimension, and the MF in the natural dimension, as well as their interrelationships. It is noted that the MF is not only an economic phenomenon, but also a societal and a natural one. In other words, there exist not only economic MFs, but also societal and natural flows. The economic MF is the core of the MF, whereas the societal and natural MFs are the basis of the MF.

Song and Xu (2009) point out that MF is a widespread phenomenon of objective things and a complex flow of substances. Thus, MF must have its substance composition. As a comprehensive substance, MF includes numerous individual substances. For example, in MF theory, commodity material flow's substance mainly comprises four substances such as storing; transportation; loading, unloading, and handling; and distribution processing, as Figure 10.2 shows.

The theory of XPMF is a theory further extended from the general MF theory (Hou et al., 2007). The XPMF concept is considered to be an extension of the MF theory. XPMF is defined as MF in which X can be singular or plural. To further elucidate that XPMF is one type of MF service model with PMF (party, material,

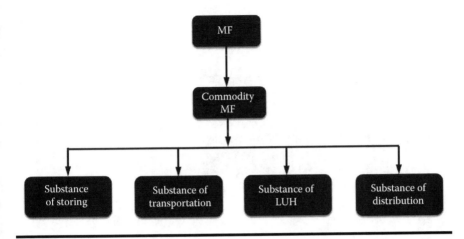

Figure 10.2 Substances in commodity material flow.

flow) fractal structure and the characteristics of XPMF, Hou et al. (2007) developed a three-pyramid synergetic operational model of XPMF, as Figure 10.3 shows. Three parameters, OIR (objective relational grade, information sharing grade, resource complementary grade), play an important role in the self-organizational process of XPMF. "Objective relational grade" (O) is the parameter of inherent unity realized through changes from disorder to order among all of the PMF subsystems from time to time. "Information sharing grade" (I) is the parameter for a PMF subsystem to realize synchronization of operations among subsystems. "Resource complementary grade" (R) is the parameter for measuring the capability that XPMF has for rapid response to the market and for making complementary resources reorganized, reusable, and expandable, dynamically. OIR is formed in the process of interactions and synergetic activities among PMF fractal units and at the same time controls and determines the formation of XPMF's new structure and its degree of being ordered. Compared to the current MF service model, the XPMF possesses the properties for MF resources to be reconfigurable, reusable, and expandable in dynamic supply chains. Figure 10.3 shows the three-pyramid synergetic operational model of XPMF. In summary, XPMF is a type of new MF theory that has applied systems theory, especially the theory of synergetics established by Hermann Haken (Lin et al., 2013).

MF theory provides deeper and extremely important insights into the future of enterprise architecture and integration, especially to the consideration of sustainable economic growth and societal development. In the context of ESs, this theory provides guidance for complex industrial information integration for the following two reasons: (1) The enterprise or business process describes the organization as a set of business processes representing the flow of materials such as goods and services and information throughout the enterprise. Based upon the new MF theory, we now have a much more comprehensive understanding of the types and quantity

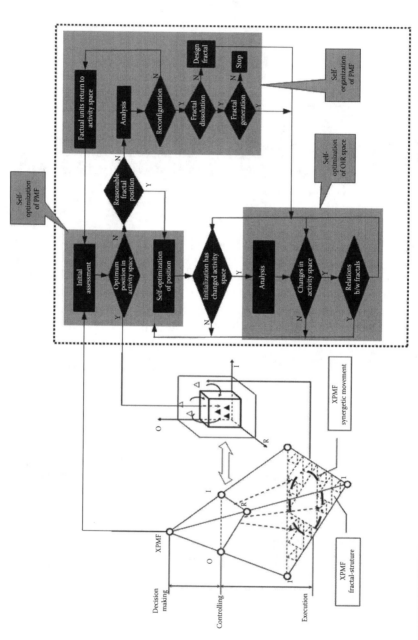

Figure 10.3 Three-pyramid synergetic operational model of XPMF. (From Hou, H. et al., *Enterp. Inform. Syst.*, 1(3), 287, 2007.)

of existing MFs. Compared to the MFs that have been represented in existing ESs, both the types and quantity revealed in the new theory are incomparable; (2) to an increasing extent, organizations are starting to discover that success will be possible only if the full range of types and quantities of MFs are considered in the design of ESs. In this particular design case, the law of requisite variety applies.

According to Warfield (1990), "the Law of Requisite Variety indicates the need for a match between the dimensionality of the Design Situation and the Target of the Design Process. This law was discovered by Ashby. This Law asserts that a Design Situation embodies a requirement for Requisite Variety in the design specifications. Every Design Situation S implicitly represents an (initially unknown) integer dimensionality Ks such that if the designer defines an integer Km number of distinct specifications (whether qualitative or quantitative or a mix of these, then: (1) If $Km < Ks$, the Target is underspecified and the behavior of the Target is outside the control of the designer; (2) if $Km > Ks$, the Target is over-specified, and the behavior of the Target cannot be compatible with the designer's wishes; (3) if $Km = Ks$, the design specification exhibits Requisite Variety, provided the designer has correctly identified and specified the dimensions; and the behavior of the design should be that which the Situation can absorb and which the designer can control, subject to the requirement that the dimensionality of the Situation is not modified by the introduction of the Target into the Situation. If the dimensionality is changed thereby, the design process can apply the Law of Requisite Variety iteratively, taking into account the dynamics of the Situation."

The increasing number of the types and the quantity of MFs that the new theory reveals implies more design options for ESs. Compared with the design options introduced in the literature (such as those of the Alberts Pattern, the Cardenas–Rivas Pattern, and the Staley Pattern), a similar pattern of design options will appear for redesigning ESs (Warfield, 2007). Figure 10.4 shows a possible design pattern for ESs. It will enable Ashby's law of requisite variety to be applied to the development of ESs.

Based upon Figure 10.4, the concept of next-generation ESs has been proposed (Xu, 2011). The next-generation ES is called entire resource planning (ERP) or CRP. In this new ES, not only have the concepts of ERP, ERPII, and ERPIII been integrated together, but the design framework has also been extended to comprehensively encompass all of the resources produced and used by enterprises, or a variety of MFs, in different industrial sectors, within the context of economic and societal development. The thinking that has guided this design is based upon the new MF theory and the law of requisite variety. In ERP, not only is the economic MF included, but the social and natural MFs are included, as well. ERP is considered a significant step forward in the evolution of ESs. Figure 10.5 shows the evolution of ESs in terms of the three stages of development efforts.

The major factors that have contributed to the birth of ERP are obvious, since one of the main challenges facing enterprises is that the expectations for sustainable economic growth and global environmental protection are rising. The limitations

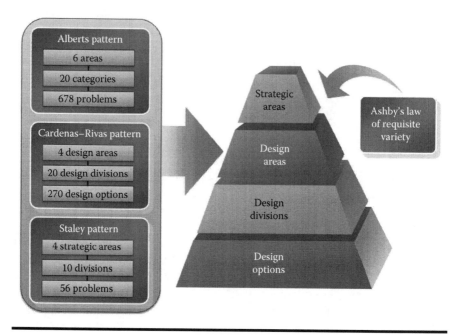

Figure 10.4   Design pattern.

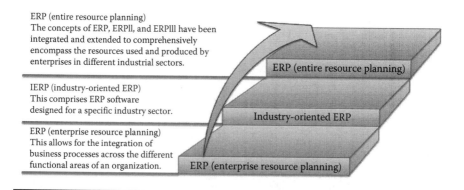

Figure 10.5   The evolution of enterprise systems.

of the existing systems restrain their flexibility to cope with such challenges, and mere modest modification of the existing ESs will not meet the requirements from the perspectives of both economic and societal development. The key characteristic of the new system is the provision of comprehensive coverage of all relevant types of resource planning, with the objective of closing the gap between the existing systems and the proposed systems.

10.3 Major New Technologies Impacting ESs

10.3.1 Internet of Things

As an emerging technology, the IoT is expected to offer promising solutions for ESs. The phrase "Internet of things" was coined at the beginning of the twenty-first century by the MIT Auto-ID Center, with special mention of Kevin Ashton (2009) and David L. Brock (2001). The IoT presents a future computing environment in which physical objects can be associated with the Internet, allowing other devices to distinguish them. The European Commission Information Society (2008) has defined the IoT as "Things having identities and virtual personalities operating in smart spaces using intelligent interfaces to connect and communicate within social, environmental, and user contexts" or "Interconnected objects having an active role in what might be called the future Internet." The term IoT is often associated with terms such as ubiquitous computing, ubiquitous network, pervasive computing, ambient intelligence, and cyber-physical systems.

As a novel paradigm, the IoT is rapidly gaining ground in wireless communications. The basic idea of the IoT is that a variety of things or objects are able to interact with each other through technologies such as RFID, sensors, and actuators (Atzori et al., 2010; Giusto et al., 2010). RFID is often seen as a prerequisite for the IoT. Through RFID, it is possible to identify, track, and monitor objects automatically, in real time, and globally if they are attached with tags (Jia et al., 2012). RFID technology comes to be one of the most popular technologies; it has been widely used in manufacturing industries, logistics, SCM, and retailing for the purpose of identification, tracking, and monitoring (Gao et al., 2010). A trend for future development is to integrate RFID technology with other technologies (such as the IoT), in order to expand its application scenarios. If objects were equipped with RFID tags, they could be identified and monitored.

As a complex cyber-physical system, the IoT is able to integrate all kinds of sensing, identification, communication, and networking devices and systems seamlessly to connect all things. As such, anything, at any time and any place, through any devices and systems, can be more efficiently accessed for any service or purpose (European Commission Information Society, 2008). "Ubiquitous" is the distinct feature of IoT technologies, so the IoT is often related to ubiquitous identification, ubiquitous sensing, ubiquitous computing, and ubiquitous intelligence.

From the perspective of the systems level, the IoT can be viewed as a highly dynamic and radically distributed networked system, composed of a very large number of smart objects that produce and consume information (Miorandi et al., 2012). The ability to interface with the physical world is achieved through devices that are able to sense physical objects and convert them into a stream of data (thereby providing information on the current context and/or environment), as well as through devices that are able to trigger actions that have an impact on physical objects through suitable actuators. From the perspective of the service level, the

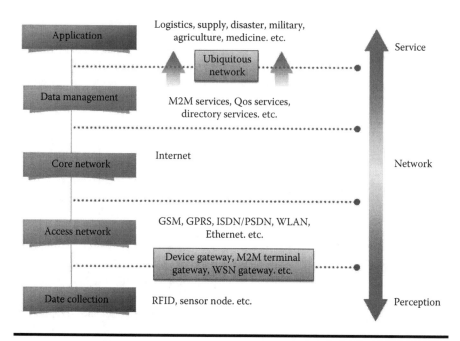

Figure 10.6   IoT system architecture. (Adapted from Jia, X. et al., RFID technology and its applications in Internet of Things (IoT), in *Proceedings of 2012 Second International Conference on In Consumer Electronics, Communications and Networks*, Yichang, China, pp. 1282–1285, 2012.)

main issues relate to how to compose and/or how to integrate the functionalities and/or resources provided by smart objects, which, in many cases, take the form of data streams generated into services (Guinard et al., 2011).

In an IoT architecture, using the cloud computing technology, millions of sensors can be associated together to acquire the massive data for the user's purposes. Objects can communicate with each other throughout the Internet, controlled by computing networks.

The IoT's system architecture can be divided into three layers (Jia et al., 2012): the perception layer, the network layer, and the service layer (or application layer), as shown in Figure 10.6.

Perception layer: This is the information origin and the core layer of IoT. All kinds of information regarding the physical world can be perceived and collected through this layer, using the technologies of sensors, WSNs, RFID, global positioning systems, intelligent terminals, etc.

Network layer: This layer, also called transport layer, which includes the access network and the core network, provides transparent data transmission capability. Using the existing mobile communication network, radio access network, WSN, and other communication equipment, such as the global system for mobile communications (GSM), general packet radio service (GPRS), worldwide interoperability for microwave access (WiMax), wireless fidelity (Wi-Fi), and the Ethernet, the information collected from the perception layer is delivered to the upper layer. At the same time, this layer is intended to provide an efficient, reliable, trusted network infrastructure platform to the upper layer and to large-scale industry applications (Lu et al., 2011).

Service layer: This layer, also called the application layer, includes the data management sublayer and the application service sublayer. The data management sublayer is intended for processing data including restructuring and cleaning and for facilitating directory service, market to market (M2M) service, quality of service (QoS), facility management, geomatics, etc., through SOA, and cloud computing technologies. The application service sublayer transforms information into content and provides a quality user interface for upper-level enterprise applications and end users, such as logistics, environmental monitoring, production operations, and natural disaster warnings.

The IoT has become the new paradigm in the evolution of ICT (Atzori et al., 2010; Miorandi et al., 2012). The IoT has the potential to reshape many existing enterprise operations and processes. The impact to be delivered by the IoT to human society will be as huge as the Internet's impact has been during the past decades, so the IoT is recognized as the next step of the Internet. The close relationship between the IoT and ESs is obvious. ESs are one of the enabling technologies of the IoT. Other IoT enabling technologies include sensors and actuators, WSNs, intelligent and interactive packaging (I2Pack), the real-time embedded system, microelectromechanical systems (MEMS), mobile internet access, cloud computing, RFID, machine-to-machine (M2M) communication, human–machine interaction (HMI), middleware, SOA, and data mining. While the IoT's technology will exhibit significant advances in a number of ICT fields, its awareness may start from existing technologies and applications and will likely follow an incremental pattern. In particular, the IoT will likely begin its expansion from the identification technologies such as RFID, which is already broadly used in numerous applications. At the same time, in its development path, the IoT will likely build on approaches that have already been introduced in a variety of relevant fields, such as wireless sensor networks as a means to collect contextual data and SOAs as the software architectural approach for extended web-based services through IoT capability.

IoT technologies will find diverse applications in many industrial sectors such as environmental monitoring, healthcare, production management (including inventory, security, and surveillance), and workplace and home support. For example, the IoT is reshaping modern food supply chains (FSCs) with promising business prospects (Pang et al., 2014). Figure 10.7 lists selected IoT application

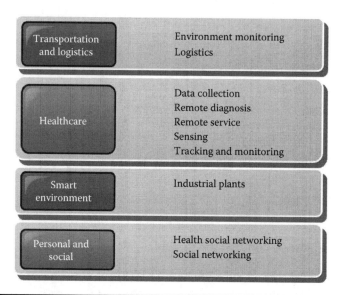

Figure 10.7 IoT application domains and relevant applications.

domains and relevant applications (Atzori et al., 2010). The following are some examples of IoT applications:

■ The IoT application in healthcare (Pang, 2013): The revolution of IoT is reshaping modern healthcare with promising economic and social prospects (Domingo, 2012). Typically, a healthcare IoT solution may include the following major functions: (1) tracking and monitoring: Powered by ubiquitous identification, sensing, and communication capacities, all of the objects in the healthcare systems (people, equipment, medicine, etc.) can be tracked and/or monitored by devices on a 24/7 basis (Pang, 2013); (2) remote service: Healthcare services such as telemedicine and remote diagnosis, emergency detection, dietary and medication management, and health social networking are expected to be enabled for remote delivery through the Internet and field devices (Pang, 2013); (3) integrated information management: Enabled by its global connectivity, all healthcare-related information (logistics, diagnosis, therapy, recovery, medication, management, finance, and even daily activity) are expected to be collected, managed, and shared efficiently (Pang, 2013); and (4) interorganizational integration: Hospital information systems are to be extended to patient's homes and are to be integrated into larger-scale healthcare networking systems (Pang, 2013). By using personal computing devices (laptop, mobile phone, tablet, etc.) and mobile Internet access (Wi-Fi, 3G, LTE, etc.), IoT-based healthcare services can be both mobile and personalized (Plaza et al., 2011). The large user base and the mature ecosystem of traditional mobile Internet service have significantly sped up the development of IoT-powered in-home healthcare (IHH) services, so-called Health-IoT. At the same time, Health-IoT extends traditional mobile Internet services into new application areas. On the other hand, the IoT is also an easy target for privacy and data breaches because of its obvious weaknesses in its early developmental stage. As a health device through which the monitor will collect data points (such as heart rate and blood sugar level), rather than delivering the data directly to the doctor's office, the data may first be routed to a local hub for managing and temporary storage. The more transfer points along the path of data travel, the higher the risk for the data to be compromised.

▪ The Food-IoT (Pang, 2013): Today's FSC can be both distributed and complex. It may have a large geographical and temporal scale, complex operation processes, and a large number of stakeholders. Such complexity may cause issues in quality control, operational efficiency, and food safety. IoT technologies offer promising potential for addressing the challenges of the traceability, visibility, and controllability. The IoT can offer the FSC so-called farm-to-plate food safety tracking—from agriculture production to food processing and preparation, storage, distribution, and consumption. Safer and more efficient FSCs are to be expected in the future. Figure 10.8 illustrates a typical IoT solution for FSC, the so-called Food-IoT. It comprises three

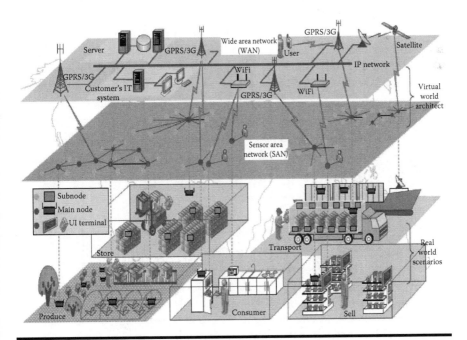

Figure 10.8   Food-IoT. (From Pang, Z. et al., *Inform. Syst. Front.*, 2014.)

parts: (1) field devices such as WSN nodes, RFID readers/tags, and user interface terminals; (2) the backbone system such as databases, servers, and many kinds of terminals connected by distributed computer networks; and (3) the communication infrastructures such as WLAN, cellular, satellite, power line, and Ethernet. As the IoT system offers ubiquitous networking capacity, all of these elements can be distributed throughout the entire FSC. Food-IoT also offers powerful but economy-sensing functionalities, since all of the environmental and event information during the life cycle of food production can be gathered on a 24/7 basis. The vast amount of raw data can be structured and organized into directly usable information for all stakeholders for the purpose of decision making.

Currently, it is broadly accepted that the technologies and applications of IoT are still in their early stages (Atzori et al., 2010; Miorandi et al., 2012). At present, only a small number of IoT applications are practically available. There are many research challenges across almost all of the aspects of the IoT solution, ranging from the enabling devices to the top-level business models. Pang (2013) proposes the scope of IoT research, as Figure 10.9 shows. Just as in the IIIE, the IoT is a complex system from the systems perspective and a typical multidisciplinary scientific subject. Atzori et al. summarize the open research issues in IoT (Atzori et al., 2010). The open issues include standard mobility support, naming transport protocol, traffic characterization and QoS support, authentication, data integrity privacy, etc. (Atzori et al., 2010).

According to Pang (2013), one of the theoretical foundations of the IoT applications in both healthcare and the food industry, as introduced earlier, is ESs and its theoretical framework-IIIE. The reason is that the information provided by the IoT system is the realization of business values; meanwhile, effective and efficient

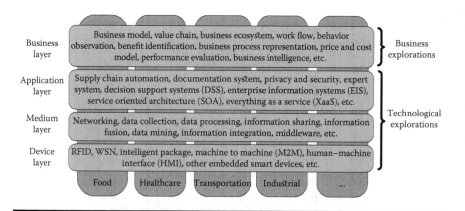

Figure 10.9 IoT research scope and multidisciplinary nature. (From Pang, Z. et al., *Inform. Syst. Front.*, 2014.)

information integration is the key for the success of an IoT solution. However, from the ESs' viewpoint, the IoT system is an enabling component of the infrastructures of future enterprises (van Sinderen and Almeida, 2011; Xu, 2011).

Pang (2013) suggests that a recommended approach for IoT development is to pay attention to business strategy at the early stage of technology development. This implies a tactical shift from technology-driven to business-technology driven, to the so-called Business-Technology Co-Design (BTCD) concept. In BTCD, a whole picture of the target business is prepared first; as such, its value proposition to drive the entire value chain can easily be discovered. Architectural trade-offs for device and service integration are based on the results from the BTDC, since only the business design can be used as the top criteria for making design trade-offs. Based upon the BTCD, Pang (2013) proposes IoT's Three Design Principles as follows.

First, an IoT solution should be modeled as a cross-boundary integration of ESs, rather than as a closed system. It will not only integrate the ESs within one business entity (intraorganizationally) but will also integrate the ESs of different entities across boundaries (interorganizationally). Therefore, interfaces at these boundaries should be explicitly presented. Correspondingly, interoperability and security should be carefully considered for these interfaces. For example, the top level of the Food-IoT solution is a Cooperative Food Cloud, in which all of the organizations in the value chain are connected through web services, as Figure 10.10 shows.

Similarly, the top level of the proposed Health-IoT solution is a cooperative health cloud (Figure 10.11). The top-level architectures represent both the formulation of value chains and the interorganizational characteristics of modern ESs.

Second, the information integration should be modeled as a cross-layer information fusion. This can be accomplished by a series of data processing at different layers ranging from the sensor node, through the gateway and server, and up to the cloud (Pang, 2013). In Figure 10.10, the information fusion of the Food-IoT solution is implemented at three layers: the on-site information fusion at sensor node level, the in-system information fusion at the gateway and server level, and the in-cloud information fusion at the cloud level.

Third, system functionality should be modeled as a cluster of services. According to Pang (2013), the basic unit of device and system integration is service, instead of a module of software or hardware. Meanwhile, attention should be paid to proper interoperability and security. For example, in Figure 10.11, an In-Home Healthcare Station (IHHS) is designed as an open platform to integrate different services from different parties. The software interfaces are merged by an open operation system and a standardized Electrical Health Record data format. The hardware interfaces are merged by a health extension based on a standard 3C (consumer, communication, and computing) terminal. Meanwhile, these architectures adapt to SOA requirements when the system is integrated into a real-world environment.

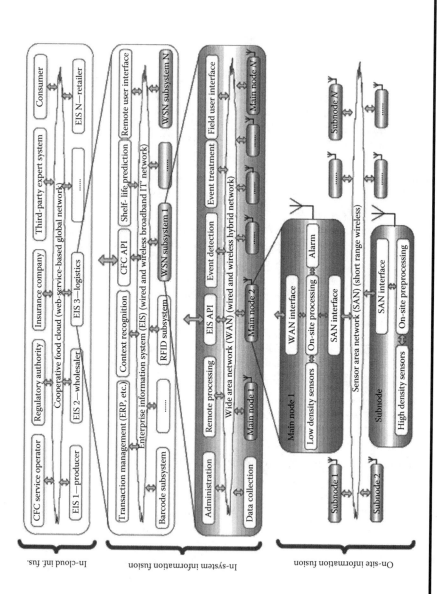

Figure 10.10   A hierarchical enterprise systems architecture for the Food-IoT solution.

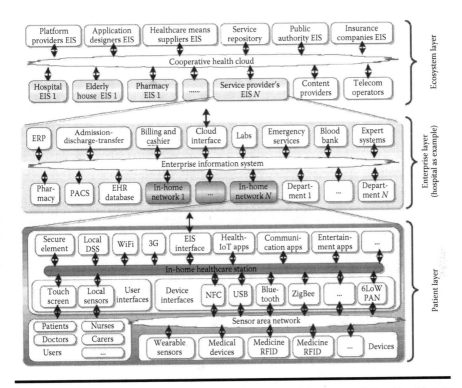

Figure 10.11 A hierarchical enterprise systems architecture of the Health-IoT solution.

As IoT has emerged as a technological revolution in the information industry, IoT is expected to become a worldwide network of interconnected objects, and its development depends on new technologies such as WSN, cloud computing, and information sensing (Li et al., 2013). In IoT-based systems, efficient data acquisition system is necessary to effectively collect and process the data, and a desirable data compression ratio is very important. Using the techniques in compressed sensing (CS) theory, data or signals can be efficiently sampled and accurately reconstructed with many fewer samples than with the Nyquist theory. Recently, research has been conducted on information acquisition in IoT and WSN with CS from the perspective of data compressed sampling, robust transmission, and accurate reconstruction to reduce the energy consumption, computation costs, and data redundancy and to increase the network capacity. In particular, the research considers distributed information sources of data and their acquisition, transmission, storage, and processing in a large-scale IoT system. The research proposes a CS framework for WSN and the IoT and introduces ways in which the framework can be utilized to reconstruct sparse or compressible data into a variety of systems involving WSN and IoT. The algorithms developed in the research are as follows:

Algorithm 10.1: Adaptive Cluster Sparsity Sensing

Input: $x \in \mathbb{R}^n$, k, $N_x \in \mathbb{R}^{n \times \tau}$, $\omega = \{w(i,t)\} \in \mathbb{R}^{n \times \tau}$, τ

Output: support set of $supp(x,k) : \Gamma$

for $i = 1 \dots n$ **do**

    for $t = 1 \cdots \tau$ **do**

        $w(i,t) = min \left\| x_i - \sum w_{ij} x_j \right\|$, $s.t. \sum w_{ij} = 1$;

    end

    $z(i) = x^2(i) + \sum_{t=1}^{\tau} w^2(i,t) N_x^2(i,t)$;

end

$\Omega \in \mathbb{R}^n$ is set as the indices corresponding to the largest k entries of z;

for $i = 1 \dots n$ **do**

    $\Gamma(i) = \Omega(i)$;

end

return $supp(\mathbf{x}, k) \leftarrow \Gamma$;

Algorithm 10.2: Adaptive Cluster Sparse Recovery Algorithm (ACSRA)

Input: $\Phi \in \mathbb{R}^{m \times n}$, $y \in \mathbb{R}^m$, $[k_{min}, k_{max}]$, Δk

Output: Recovered signal x^*.

Initialization: $l = 0$, residual $y^0 = y$, $\Gamma = supp(x) = \varnothing$, $x = 0$, $k = k_{min}$;

repeat

    Apply Algorithm 10.1 to find support set Γ^l

    with sparsity number k;

    update Γ; $x^l = \Phi_\Gamma^\dagger y^l$;

    $y^{l+1} = y^l$;

    $\Gamma^{l+1} = \Gamma^l$;

    $k = k + \Delta k$;

    $l = l + 1$;

until $x^{l+1} - x^l \leq \epsilon$;

$x^* = \Phi_\Gamma^\dagger y^{l+1}$

Figure 10.12 shows the CS scheme over the IoT, while Figure 10.13 illustrates an example. The nonsparse raw data in Figure 10.13a can be sparsely represented over a wavelet basis, as shown in Figure 10.13b.

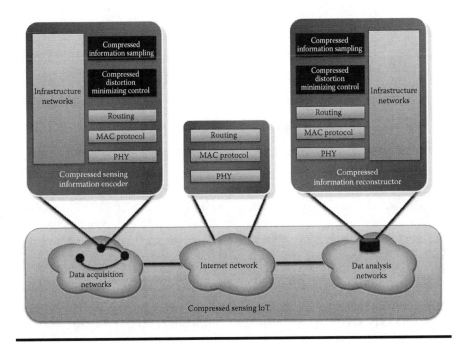

Figure 10.12   Compressed sensing scheme over the IoT.

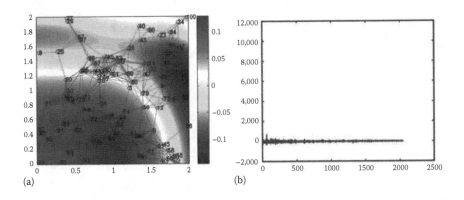

Figure 10.13   An illustration of the compressibility of the network. (a) Monitoring scenario and (b) sparse representation of monitoring data.

In summary, as a complex cyber-physical system, the IoT integrates all kinds of sensing, identification, communication, networking, and information management devices and systems and is able to seamlessly link all things according to the interests. The emerging technology breakthrough of the IoT is expected to offer promising solutions across many application areas. IoT-based technical architectures that support interoperability, security, and information system integration are being tested and improved by both researchers and practitioners. It is expected that ESs, services, and solutions based on the IoT will resolve many challenges facing today's enterprises and will be increasingly used to integrate and extend business functions and processes across the boundaries at both intra- and interorganizational levels.

10.3.2 Cloud Computing

As a new computing paradigm, cloud computing has received a lot of attention from enterprises and has been integrated or applied to create enterprise architectures for ESs (Wang et al., 2012). There are many definitions for cloud computing. A widely accepted definition is given by Mell and Grance (2011): "a model for enabling convenient, on-demand network access to a shared pool of configurable computing resources (e.g., networks, servers, storage, applications, and services) that can be rapidly supplied and released with minimal management effort or service provider interaction."

Wang et al. (2012) provide a review on the enterprise cloud service architecture based on recent literature. The cloud computing architecture is composed of three layers: resource, platform, and application. The resource layer is the infrastructure layer, which includes physical and virtualized computing, storage, and networking resources (Qian et al., 2009). The platform layer includes components such as the web server, the application server, and the enterprise service bus. The application layer serves the user and is mainly used for transaction processing and interaction (Qian et al., 2009).

According to Mell and Grance (2011), cloud computing includes three major delivery models (Wang et al., 2012):

1. Software as a service (SaaS): The consumer is able to use an application to meet specific needs. However, the consumer does not have control over the hardware, network infrastructure, and operating systems on which the application is running. Salesforce.com and Google Apps are well-known examples of SaaS.
2. Platform as a service (PaaS): The consumer uses a hosting environment for application development. The consumer has control over the applications and possibly has partial control over the hosting environment. But the consumer does not have control over the hardware, network infrastructure, and operating systems. An example of PaaS is Microsoft Azure.
3. Infrastructure as a service (IaaS): The consumer has greater access to computing resources including processing power, storage, networking components, and middleware.

In cloud computing, everything is treated as a service (Xu, 2012). These services define a layered system structure for cloud computing, as Figure 10.14 shows (Xu, 2012). At the infrastructure layer, processing, storage, networks, and other

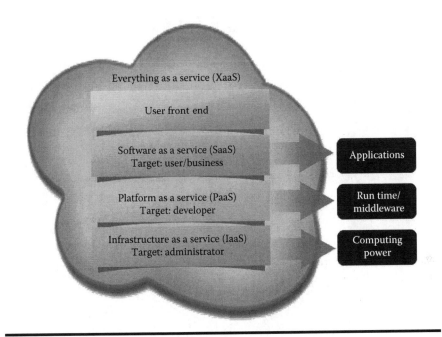

Figure 10.14   A layered system structure for cloud computing. (Adapted from Xu, X., *Robot. Comp. Int. Manuf.*, 28(1), 75, 2012.)

fundamental computing resources are defined as standardized services over the network. Cloud providers' clients can deploy and run operating systems and software for their underlying infrastructures. The middle layer, that is, PaaS, provides abstractions and services for developing, testing, deploying, hosting, and maintaining applications in the integrated development environment. The application layer provides a complete application set of SaaS. The user interface layer at the top enables seamless interaction with all of the underlying layers (Xu, 2012).

In general, the consumer has control over the operating system, storage, deployed applications, and possibly networking components. However, the consumer cannot control the underlying cloud infrastructure. Enterprises can upscale or downscale the resources according to their needs and will only pay for the capacity and space that will be used.

According to Mell and Grance (2011), cloud-based services can be deployed via four methods:

1. Public cloud: A public cloud is characterized as being available to the general public from a third-party cloud service provider such as Google, Amazon, Microsoft, and others that run public clouds. A public cloud does not mean that a user's data are publicly visible or not secure; public cloud vendors typically provide an access control or security mechanism for their users (Wang et al., 2012).

2. Private cloud: Cloud services can be offered on private networks. Private clouds are typically designed and managed by an IT department within an organization. In a private cloud service, data and processes are managed within the organization without the restrictions of network bandwidth or security exposures, and without the legal requirements that using the public cloud services might entail (Wang et al., 2012).

3. Community cloud: Using a community cloud, the cloud infrastructure is shared by several organizations and supports a specific community that has shared concerns, for example, missions, security requirements, policies, and/or compliance considerations. It can be managed by the organization or by a third party, and it may exist either on-premises or off-premises (Wang et al., 2012).

4. Hybrid cloud: A hybrid cloud is a combination of a public, private, and/or community cloud that interoperates. In this model, users typically outsource non-business-critical information and processing to the public cloud, while keeping business-critical services and data in the private cloud (Wang et al., 2012).

As a new paradigm for user services, cloud computing is a powerful way to transform existing ESs toward being more cloud computing oriented. Cloud computing offers many potential benefits such as agility, flexibility, scalability, virtualization, and cost saving to ESs by making their services available to enterprise users anytime and anywhere on the Internet. Thus, a growing trend is to move ESs to the cloud. Several approaches for designing enterprise cloud service architecture have been proposed and show ways to move ESs to the cloud. A common approach for designing enterprise cloud service architecture is based on the layered approach used in enterprise service-oriented architectures (ESOAs) (Zhang et al., 2010). The service management, application management, identity management, network monitoring, and governance used in ESOA have been reused in the design of new enterprise cloud service architectures in recent years (Zhang et al., 2010). Based on the layered approach, enterprises can eventually develop or customize an enterprise cloud service architecture that works for them by continuous refinement. Thus, the enterprise cloud service architecture has great potential to be used for the development and maintenance of ESs and will make ESs more flexible and powerful. However, there are still some research challenges (such as compatibility and interoperability issues) associated with the design of enterprise cloud service architecture and its applications. Other issues (such as application integration in the cloud environment, lack of a widely accepted and used standard, monitoring of quality of services in the cloud environment, and the security of the cloud platform and of data in transmission) need to be considered when enterprises move their ESs to the cloud. To encourage wide adoption of cloud solutions, these challenges will have to be addressed, in order to make cloud-oriented ESs more secure and reliable.

A cloud architecture of an integrated ES that maps the integration of an ES into a private cloud has been proposed (Li et al., 2012). It can provide various cloud services. Figure 10.15 shows that the system has a three-layer architecture: the front-end layer, the middle layer, and the back-end layer, which are connected by hybrid wireless networks that involve the ES. In the front-end layer, IaaS is built on top of virtualized computing storage, with network resource PaaS at the middle layer and SaaS at the user application's back-end layer. The application module in the system can provide services, data analysis, web applications, and security protection. The data/information module provides data loss protection, service log file, service activity, monitoring, and data processing. The management provides services such as a remote management interface, and networking is able to provide network services and protection. The computing storage can provide cloud computing resource, cloud data storage, and integrity, encryption, and masking.

A typical cloud application of an ES relates to cloud manufacturing. The manufacturing industry is exploring cloud computing in order to improve existing manufacturing structures and ESs, to share and provide on-demand networked manufacturing services, and to better satisfy specific manufacturing enterprise needs (Huynh and Quan, 2008). During the past few years, cloud manufacturing has become an emerging research topic in the manufacturing industry. Researchers have proposed a number of new approaches and techniques to encapsulate various virtualized manufacturing resources and capabilities as cloud-based services and to implement enterprise cloud manufacturing service frameworks and platforms (Tao et al., 2011; Xu, 2012; Zhang et al., 2010).

Currently, many modern manufacturing enterprises resort to building ESs (Xu, 2011) and production-oriented networked manufacturing environments in order to realize the integration of distributed manufacturing resources. However, there are many issues such as agility, interoperability, and scalability with networked manufacturing environments in manufacturing enterprises (Xu, 2012) that need to be considered. The research topics about how to realize full-scale sharing and on-demand use of various manufacturing resources and capabilities in networked manufacturing have not yet been effectively addressed. The flexibility of ESs has actually not yet been fully developed either, because of the limitations of services and the low efficiency in the decision and management process. In an effort to address these issues that are inherent in networked manufacturing and to provide effective support for the integration of large-scale, heterogeneous, and distributed manufacturing resources, researchers have been developing alternative approaches and methods to support more effective integration of distributed manufacturing resources within and across enterprises. Many manufacturing models and technologies, such as agile manufacturing, virtual manufacturing, application service provider, the manufacturing grid, and others, have been proposed. Some of them have been widely used by manufacturing enterprises. For example, grid computing technology has been widely investigated and applied in manufacturing during the

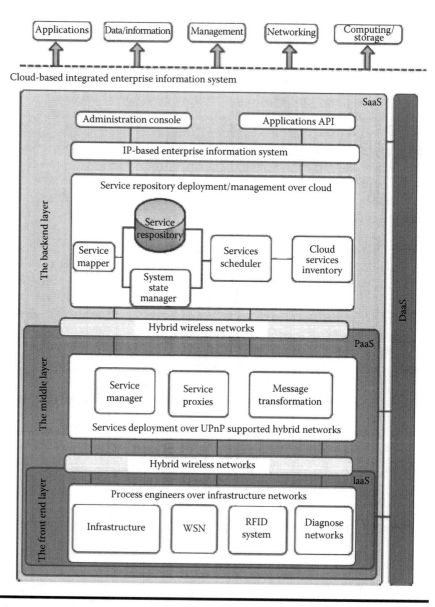

Figure 10.15 Architecture of cloud-based integrated enterprise system. (From Li, S. et al., *Enterp. Inform. Syst.***, 6(2), 165, 2012.)**

past 10 years. Researchers have developed many applications based on a manufacturing grid and have used them to support various manufacturing tasks or activities including product design, manufacturing resource integration and sharing, enterprise management, enterprise collaboration, resource allocation, and scheduling. The manufacturing grid encapsulates heterogeneous manufacturing resources in various locations and provides these resources on the enterprise computing network for online access. As a result, manufacturing users can access the remote resources deployed on the manufacturing grid as needed. Research on the architecture, application prototype, resource service composition, and other aspects of manufacturing grid have also made a fundamental contribution to the subsequent development of cloud manufacturing (Tao et al., 2012).

To transform production-oriented manufacturing to service-oriented manufacturing and to construct complete and intelligent manufacturing ESs, a computing and service-oriented manufacturing paradigm, that is, cloud manufacturing (*CMfg*), is proposed; it offers a combination of advanced computing technologies, such as cloud computing, high-performance computing, service-oriented technologies, and the IoT (Tao et al., 2011; Xu, 2012). Figure 10.16 shows the basic idea of the *CMfg* model (Cheng et al., 2013). There are

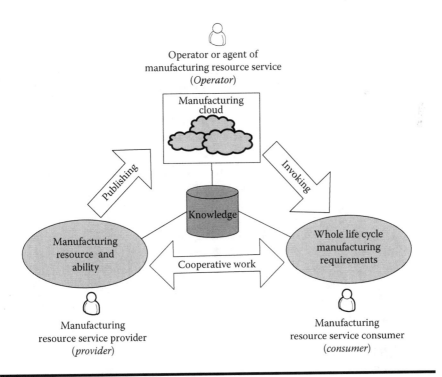

Figure 10.16 *CMfg* model. (From Cheng, Y. et al., *Proc. Instit. Mech. Eng. Part B J. Eng. Manuf.*, 227(12), 1900, 2013.)

primarily three classes of users in the *CMfg* system: resource service "providers," "consumers," and "operators."

1. A "provider" owns various manufacturing capabilities and resources and is involved with the entire life cycle of the manufacturing activity. It publishes and registers its idle resource, product, and manufacturing capability to the *CMfg* platform and provides manufacturing resources and capability service, which consumers may request. The provider can be an enterprise, an organization, or a third party.
2. A "consumer" is the subscriber of the available resource service and capability in the *CMfg* platform. It searches the optimal manufacturing resource and capability service and requests the use of services based on needs.
3. An "operator" operates the entire *CMfg* platform regarding services, capabilities, and functions for "consumers," "providers," and the third parties. Operators deal with the organizing, acquiring, and licensing of manufacturing resource and capability services and provide, update, and maintain the technologies and services involved in the operation of the platform.

There is a variety of existing research on manufacturing resource allocation and scheduling for advanced manufacturing systems. An interesting resource service scheduling problem in *CMfg* platform that is introduced by Cheng et al. (2013) is a combinational optimization problem in allocating services to providers for certain manufacturing tasks. This scheduling problem consists of multiple providers (n providers), one operator, and multiple consumers (m consumers), as shown in Figure 10.17. While one handles the tasks with the same service function requirements submitted by different consumers, there can be many services from different providers. At a certain time, let the service requirements submitted by consumer j to the platform be k_j ($k_j = 1, 2, ..., Q_j$), and the services provided by provider i be k_i, of which n_i services can satisfy the functional requirements of k_j. The problem of resource service scheduling that needs to be investigated is this: an allocation solution, that is, a mapping between k_i and k_j, to maximize the comprehensive utility of the entire system under multiple objectives (e.g., minimizing cost, energy consumption and risk, and maximizing reliability) and multiple constraints must be defined.

Cloud computing is a core enabling technology for cloud manufacturing. However, the resources involved in cloud computing are primarily computational resources (e.g., server, storage, network, and software), and they are primarily provided as services for users in the following three aspects: "IaaS," "PaaS," and "SaaS," as introduced in Figure 10.15. While in a *CMfg* system, in addition to these IT resources, various manufacturing resources and capabilities in the entire life cycle of manufacturing are connected and encapsulated as manufacturing cloud services (MCSs) and provide the users with different service models based on IaaS, PaaS, and SaaS. The service modes in *CMfg* include (1) "design as a service" (DaaS), in which the design resource and capability are provided as a service;

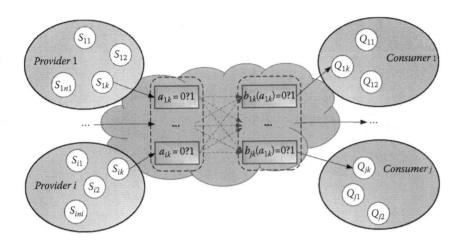

Figure 10.17 A resource service scheduling model in *CMfg*. (From Cheng, Y. et al., *Proc. Instit. Mech. Eng. Part B J. Eng. Manuf.*, 227(12), 1900, 2013.)

(2) "manufacturing as a service" (MFGaaS), in which the manufacturing resource and capability are offered as a service; (3) "experimentation as a service" (EaaS), in which experimentation resource and capability are provided as a service; (4) "simulation as a service" (SIMaaS), in which the simulation resource and capability are provided as a service; (5) "management as a service" (MaaS), in which the management resource and capability are provided as a service; (6) "maintaining as a service" (MAaaS), in which the maintaining resource and capability are provided as a service; and (7) "integration as a service" (INTaaS), in which the integrated resource and ability, information system, and platform are provided as a service (Tao et al., 2013).

In addition, researchers have proposed different ways to create multilayer architectures for implementing cloud manufacturing systems. Table 10.1 summarizes the layers proposed by different researchers to build the software architecture of enterprise cloud manufacturing systems.

Figure 10.18 shows an example of cloud manufacturing proposed by Tao et al. (2011) for the cloud manufacturing system named *CMfg*. The architecture consists of layers: the resource layer, the perception layer, the virtual resource layer, the core cloud service layer, the application layer, the portal layer, the enterprise cooperation application layer, and other layers, such as the knowledge layer, the cloud security layer, and the wider Internet layer.

Cloud manufacturing is an application that is constructed on the basis of the cloud-oriented manufacturing domain. Cloud manufacturing is composed of common functional components such as the management of cloud services, the management and access of data, and security management. The most important

Table 10.1 Proposed Layers in Cloud Manufacturing Platform Architectures

Proposed Layers	Authors
Resource layer, perception layer, virtual resource layer, core cloud service layer, application layer, portal layer, enterprise cooperation application layer, knowledge layer, cloud security layer and wider internet layer	Tao et al. (2011)
Infrastructure layer, manufacturing resources layer, business unit layer, business cloud and resource cloud, model layer for cloud manufacturing process, manufacturing cloud layer, and ontology layer	Wang et al. (2012)
Interface layer, modular interpretation layer, store/retrieve layer, and XML layer	Valilai and Houshmand (2013)
Application layer, global service layer, virtual service layer, and manufacturing resource layer	Xu (2012)
Application layer, portal layer, basement layer, access layer, and functional layer	Jiang et al. (2012)
Physical layer, connection layer, virtual layer, and service application layer	Lv (2012)
Manufacturing resource layer, integrated operation environment layer, basic supporting layer, engine layer, tool layer, service component layer, service module layer, business model layer, transaction layer, enterprise service bus (ESB) layer, user layer	Huang et al. (2012)
Physical layer, resource-oriented interface layer, virtual resource layer, core services layer, service-oriented interface layer, application layer	Ning et al. (2011)
Resource layer, resource-perception layer, resource virtual access layer, manufacturing cloud core service layer, transmission network layer, terminal application layer	Zhang et al. (2014)

component of the cloud manufacturing architecture is the management function of manufacturing resources and manufacturing tasks.

Currently, the application of cloud manufacturing is still in the initial experimental and testing stage. A lot of research and system development work is needed to move cloud manufacturing to a higher level with a more mature manufacturing model for modern manufacturing enterprises. With the support of the research on extended enterprises in the framework of IIIE, a future

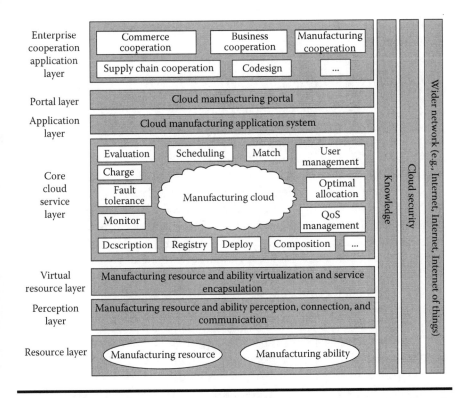

Figure 10.18 Architecture of a cloud manufacturing system. (From Tao, F. et al., *Proc. Instit. Mech. Eng. Part B J. Eng. Manuf.*, 225(10), 1969, 2011.)

research trend in cloud computing is to develop extended ES models based on the characteristics of cloud computing. Eventually, various ESs are expected to offer the function of virtualized services on the cloud and to be able to dynamically assemble solutions to support the sharing of manufacturing resources and capabilities among geographically dispersed enterprises using the manufacturing cloud. Meanwhile, the integration of existing ESs such as enterprise resource planning systems, SCM systems, customer relationship management systems, and product data management systems with cloud manufacturing platforms are expected to appear. To construct cloud-based manufacturing ESs, research is required to improve the architecture, resource sharing, management, and cooperative mechanisms in various ESs.

In the past years, various enterprise cloud computing platforms have emerged. As consumers and enterprises have gradually realized the potential benefits of on-demand SaaS, PaaS, and IaaS, there has been a substantial increase in the number of cloud computing platforms (Buyya et al., 2009; Wang et al., 2012). In general, each cloud computing platform has its own characteristics, strengths, and weaknesses (Wang et al., 2012). The following are some emerging cloud computing platforms:

- The Abiquo enterprise cloud is designed to build, integrate, and manage the public, private, and hybrid clouds. Abiquo is characterized by features that provide users with full control over server, storage, and network resources. It also allows enterprises to deploy and manage cloud resources using an industry-standard API (Peng et al., 2009).
- Aneka is a cloud development and management platform with rapid application development and workload distribution capabilities (Buyya et al., 2009). As a .NET-based service-oriented resource management platform, Aneka is designed to support multiple application models, persistence and security solutions, and communication protocols in private and public network environments.
- The Amazon Elastic Compute Cloud provides users with a virtual environment to run Linux-based applications. The users need to create a new Amazon Machine Image (AMI) or use an existing AMI from a global library. Then the user could submit AMIs to Amazon Simple Storage Service. Afterward, the users could start, stop, and monitor instances of the uploaded AMIs (Buyya et al., 2008).
- Eucalyptus is a cloud computing software platform for the on-premise (private) IaaS cloud. It is an open-source implementation of Amazon EC2 and is fully compatible with the popular Amazon Web Services including the Amazon Elastic Compute Cloud and Amazon Simple Storage Service (Peng et al., 2009).
- The Google App Engine allows the running of web-based applications written using the Python programming language. The Google App Engine supports APIs for data storage, Google Accounts, URL fetch, image manipulation, and e-mail services (Buyya et al., 2008). A web-based Administration Console is offered to the user to manage web applications, as well.
- Microsoft Azure is designed to allow software developers to easily create, host, manage, and scale both web and nonweb applications through Microsoft data centers. Microsoft Azure supports many proprietary development tools and protocols (Buyya et al., 2009).
- The Nimbus Platform is a cloud computing platform for science, which provides an integrated set of tools that deliver the versatility of infrastructure clouds to scientific users. Nimbus is also compatible with the Amazon Elastic Compute Cloud and Amazon Simple Storage Service. Nimbus allows users to lease remote resources and build the required computing environment by deploying virtual machines (Peng et al., 2009).
- OpenNebula is an open-source cloud computing product for data center virtualization. OpenNebula provides customizable solutions for users to enable them to build virtualized enterprise data centers and private cloud infrastructures and allows users to deploy and manage virtual machines on physical resources (Peng et al., 2009).

These emerging cloud computing platforms pave the way for the efficient deployment and provision of cloud services and applications to enterprises at different levels. However, each cloud computing platform has certain unresolved issues (Peng et al., 2009). Enterprises need to be aware of each platform's characteristics, features, underlying technologies, and strengths and weaknesses before a cloud computing platform is selected for enterprise deployment and implementation. At present, many enterprises have built successful applications (such as business applications and manufacturing applications) on top of cloud computing platforms. Meanwhile, more and more enterprise applications are moving to the cloud as more enterprises adopt cloud solutions. At the same time, enterprise cloud applications are becoming more configurable, scalable, and multi-tenant-aware as more researchers are using multi-tenancy patterns to develop and deploy service-oriented enterprise cloud applications (Mietzner et al., 2011). To achieve high-quality assurance in enterprise cloud services, service level agreement (SLA)–aware enterprise cloud computing architecture will need to be increasingly adopted in enterprise service computing (Wang et al., 2012). The methodology and design principles of SLA and QoS will be increasingly emphasized in designing enterprise cloud service architectures. As more ESs are moving to the cloud, more empirical research is required to study the factors involved in enterprise cloud migration. Numerous factors have been identified that have an effect on the migration of enterprise applications to cloud platforms. Such factors can be used to determine when and how enterprises should migrate applications to the cloud (Leavitt, 2009; Rimal et al., 2011). These factors include (Wang et al., 2012) the following:

- Mechanism of enterprise applications migrating to the cloud
- Procedures to extend enterprises' policies and governance to cloud deployment
- Designing of public clouds for meeting ESs' scope
- Management's attitude and perception of possible impacts and changes induced by cloud migration in the existing environment
- Implementation techniques, domain knowledge, and cost concerns
- Organizations' positions and competitive strategies, the influence of internal and external parties on the adoption decision process, and organizational readiness

In summary, the cloud is a new technology that will have a significant impact on ESs. Implementing cloud computing is challenging for enterprises that have large-scale and complex ESs. Integrating cloud computing with ESs requires enterprises to understand the requirements of diverse stakeholders, to understand how organizational processes have to change to make effective use of the new systems, to understand how new systems can be interoperable with a range of existing systems, and to understand the need to build resilience into the systems so that essential services can be provided with a high level of availability.

10.4 Other Methods and Techniques Impacting ESs

10.4.1 Software Architecture Methods

ESs have emerged as a strategic and effective tool for integrating and extending business processes and functions across supply chains at both intra- and interorganizational levels. ESs have been adopted by numerous industries. As a result, designing appropriate software architectures has become important for ESs. Software architecture consists of a set of structures and relationships for reasoning about the functionalities, behaviors, and qualities of the system. Since software architecture includes a set of system components as well as their topological relations in an ES, software architecture can be used as a bridge between the stakeholders' goals and the implementation of business functions. The decisions made during architectural analyses, therefore, not only shape what the ES will actually deliver but also determine the priority in which the requirements should be fulfilled. Determining architecturally significant requirements is important in designing ESs, since the underlying architectures impose shaping limits regarding how the ES will evolve throughout its life span. While changes conforming to the software architecture can be readily accommodated, the ones against the architectural blueprint are unlikely to be addressed adequately without serious (and sometimes prohibitive) reengineering efforts.

Due to the importance of software architecture, research on software architectures in ESs has received increasing attention. A number of different software architectures have been proposed to develop ESs for supporting a wide range of industrial applications. While these proposals and the corresponding methodologies are helpful in determining the appropriate architectures for developing ESs, systems methodologies for evaluating software architectures are scarce. Specifically, the research on software architectures for ESs remains insufficient to deal with today's IT-driven industrial automation. Thus, there is a critical need to evaluate software architecture design methods and then to select most appropriate ones to fulfill the industrial application goals.

Requirement analysis is critical to the design and development of ESs. Some of the requirements may have a global impact on the underlying software infrastructure and therefore need to be thoroughly examined. Requirements modeling is an important activity in the process of designing and managing enterprise architectures (Engelsman et al., 2011). Requirements modeling captures the motivation and rationale behind enterprise architectures in terms of goals and requirements. Through these goals and requirements, architecture artifacts (such as business services, processes, and supporting software applications) can be related to the business goals as defined by the business strategy at a higher level. This forms the basis for the assessment of the contribution of architecture artifacts to business goals (Quartel et al., 2012). Business requirements modeling will become increasingly important in designing enterprise architectures in the future.

Despite the increasing number of proposals for designing ESs, little is known about analyzing software architecture trade-offs to tackle the dependencies, interactions, and interplays of architecturally significant requirements. Lack of this knowledge makes it difficult (or even impossible) to create systems methods that can help designers to reason, analyze, and determine a suitable software infrastructure for ESs. On the other hand, architecturally significant requirements do not always manifest themselves in an explicit way. Architecturally significant requirements often represent only a small subset of all the requirements. When they are incorrect, incomplete, inaccurate, or lack details, an enterprise architecture based on them will be problematic. Furthermore, while mapping the requirements to the architectural components, the following issues require to be carefully considered (Rimal et al., 2011):

■ What kind of architectural components are frequently developed in building ESs?
■ How do architectural components relate to system requirements?
■ How to extract the generic and reusable model to classify relations between requirements and architectural components/patterns?
■ How to abstract key architectural assessment made in existing ES?

In summary, a systems approach is required to identify implicit yet significant requirements and to relate them to the architectural components. In particular, when designing an ES's software architecture in practice, designers often encounter multiple requirements as well as conflicting requirements. For instance, a trade-off exists between the planned downtime required to address ES's availability and the maintainability of the system. In other words, selecting certain requirements to be architecturally significant simultaneously excludes some other requirements being considered as the driving forces for the development of the ES. So far, little research has been conducted on the analysis of architecture interactions and trade-offs. Niu et al. (2013a,b) propose a framework consisting of an integrated set of activities to help tackle requirements analysis in practice. Specifically, they leverage quality attribute scenarios to elicit implicit yet significant requirements, to model requirements interplays, to manage terminological interferences, and to determine change impacts. They also apply the proposed framework to a customer relationship management software system. The results show that the framework offers concrete insights and can be incorporated into an organization's ES practice.

Furthermore, when making architectural decisions, taking the key driving factors into account is indispensable. Among these factors, business goals and stakeholder needs that have a broad scope of impact need to be treated as architecturally significant. For example, for a logistics management software solution, accuracy and performance are extremely crucial in delivering high-quality routing, scheduling, and dispatching services. In contrast, requirements like persistence, though relevant, may require only a secondary and thus relatively flexible treatment during software implementation. It is interesting to note the relationship between architecturally significant requirements and nonfunctional requirements. In software engineering, functional requirements describe what the system does, and nonfunctional requirements describe how well a system's functions are accomplished. In designing and implementing an ES, software architecture must support the key business drivers. These drivers are also referred to as quality attributes or NFRs. Currently, the majority of software architectures support only a single NFR. To evaluate software architecture, Niu et al. (2013a,b) propose a scenario-based method to assess how software architecture affects the fulfillment of business requirements. As Figure 10.19 shows, the scenarios play two important roles in the evaluation. First, they allow the abstract NFRs to be concretely defined, operationally measured, and meaningfully communicated among the stakeholders. Second, they link architecture choices to the satisfaction of the ES drivers, which helps management to make an informed decision about the system—a decision that is best suited to their needs. An empirical evaluation of the selection of a supply chain software tool has shown that their developed method offers remarkable insights for software development.

Currently, the research on software architecture for ESs remains insufficient. More systems methods are expected to compare and evaluate the software architecture of ESs.

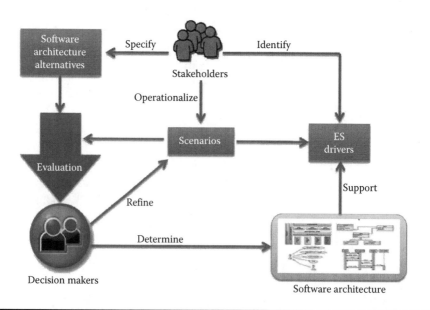

Figure 10.19   A framework for scenario-based enterprise systems software architecture analysis.

10.4.2 Networking

Networking is critical to data communication in ESs. The emerging hybrid wireless networks such as WLAN (Wireless LAN), WMN, WSN, Bluetooth-based Piconet, RFID-based information systems, and a mixture of wireless-based networks make it possible to use wireless at a large scale in manufacturing processes and offer opportunities to enterprises for using wireless in many processes and activities (Li et al., 2012). In recent years, the emergence of technologies such as cloud computing, WSN, RFID, Web 2.0, and other cutting-edge technologies further leads to the development of integrated ESs that can integrate hybrid wireless networks and cloud computing–based technologies for extended enterprises in order to achieve efficiency, competency, and competitiveness (Marston et al., 2011).

ESs benefit tremendously from emerging networking technologies, which are increasingly deployed. In the past decade, the rapid development of wireless technologies and the progress made in manufacturing low-cost and low-power devices have led to the massive deployment of WSN and Wireless Sensor Actuator Networks (WSAN) in many industrial sectors. Many industrial applications based on WSN and WSAN have been developed for specific purposes and functions, such as data collection and target tracking (Marin-Perianu et al., 2007). In a networked service of sensors, data gathered from the sensors are transferred to the Internet. Then users can use the data through web services. For example, a wireless industrial sensor network can be used to collect data from a machine equipped with sensor nodes, and data are then transmitted to the sink node. Next, a control system that is connected to the sink node will obtain the data. Driven by new emerging wireless networks–based technologies, a hybrid wireless networks–based information exchange in ESs is considered to be moving toward one of the mainstream technologies for few reasons: (1) The scalability and robustness of ESs: The IP network–based ESs can be effectively developed for different enterprises. However, the scalability and robustness of the existing ESs may cause efficiency problems as new wireless technologies are applied. Hybrid wireless networks can effectively and wirelessly connect the different subsystems in ESs (such as data centers, CAD, and CAM) that are equipped with wireless communication technologies, (2) The collaborative environment supported by wireless ESs; In ESs, the peer-to-peer collaboration paradigm can change the traditional passive exchange pattern. Integrating ESs with hybrid wireless networks can effectively connect different systems in large enterprises. Hybrid wireless network-based collaborative approaches can significantly improve the collaborative environment, (3) Effective cooperation in wireless ESs: In a hybrid wireless networking ES, the components can work properly, without interfering with each other.

With the emerging wireless networks, it is necessary to develop a scheme that can seamlessly integrate the WSN and WSAN into existing ESs. Li et al. (2012) propose a framework to integrate the hybrid wireless networks and cloud computing with ESs. This framework includes the front-end layer, middle layer, and back-end layers connecting to ESs. The proposed approach integrates access control functionalities within the hybrid framework that provide users with filtered views on available cloud services. However, there is a reliability issue when integrating wireless networks with ESs. Since a wireless sensor network can have thousands of sensor nodes, it is important to keep the network scalable and reliable so that the network can tolerate dynamic network changes. For example, sensing data from sensor nodes must be transmitted to the sink node reliably and on time. Delayed or lost data may cause industrial applications to malfunction (Heo et al., 2009). To realize real-time and reliable communication in wireless industrial sensor networks, Heo et al. (2009) propose EARQ, which is a novel routing protocol for wireless industrial sensor networks. In EARQ, a node estimates the energy cost, delay, and reliability of the path to the sink node, based only on information from the neighboring nodes. Then it calculates how to select a path using the estimates. A path with lower energy cost is to be selected. To achieve real-time delivery, only those paths that will deliver a packet in time are selected. To achieve reliability, the system may send a redundant packet via an alternate path. Experimental results show that EARQ is suitable for industrial applications, due to its capacity for energy efficient, real-time, and reliable communications.

Wireless sensors are made by many different manufacturers. Integrating such devices from different manufacturers into ESs can be challenging. As Internet technology becomes the basic carrier for interconnecting most of the other components of an ES, a common approach is to map the wireless sensor network to the Internet and then to provide these devices with the same interoperability as the other components of the ES. SOA is often applied to such mapping. Researchers have proposed several device-level SOA technologies such as Jini, UPnP, and the Device Profile for Web Services (DPWS). Using all of these technologies, DPWS is fully compatible with web services technology and is compatible with a large base of the other components of an ES that uses web services. Samaras et al. (2009) propose an advanced middleware solution to assist in the integration of a WSN into an ES at a high abstraction level. The proposed middleware is based on the DPWS, which is an SOA technology at the device level. By utilizing the proposed middleware, the process for connecting a WSN to the Internet to achieve integration into ESs in which all its components conform to an SOA standard has been demonstrated.

Recent advances in wireless sensor networks have led to pervasive health monitoring for both homecare and hospital environments. The concept of body sensor networks (BSNs) has been introduced recently. In BSN, miniaturized wearable or implantable wireless sensors are used for the continuous monitoring of patients. However, existing BSNs are still unable to support long-term monitoring in healthcare or to provide an energy-efficient, low-communication-burden, and inexpensive scheme. The emerging CS holds considerable promise for continuously acquiring biomedical signals in BSNs, which will enable nodes to employ a much lower sampling rate than the Nyquist, while still being able to accurately reconstruct signals.

CS-based BSNs are expected to significantly enhance the quality of healthcare and to improve the ability of prevention, early diagnosis, and treatment of chronic diseases.

Capitalizing on the sparsity of biomedical signals in transfer domains, Li et al. (2013) developed a continuous biomedical signal acquisition system that explores a sparsification model to find the sparse representation of biomedical signals. The sparsified measurements of signals are wirelessly transmitted to a fusion center through BSNs. Meanwhile, a weighted group sparse reconstruction algorithm was proposed to accurately reconstruct the signals at the fusion center. Simulation results show that on a random sampling over BSN, the proposed group sparse algorithm shows high efficiency, strong stability, and robustness.

As mentioned earlier, WSN is the key enabling technology of the IoT. When the WSN is integrated with the IoT, it is extended to the Ubiquitous Sensor Networks (USN). The main components or layers of USN are (Pang, 2013) (1) sensor networking, also called the sensor area network (SAN). The sensors are used to collect and transmit information; (2) access networking, also called WAN, which collects information from a group of sensors and facilitates communication with a control center or external entities; (3) network infrastructure; (4) middleware; and (5) an application platform. For real-time sensing, monitoring, and tracking, as a challenge to the enabling technology, the mobility and deployment area is mainly limited by the WAN and the power supply of the WSN gateway. For enhanced mobility, Munir et al. (2007) propose a three-tiered mobile WSN architecture. Inspired by this, a gateway architecture called WAN–SAN Coherent Architecture was proposed by Pang (2013), as shown in Figure 10.20. With this architecture, mobility can be enhanced significantly. It also supports more sensor types, higher-performance data processing, and lower power consumption. In addition, both system complexity and deployment cost can be reduced.

More research efforts have been directed toward integrating emerging networking technologies with ESs. As a result, ESs will be advanced with more functionality for pervasive services. We expect that such research efforts will lead to a new generation of ESs in the future.

10.5 Summary and Challenges

The emerging interest in ESs and industrial information integration has increased in recent years. At present, many ES implementations have not yet employed the advanced technology and the new techniques developed under the umbrella of IIIE. In other words, some technologies and applications introduced in this book are not yet fully applied in ESs, although they have great potential to play a major role in the near future. Efforts are now focusing on blending the capabilities of the existing ESs with new emerging technologies. Such a blending will enable the harnessing of the power of current and emerging technologies and is likely to

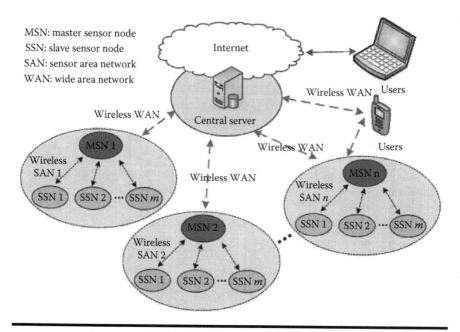

Figure 10.20 WAN-SAN Coherent Architecture of WSN. (From Pang, Z., Technologies and architectures of the internet-of-things (IoT) for health and well-being, PhD thesis, Royal Institute of Technology, Stockholm, Sweden, 2013.)

dramatically improve enterprise performance through the adoption of new innovative ESs technologies.

In addition to improving enterprise performance by adopting new innovative ESs technologies, the performance of SCM can also be improved through proper integration of the existing ESs and new technologies. One of the approaches to SCM is to develop cooperative extended ESs that are capable of promoting cooperation between the participating enterprises in the supply chain. The successful ESs for SCM rely more upon sophisticated technologies than those that are currently available. With the coming of more advanced technologies, the overall performance of SCM will improve. Research and practice have proven that ESs have been and will continue to be a basic enabler for successful SCM.

There are many challenges and issues that need to be resolved in order for ESs to become more successful. Designing ESs involves complexity that mainly stems from their high dimensionality and complexity. In recent years, there have been significant developments in this newly emerging technology, as well as actual and potential applications in various industrial sectors; however, the development of advanced methodologies, especially formal methods and a systems approach, has to be synced with the rapid technological developments. Even in the designing of manufacturing systems, a component of ESs, there exists a gap between the level of complexity inherent in manufacturing systems and the rich set of formal methods that could potentially contribute to the design of advanced manufacturing systems. Despite advancements in the field of ESs, both in academia and industry, significant challenges remain. They need to be dealt with in order to fully realize the potential of ESs. IIIE will continue to embrace cutting-edge ESs technology and techniques and will open up new applications that will impact industrial sectors. ESs can and will contribute to the success of this endeavor.

References

Ashton, K. 2009. That 'Internet of Things' thing. *RFID Journal*, June 22, 2009.

Atzori, L., Iera, A., and Morabito, M. 2010. The internet of things: A survey. *Computer Networks*, 54(15), 2787–2805.

Brock, D. 2001. *The Electronic Product Code: A Naming Scheme for Physical Objects.* Auto-ID Publication, MIT-AUTOID-WH-002, Cambridge, MA.

Buyya, R., Yeo, C. S., and Venugopal, S. 2008. Market-oriented cloud computing: Vision, hype, and reality for delivering IT services as computing utilities. *Proceedings of 10th IEEE International Conference on High Performance Computing and Communications*, Dalian, China, pp. 5–13.

Buyya, R., Yeo, C. S., Venugopal, S., Broberg, J., and Brandic, I. 2009. Cloud computing and emerging IT platforms: Vision, hype, and reality for delivering computing as the 5th utility. *Future Generation Computer Systems*, 25(6), 599–616.

Cheng, Y., Tao, F., Liu, Y., Zhao, D., Zhang, L., and Xu, L. 2013. Energy-aware resource service scheduling based on utility evaluation in cloud manufacturing system. *Proceedings of the Institution of Mechanical Engineers, Part B: Journal of Engineering Manufacture*, 227(12), 1900–1914.

Engelsman, W., Quartel, D., Jonkers, H., and van Sinderen, M. 2011. Extending enterprise architecture modelling with business goals and requirements. *Enterprise Information Systems*, 5(1), 9–36.

European Commission Information Society. 2008. *Internet of Things in 2020: A Roadmap for the Future*, Workshop Report, Prague, May 10, 2009.

Finzen, J., Riedl, C., May, N., and Stathel, S. 2010. Innovation in the Internet of Services. *Proceedings of XX International RESER Conference*, Gothenburg, Sweden, pp. 1–21.

Gadea, C., Ionescu, B., and Ionescu, D. 2010. Real-time collaborative intelligent services for sensor networks. *Proceedings of 2010 International Joint Conference on Computational Cybernetics and Technical Informatics*, Timisoara, Romania, pp. 511–516.

Gao, J., Pang, Z., Chen, Q., and Zheng, L. 2010. Interactive packaging solutions based on RFID technology and controlled delamination material. *Proceedings of 2010 IEEE International Conference on RFID*, Orlando, FL, pp. 158–165.

Giusto, D., Iera, A., Morabito, G., and Atzori, L. 2010. *The Internet of Things*. Springer, Springer Heidelberg, ISBN: 978-1-4419-1673-0.

Guinard, D., Trifa, V., Mattern, F., and Wilde, E. 2011. From the Internet of Things to the web of things: Resource oriented architecture and best practices. In Uckemann, D. et al. (Eds.), *Architecting the Internet of Things*, Springer Heidelberg.

He, W. and Xu, L. 2013. Integration of distributed enterprise applications: A survey. *IEEE Transactions on Industrial Informatics*, 10(1), 35–42, 2014.

Heo, J., Hong, J., and Cho, Y. 2009. EARQ: Energy aware routing for real-time and reliable communication in wireless industrial sensor networks. *IEEE Transactions on Industrial Informatics*, 5(1), 3–11.

Hou, H., Xu, S., and Wang, H. 2007. A study on X party material flow: The theory and applications. *Enterprise Information Systems*, 1(3), 287–299.

Huang, B., Li, C., Yin, C., and Zhao, X. 2012. Cloud manufacturing service platform for small- and medium-sized enterprises. *International Journal of Advanced Manufacturing Technology*, 65(9–12), 1261–1272, 2013.

Huynh, S. and Quan, D. 2008. Cloud computing in manufacturing environment. *Proceedings of 2008 Spring Meeting & Fourth Global Congress on Process Safety*, New Orleans, LA.

Jia, X., Feng, Q., Fan, T., and Lei, Q. 2012. RFID technology and its applications in Internet of Things (IoT). *Proceedings of 2012 Second International Conference on In Consumer Electronics, Communications and Networks*, Yichang, China, pp. 1282–1285.

Jiang, W., Ma, J., Zhang, X., and Xie, H. 2012. Research on cloud manufacturing resource integrating service modeling based on cloud-agent. *Proceedings of 2012 IEEE Third International Conference on Software Engineering and Service Science*, Beijing, China, pp. 395–398.

Leavitt, N. 2009. Is cloud computing really ready for prime time? *Computer*, 42(1), 15–20.

Li, L. 2007. *Supply Chain Management: Concepts, Techniques and Practices*. Hackensack, NJ: World Scientific.

Li, S., Xu, L., and Wang, X. 2013. A continuous biomedical signal acquisition system based on compressed sensing in body sensor networks. *IEEE Transactions on Industrial Informatics*, 9(3), 1764–1771.

Li, S., Xu, L., Wang, X., and Wang, J. 2012. Integration of hybrid wireless networks in cloud services oriented enterprise information systems. *Enterprise Information Systems*, 6(2), 165–187.

Lin, Y., Duan, X., Zhao, C., and Xu, L. 2013. *Systems Science Methodological Approaches*. Boca Raton, FL: CRC Press.

Lu, Y., Chen, T., and Meng, Y. 2011. Evaluation guiding system and intelligent evaluation process on the Internet of Things. *American Journal of Engineering and Technology Research*, 11(9), 537–541.

Lv, B. 2012. A multi-view model study for the architecture of cloud manufacturing. *Proceedings of 2012 Third International Conference on Digital Manufacturing and Automation*, Guilin, China, pp. 93–97.

Marin-Perianu, M., Meratnia, N., Havinga, P., de Souza, L., Muller, J. Spiess, P., Haller, S., Riedel, T., Decker, C., and Stromberg, G. 2007. Decentralized enterprise systems: A multiplatform wireless sensor network approach. *IEEE Wireless Communications*, 14(6), 57–66.

Marston, S., Li, Z., Bandyopadhyay, S., Zhang, J., and Ghalsasi, A. 2011. Cloud computing-the business perspective. *Decision Support Systems*, 51(1), 176–189.

Mell, P. and Grance, T. 2011. The NIST definition of cloud computing. National Institute of Standards and Technology. Gaithersburg, MD. Retrieved from http://csrc.nist.gov/groups/SNS/cloud-computing/.

Mietzner, R., Leymann, F., and Unger, T. 2011. Horizontal and vertical combination of multi-tenancy patterns in service-oriented applications. *Enterprise Information Systems*, 5(1), 59–77.

Miorandi, D., Sicari, S., De Pellegrini, F., and Chlamtac, I. 2012. Internet of things: Vision, applications and research challenges. *Ad Hoc Networks*, 10(7), 1497–1516.

Munir, S., Ren, B., Jiao, W., Wang, B., Xie, D., and Ma, M. 2007. Mobile wireless sensor network: Architecture and enabling technologies for ubiquitous computing. *Proceedings of 21st International Conference on Advanced Information Networking and Applications Workshops*, Niagara Falls, Canada, pp. 113–120.

Ning, F., Zhou, W., Zhang, F., Yin, Q., and Ni, X. 2011. The architecture of cloud manufacturing and its key technologies research. *Proceedings of 2011 IEEE International Conference on Cloud Computing and Intelligence Systems*, Beijing, China, pp. 259–263.

Niu, N., Xu, L., and Bi, Z. 2013a. Enterprise information systems architecture-analysis and evaluation, *IEEE Transactions on Industrial Informatics*, 9(4), 2147–2154.

Niu, N., Xu, L., Cheng, C., and Niu, Z. 2013b. Analysis of architecturally significant requirements for enterprise systems. *IEEE Systems Journal*, DOI: 10.1109/JSYST.2013.2249892.

Pang, Z. 2013. Technologies and architectures of the internet-of-things (IoT) for health and well-being, PhD thesis, Royal Institute of Technology, Stockholm, Sweden.

Pang, Z., Chen, Q., Han, W., and Zheng, L. 2014. Value-centric design of the internet-of-things solution for food supply chain: Value creation, sensor portfolio and information fusion. *Information Systems Frontiers*.

Peng, J., Zhang, X., Lei, Z., Zhang, B., Zhang, W., and Li, Q. 2009. Comparison of several cloud computing platforms. *Proceedings of Second International Symposium on Information Science and Engineering*, Shanghai, China, pp. 23–27.

Plaza, I., Martin, L., Martin, S., and Medrano, C. 2011. Mobile applications in an aging society: Status and trends. *Journal of Systems and Software*, 84(11), 1977–1988.

Qian, L., Luo, Z., Du, Y., and Guo, L. 2009. Cloud computing: An overview. *Lecture Notes in Computer Science*, Springer Heidelberg, Vol. 5931, pp. 626–631.

Quartel, D., Steen, M., and Lankhorst, M. 2012. Application and project portfolio valuation using enterprise architecture and business requirements modeling. *Enterprise Information Systems*, 6(2), 189–213.

Rimal, B., Jukan, A., Katsaros, D., and Goeleven, Y. 2011. Architectural requirements for cloud computing systems: An enterprise cloud approach. *Journal of Grid Computing*, 9(1), 3–26.

Samaras, I., Gialelis, G., and Hassapis, G. 2009. Integrating wireless sensor networks into enterprise information systems by using web services. *Proceedings of Third International Conference on Sensor Technologies and Applications*, Athens, Greece, pp. 580–587.

Song, B. and Xu, S. 2009. The theory of material flow substance. *Systems Research and Behavioral Science*, 26(2), 251–258.

Tao, F., LaiLi, Y., Xu, L., and Zhang, L. 2013. FC-PACO-RM: A parallel method for service composition optimal-selection in cloud manufacturing system. *IEEE Transactions on Industrial Informatics*, 9(4), 2023–2033, 2013.

Tao, F., Zhang, L., Lu, K., and Zhao, D. 2012. Research on manufacturing grid resource service optimal-selection and composition framework. *Enterprise Systems*, 6(2), 237–264.

Tao, F., Zhang, L., Venkatesh, V. C., Luo, Y., and Cheng, Y. 2011. Cloud manufacturing: A computing and service-oriented manufacturing model. *Proceedings of the Institution of Mechanical Engineers, Part B: Journal of Engineering Manufacture*, 225(10), 1969–1976.

Valilai, O. and Houshmand, M. 2013. A collaborative and integrated platform to support distributed manufacturing system using a service-oriented approach based on cloud computing paradigm. *Journal of Robotics and Computer-Integrated Manufacturing,* 29(1), 110–127.

van Sinderen, M. and Almeida, J. 2011. Empowering enterprises through next-generation enterprise computing. *Enterprise Information Systems,* 5(1), 1–8.

Wang, H., He, W., and Wang, F. 2012. Enterprise cloud service architecture. *Information Technology and Management,* 13(4), 445–454.

Warfield, J. 1990. *A Science of Generic Design Managing Complexity through Systems Design.* Salinas, CA: Intersystems Publications.

Warfield, J. 2007. Systems science serves enterprise integration: A tutorial. *Enterprise Information Systems,* 1(2), 235–254.

Xu, L. 2011. Enterprise systems: State-of-the-art and future trends. *IEEE Transactions on Industrial Informatics,* 7(4), 630–640.

Xu, S. 2008. The concept and theory of material flow. *Information Systems Frontiers,* 10(5), 601–609.

Xu, X. 2012. From cloud computing to cloud manufacturing. *Robotics and Computer-Integrated Manufacturing,* 28(1), 75–86.

Zhang, L., Luo, Y., Tao, F., Li, B., Ren, L., Zhang, X., Guo, H., Cheng, Y., Hu, A., and Liu, Y. 2014. Cloud manufacturing: A new manufacturing paradigm. *Enterprise Information Systems,* 8(2), 167–187, 2014.

Zhang, Q., Cheng, L., and Boutaba, R. 2010. Cloud computing: State-of-the-art and research challenges. *Journal of Internet Services and Applications,* 1(1), 7–18.

Index